012437352

Surveying for Construction
Third Edition

Surveying
for
Construction
Third Edition

William Irvine, FRICS

Lecturer in Land Surveying
Glasgow College of Building and Printing

McGRAW-HILL BOOK COMPANY

London · New York · St Louis · San Francisco · Auckland · Bogotá · Caracas · Hamburg
Lisbon · Madrid · Mexico · Milan · Montreal · New Delhi · Panama · Paris · San Juan
São Paulo · Singapore · Sydney · Tokyo · Toronto

Published by

McGRAW-HILL Book Company Europe

SHOPPENHANGERS ROAD · MAIDENHEAD · BERKSHIRE · ENGLAND

TEL: 0628-23432; FAX: 0628-770224

British Library Cataloguing in Publication Data
Irvine, William
  Surveying for construction.——3rd ed.
  1. Surveying
  I. Title
  526.9'024624   TA549

ISBN 0-07-707041-0

56 AP 92

Typeset by STYLESET LIMITED · Warminster · Wiltshire

Printed and bound in Great Britain by The Alden Press · Oxford

Contents

Preface

This is the third edition of *Surveying for Construction*. It has been written primarily for the non-practising surveying student; that is, the student who does not work regularly on surveying but who is required to show, during the course of his work or in examinations, that he understands the basic principles of the subject and can apply these principles to a variety of situations as they arise.

This third edition is necessitated by the fact that surveying has harnessed the new technologies of electronics and computers, and even in the most modest surveying courses the impact of the new developments is being felt.

The age-old question concerning the depth of knowledge that the surveying technician should possess then becomes daily more complex.

The author's opinion is that the surveying technician should possess a sound knowledge of the basic surveying procedures and methods; should understand the mathematical principles of calculating a survey; be able to write a simple computer program; and be aware of the most recent developments in surveying technology.

In an attempt to cover these aspects, this third edition has been completely revised, enlarged and rewritten wherever necessary.

New sections have been added on automatic levels, theodolite axis systems, electromagnetic distance measurement, the use of laser lining equipment in setting out, and the radiation method of detail surveying.

The opportunity has been taken to revise the topics of levelling reduction methods, sections, errors in linear measurement, calculation of coordinates, areas, volumes, and vertical control in setting out, all of which required some elucidation in the second edition.

In response to some valid criticisms, more examples have been introduced into the text and the test papers have been enlarged and updated.

A new chapter has been included on the use of computers and spreadsheets to reflect the current trends in surveying courses. It should be emphasized at this point that the programs have been written for the novice computer operator and cannot be compared with the long, sophisticated and usually very expensive commercial software packages currently available.

The use of the scientific calculator in coordinate calculations has been fully explained in the traversing chapters as it is the author's experience that students fail to realise the full potential of these instruments.

The book covers the syllabuses of the various bodies connected with the construction industry. Consequently it is suitable for:

1. Professional examinations of
(a) The Royal Institution of Chartered Surveyors (General Practice, Quantities and Building Surveying sections)
(b) The Chartered Institute of Building.
(c) The Incorporated Association of Architects and Surveyors.
2. Examinations of
(a) The Construction Surveyors Institute.
(b) Institute of Clerks of Works.
3. Certificates and Diplomas of B.T.E.C. in Building, Engineering and Topographic Studies.
4. Certificates and Diplomas of the Scottish Vocational Education Council in Building, Engineering, Mining and Topographic Studies and the Scottish National Certificate (Modular).
5. City and Guilds of London Institute.
6. Council for National Academic Awards courses in Building, Engineering and Surveying.
7. Polytechnic and College Diplomas in Building Surveying and Quantity Surveying.
8. National Certificates and Diplomas in Building and Engineering.
9. Regional Examining Bodies awards in Engineering, Construction and Mining.

Acknowledgements

My sincere thanks are due to the following individuals and organizations:

The Director General of the Ordnance Survey for permission to reproduce Ordnance Survey maps.
British Standards Institution, 2 Park Street, London W1A 2BS, with whose permission the drawings from *BS 4484: Measuring Instruments for Constructional Works: Part 1 1969; Metric Graduation and Figuring of Instruments for Linear Measurements* and adaptations of material from *BS 1192: Recommendations for Building Drawing Practice* are reproduced.
Clarkson Lord & Co Ltd.
D. T. F. Munsey FRICS and the Technical Press Limited for allowing the reproduction of tacheometric tables from the book *Tacheometric Tables for the Metric User.*
E. Mason (Editor) and Virtue and Co Ltd for permission to adapt material from *Surveying* by J. L. Holland, K. Wardell, and A. G. Webster.
AGA Geodimeters; Hilger & Watts (now Rank Precision Industries); Kern & Co. Ltd; Laser Alignment Inc.; Microsoft Corporation; Projectina Co Ltd; Rabone Chesterman Ltd; Sokkisha & Co Ltd.; Surveying & General Instrument Co Ltd; Topcon Instrument Corporation of America; Vickers Instruments Ltd; Wild Heerbrugg (UK) Ltd; W. F. Stanley & Co Ltd; for permission to photograph instruments and/or to adapt illustrations, etc., from their technical literature.
The City and Guilds of London Institute; The Institute of Building; The Royal Institution of Chartered Surveyors; The Scottish Vocational Education Council; The Technical Education Council; for permission to use questions from their examination papers.
The Principal, David McEwan BSc, CEng, MICE, FCIQB, and members of staff of Glasgow College of Building and Printing for their valued advice.
My long suffering wife for her endless patience and understanding.

I wish to make it clear that the responsibility for the drawing and accuracy of all diagrams, the compilation and solution of all test questions, the model answers to all examination questions, and the compilation of all computer programs and spreadsheets is entirely my own.

If I have inadvertently used material without permission or acknowledgement, I sincerely apologize and hope that any oversight will be excused.

1. Basic principles of surveying

Exactly what is meant by the term 'surveying' in the construction industry?

Perhaps it is less difficult to answer this question if an examination is made of the duties of a land surveyor employed in the industry. Among other things he is called upon to determine heights and distances; to set out buildings, sewers, drains and roadways; to determine areas and volumes of regular and irregular figures; and to prepare and finish detailed drawings.

These operations involve the application of practical mathematics, the use, handling, and understanding of several types of instrument, the calculation of the observations, and finally the presentation of the work in the form of prepared plans.

The surveyor must as a consequence be a practical mathematician knowledgeable in geometry and trigonometry and possess a fair ability for drawing and painting.

In the light of the above paragraphs it might be fairly said that land surveying could be defined in its simplest terms as the art of measuring an area on the earth and representing it to some suitable scale on paper.

All the physical measurements and scaling are now done on the metric system in Great Britain, as in the majority of the world's nations, and perhaps it is wise to begin a study of surveying with a revision of the units of measurement commonly used.

The metric system of measurement

The metric system was officially introduced in France on 22 June 1799. The fundamental unit of measurement is the metre and at the time of its adoption, two marks were etched on a bar of platinum at a distance apart of 1/10 000 000 part of the earth's quadrant, that is, the distance from the Equator to the North Pole measured on the meridian of Paris.

In 1872, at a meeting of the interested nations in Paris, it was decided that the bar should be made of 90 per cent platinum and 10 per cent iridium and to this day, it is preserved in Sèvres in France.

In 1960, however, the metre was finally defined as being equal to 1 650 763.73 wavelengths in vacuum of the orange radiation $(2P_{10} - 5d_5)$ of a krypton atom of mass 86.

The metric system was standardized in 1964 and the Système Internationale (SI) sets out the basic and derived units which have been agreed internationally.

The following units are most important to the surveyor:

Quantity	Unit	Symbol
Length	Metre	m
Area	Square metre	m^2
Volume	Cubic metre	m^3
Mass	Kilogramme	kg
Capacity	Litre	l

Taking any one of the above-listed quantities as the basic unit, a table of multiples and sub-multiples is derived by prefixing the basic unit as follows:

Prefix	Multiplication factor	Derived unit	SI recommended unit
kilo	1000	kilometre	kilometre (km)
hecto	100	hectometre	
deca	10	decametre	
		metre	metre (m)
deci	0.10	decimetre	
centi	0.01	centimetre	
milli	0.001	millimetre	millimetre (mm)

Table 1.1

In Table 1.1 it should be noticed that only three units are recommended for general use. This is true for other quantities also and Table 1.2 shows the small selection of the units included in the SI system which are now in common use.

Finally it is necessary to see exactly how mass, volume, and capacity are related. Table 1.3 shows the basic relationship from which others may be deduced.

Quantity	Recommended SI unit	Other units which may be used
Length	kilometre (km) metre (m) millimetre (mm)	centimetre (cm)
Area	square metre (m²) square millimetre (mm²)	square centimetre (cm²) hectare (100 × 100 m) (ha)
Volume	cubic metre (m³) cubic millimetre (mm³)	cubic decimetre (dm³) cubic centimetre (cm³)
Mass	kilogramme (kg) gramme (g) milligramme (mg)	
Capacity	cubic metre (m³) cubic millimetre (mm³)	litre (l) millilitre (ml)

Table 1.2

Volume	Mass	Capacity
1 cubic metre	1000 kilogrammes	1000 litres
1 cubic decimetre	1 kilogramme	1 litre
1 cubic centimetre	1 gramme	1 millilitre

Table 1.3

Surveying mathematics

It has already been stated that the surveyor must be a practical mathematician. The branch of mathematics in which he must be most proficient is trigonometry.

The following formulae are given because of their general usefulness in everyday situations. No attempt is made to prove them. Should any proof be required or any further formulae needed a good textbook must be consulted.

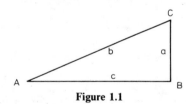

Figure 1.1

Fundamental trigonometrical ratios
In Fig. 1.1 angle B is a right angle. Sides *a, b,* and *c* lie opposite the angles A, B, and C.

1. $\text{Sin A} = \dfrac{a}{b}$ 4. $\text{Cosec A} = \dfrac{b}{a} = \dfrac{1}{\text{Sin A}}$

2. $\text{Cos A} = \dfrac{c}{b}$ 5. $\text{Sec A} = \dfrac{b}{c} = \dfrac{1}{\text{Cos A}}$

3. $\text{Tan A} = \dfrac{a}{c}$ 6. $\text{Cot A} = \dfrac{c}{a} = \dfrac{1}{\text{Tan A}}$

In Fig. 1.2 all of the angles are acute while in Fig. 1.3 angle A is obtuse.

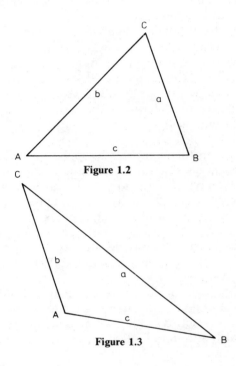

Figure 1.2

Figure 1.3

In both triangles the following formulae apply.

7. *Sine Rule*

$$\frac{a}{\sin A} = \frac{b}{\sin B} = \frac{b}{\sin C} = 2R$$

(where *R* is the radius of the circumscribing circle)

8. *Cosine Rule*

$$a^2 = b^2 + c^2 - 2bc \cos A$$

(Note: where A is obtuse cos A =
$$-\cos (180 - A)$$

9. *Area Rule*
   Area of triangle ABC = $ab \sin C$

10. Tangent half angle formula

$$\tan \frac{(A - B)}{2} = \frac{(a - b)}{(a + b)} \cot \frac{C}{2}$$

11. The 's' formula
Area of triangle ABC =
$$\sqrt{(s)\,(s-a)\,(s-b)\,(s-c)}$$

$$\cos\frac{A}{2} = \sqrt{(s)\,\frac{(s-a)}{bc}}$$

where $s = \dfrac{(a+b+c)}{2}$

Any other trigonometric formula and all geometrical theorems will be given as the need arises.

Basic principles of surveying

In the introduction to this book it was stated that surveys are represented to scale on paper. These pieces of paper are called maps or plans depending on the scale to which they are drawn.

Since the flat surface of a piece of paper can truly represent only the horizontal plane, and since the surface of the earth is known to be curved, it follows that the earth's surface cannot be exactly represented on a map, without recourse to some special form of surveying known as Geodetic Surveying.

Over large tracts of land, for example the whole of Great Britain, geodetic techniques are employed to provide the basic framework or backbone of the survey.

Closely allied to geodetic surveying is the science of aerial surveying and nowadays, whenever topographic maps of a large area are to be compiled, the details are obtained from aerial surveys.

Both of these forms of surveying are subjects in their own right and in general are beyond the scope of construction surveying. For a study of these more advanced techniques, readers are referred to the numerous textbooks available on the subjects.

This book deals with the surveys of relatively small tracts of land and uses the techniques of Plane Surveying in which the curvature of the earth's surface is ignored.

Beginning with the survey of a simple living room or classroom, it will quickly be appreciated that geometric design and order are fundamental if an accurate scale drawing of the room is to be produced.

In order to make a survey of the room shown in Fig. 1.4, the dimensions AB, BC, CD, DA, and AC are measured using a tape and plotted to scale using a scale rule and compasses. The dimension BD is also measured to provide a check on the work. It is not permissible to assume that any angle is right angled, hence the check BD is required.

In even this simplest of surveys, the control points of the room are grouped in sets of three such that one

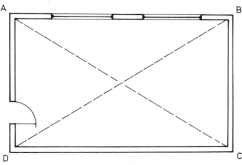

Figure 1.4

unknown position, e.g., point C, is fixed from the two known positions A and B. Unknown point D is then fixed from the two known points A and C.

In each of the Figs. 1.5 to 1.8, points X, Y and Z lie on a horizontal plane. Their relative positions can be fixed in a number of ways.

1. *Using linear measurement only*
   (a) *Trilateration* The word means measurement of three sides. When the principle is applied to Fig. 1.5, the lengths XY, YZ, and XZ are all measured in the field. Length XY is then drawn to scale on paper, and arcs representing the lengths YZ and XZ are drawn using compasses to intersect in the point Z.

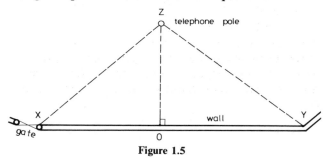

Figure 1.5

   (b) *Lines at right angles* In the field, lengths XO and OY are measured along line XY, and line OZ is measured exactly at right angles to XY (Fig. 1.5). Using a set-square and scale rule, point Z can again be plotted in its correct relationship with X and Y. This method of surveying is known as offsetting.

2. *Using linear and angular measurements*
   (a) *Triangulation* In the field, line XY known as the base line, is measured using a tape, and angles ZXY and XYZ are measured using some angular measuring instrument (Fig. 1.6). The survey is then drawn to scale on paper using a scale rule and protractor. Alternatively, the lengths XZ and YZ could be calculated and plotted by trilateration.
   (b) *Polar coordinates* In the field, lines XY, XZ, and angle ZXY are measured as before (Fig. 1.7). Using a scale rule and protractor, the survey is plotted to scale. Where two lines and the included angle are

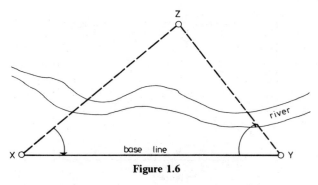

Figure 1.6

measured, the method of surveying is known as traversing.

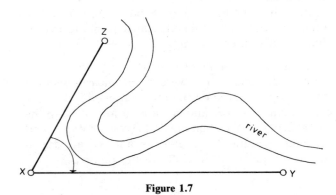

Figure 1.7

All of the above principles are used in building surveys in some form or other and indeed a survey of a relatively small building site may combine all four methods.

Example 1 The following four sets of information illustrate the principles of plane surveying which have been used to survey the relative positions of points A, B, and C. Plot the information to scale 1:2500 (i.e., 1 mm = 2.5 m).

1. Trilateration: line AB = 273.2 m
   line AC = 200.0 m
   line CB = 244.9 m

2. Triangulation: line AB = 273.2 m
   angle BAC = 60° 00′
   angle CBA = 45° 00′

3. Polar coordinates: line AC = 200.0 m
   line AB = 273.2 m
   angle BAC = 60° 00′

4. Lines at right angles: line AB = 273.2 m
   line AO = 100.0 m
   line OC = 173.2 m

Solution The points are plotted to scale in Fig. 1.8 and all of the measurements relate to the same points.

Application of principles
(framework and detail surveys)

All of the principles previously outlined are used in some form in construction surveys. The following cases demonstrate the use of the principles in various circumstances.

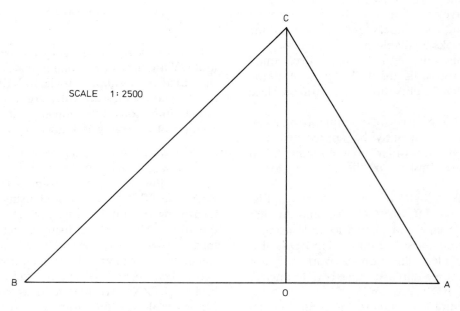

SCALE 1 : 2500

Figure 1.8

1. Trilateration and offsetting

Figure 1.9 shows a house and garage, standing in about 1 hectare of ground. A plan of the house and grounds is required.

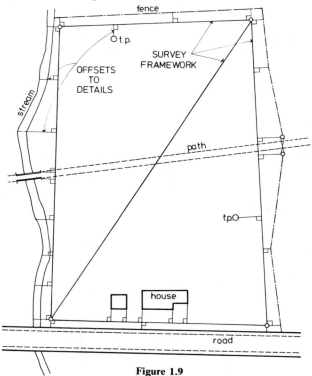

Figure 1.9

This survey and all other surveys are carried out in two parts.

(a) *Framework survey* A framework is established over the whole site to form a sound geometrical figure which can be readily plotted. It should be noted that, even though the path neatly bisects the enclosure, the framework covers the site in its entirety. It is quite wrong to survey both portions independently and then attempt to join them. Chapter 4 deals fully with the procedures in making a linear survey. At this stage, it is sufficient to know that the area is divided into triangles forming a trilateration framework onto which the detail survey is built. Thus, the fundamental surveying rule of 'working from the whole to the part' is complied with.

(b) *Detail survey* The fences, stream, buildings, and path are the details which are added to the framework by measuring short lengths perpendicular to the main lines. These short distances are called offsets.

2. Triangulation and traversing

Figure 1.10 shows a part of the major triangulation survey of Great Britain. The various triangles and quadrilaterals form the basic framework. All angles shown in the figure were measured with a very high degree of precision, but only one length was measured. Very great care was taken to ensure that the measurement was made as accurately as possible, because the positions of the other points were calculated using the length of this baseline. The techniques of triangulation are more fully covered in Chapter 18. At this stage, it is sufficient to know that this basic framework is calculated in rectangular coordinates, and all of the coordinated points are permanently marked on the ground by triangulation pillars.

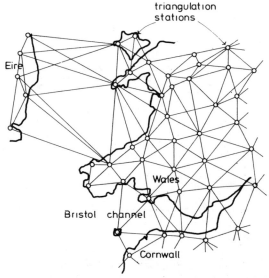

Figure 1.10

Figure 1.11 shows three of these triangulation pillars forming the framework for traverse surveys along the river and roadway. It should be remembered from the basic principles that a traverse is the measurement of two lines and the angle contained between them, therefore stations B, A, and P form the first section of the traverse BAPQRSC. The traverses form secondary frameworks from which the necessary offsets are taken to the various points of detail. Once again the basic 'whole to part' rule has been followed.

The techniques of traversing and the subsequent calculations are fully dealt with in Chapters 9–13. It must be pointed out that the applications outlined in this section contain a few oversimplifications which are necessary at this stage.

It is very important that the surveyor and engineer have a sound knowledge of basic mathematics. The following examples indicate some situations in which the basic principles are used.

Example 2 In a simple triangulation scheme, several triangles have been linked to produce the Fig. 1.12. The base line AB has been measured by

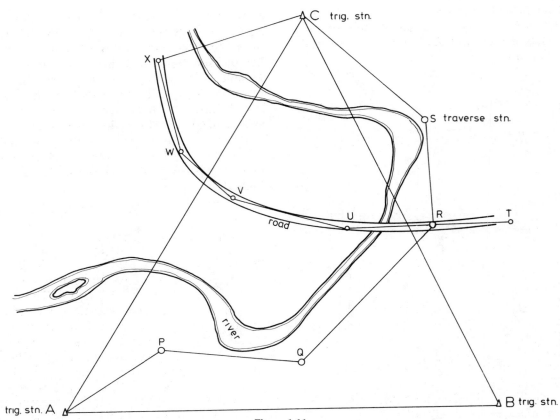

Figure 1.11

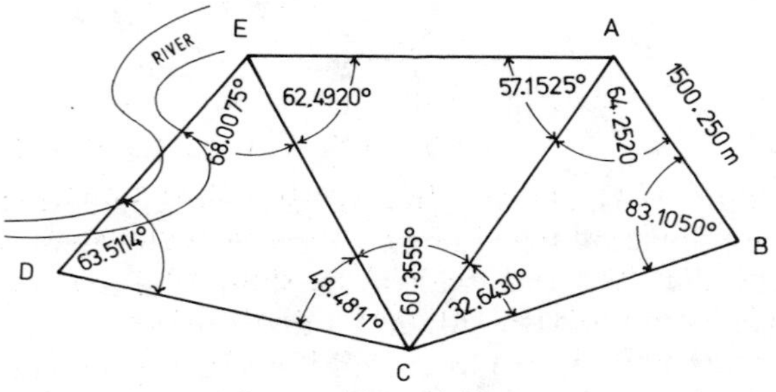

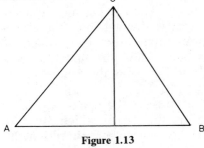

Figure 1.12

tape and the angles of the various triangles by a theodolite. The distance DE across the river is required.

Solution

In ABC

$$\frac{AC}{\sin 83.1050} = \frac{AB}{\sin 32.6430}$$

$$AC = \frac{1500.25 \times \sin 83.1050}{\sin 32.6430} = 2761.201 \text{ m}$$

In ACE

$$\frac{CE}{\sin 57.1525} = \frac{AC}{\sin 62.4920}$$

$$CE = \frac{2761.201 \times \sin 57.1525}{\sin 62.4920} = 2615.415 \text{ m}$$

In DEC

$$\frac{DE}{\sin 48.4811} = \frac{CE}{\sin 63.5114}$$

$$DE = \frac{2615.415 \times \sin 48.4811}{\sin 63.5114} = 2187.942 \text{ m}$$

Example 3 A triangular plot of ground (Fig. 1.13) has been surveyed by measuring the line AB = 231 m and the altitude 164.0 m. Calculate the area of the plot in hectares.

Figure 1.13

Solution

Area of ABC = ½ base × altitude
= 115.5 × 164 m²
= 18 942 m²
= 1.894 hectares

2. Plan scales

Most often, the purpose of making a survey is to produce some kind of plan or map. What kind of plan will result depends to a large extent on the scale to which it is drawn.

The scale of a map is the ratio between any distance on the map and the same distance on the ground. If 10 millimetres on a plan represent a ground distance of 10 kilometres, the scale would be very small and there would be little detail shown, rather like a page in an atlas. If, however, the same 10 millimetres represented only a distance of 1 metre on the ground, the scale would be large and even small details could be shown.

The scale of a map depends upon the purpose for which the map is required. When going on a motoring holiday it is desirable to be able to handle the map easily in the car and at the same time have a fair distance represented on it. A map 1 metre square might represent 200 kilometres square, which would be a fairly small scale but sufficiently large to enable the motorist to find his way around the countryside. An architect, on the other hand, when planning a house wants to be able to draw details of doors, windows, etc.; therefore, a window 1 metre wide might be shown on the plan 10 millimetres wide.

Methods of showing scale

The scale of a map or plan can be shown in three ways.

1. It may simply be expressed in words
For example, 1 centimetre represents 1 metre. By definition of scale this simply means that one centimetre on the plan represents 1 metre on the ground.

2. By a drawn scale
A line is drawn on the plan and is divided into convenient intervals such that distances on the map can be easily obtained from it. If the scale of 1 centimetre represents 1 metre is used, the scale drawn in Fig. 2.1 would be obtained.

Figure 2.1

Figure 2.1 is an example of an open divided scale in which the primary divisions (1.0 metre) are shown on the right of the zero. The zero is positioned one unit from the left of the scale and this unit is subdivided into secondary divisions.

An alternative method of showing a drawn scale is to fill in the divisions thus making a filled line scale, an example of which is shown in Fig. 2.2.

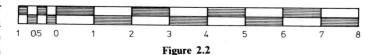
Figure 2.2

3. By a representative fraction
With this method of showing scale, a fraction is used in which the numerator represents the number of units on the map (always 1) and the denominator represents the number of the same units on the ground. With a scale of 1 centimetre represents 1 metre, the representative fraction will be 1/100, sometimes shown as 1:100, since there are 100 centimetres in 1 metre.

A representative fraction is the international way of showing scale. Any person looking at the RF on a map thinks of the scale in the units to which he is accustomed. A scale of 1:129 600 would mean to an American that 1 inch equals 129 600 inches or 2 miles; to a Japanese 1 sun equals 129 600 sun or 1 ri; and to a Continental 1 millimetre equals 129 600 millimetres or 129.6 metres.

Conversion of scales to representative fractions

Example 1 If 1 centimetre on a map represents 10 metres on the ground, the RF will be:

$$\frac{\text{Map units}}{\text{Ground units}} = \frac{1 \text{ centimetre}}{10 \times 100 \text{ centimetres}} = 1:1000$$

Example 2 If the RF of a map is 1:120, how many units on the ground do 2 map units represent?

$$RF = 1:120$$
$$\text{i.e., } 1 \text{ unit} = 120 \text{ units}$$
$$\text{therefore } 2 \text{ units} = 2 \times 120 \text{ units}$$
$$= 240 \text{ units}$$

Example 3 If the RF on a map is 1:1200 how far apart would two points be on the map if their actual ground distance is 360 metres?

The problem can be solved by using the following reasoning:

$$RF = 1:1200$$
i.e., 1 m on a map equals 1200 m on the ground. x m on a map equal 360 m on the ground or

$$\frac{1}{1200} = \frac{x}{360} \text{ metres}$$

$$\text{therefore } x = \frac{360}{1200} \text{ metres}$$

$$= 0.3 \text{ metres}$$

$$= 300 \text{ mm}$$

Conversion of areas by representative fractions

If the RF on a plan is a very large one, say $\frac{1}{4}$, the meaning assigned to the fraction, as already discovered, is that 1 unit on the plan represents 4 units on the ground. A square of 1 unit on the plan will therefore represent a ground area of (4 units × 4 units). From these facts emerges a simple formula:

$$\text{Plan scale} = 1:4$$

$$\text{Plan area} = 1 \times 1 \text{ sq. units}$$

$$\text{therefore Ground area} = (1 \times 4) \times (1 \times 4) \text{ sq. units}$$

$$= 1 \times (4 \times 4) \text{ sq. units}$$

$$= 1 \times 4^2 \text{ sq. units}$$

$$\text{i.e., Ground Area} = \text{Plan Area} \times 4^2$$
$$= \text{Plan area} \times (\text{Scale factor})^2$$

$$\text{therefore Plan area} = \frac{\text{Ground area}}{4^2}$$

$$\text{therefore } \underline{\text{Plan area} = \text{Ground area} \times (RF)^2}$$

Example 4 The scale of a plan is 1:4. If a square on the plan measures 3 by 3 units, what is the corresponding ground area?

Plan scale $= 1:4$
Plan area $= 3 \times 3 = 9$ sq. units
Ground area $= (3 \times 4) \times (3 \times 4) = 144$ sq. units

By the formula

$$\text{Plan area} = \text{Ground area} \times (RF)^2$$
$$9 = x \times \tfrac{1}{16} \text{ sq. units}$$
$$\text{therefore } x = 144 \text{ sq. units}$$

Example 5 An area was measured on a plan by a rule as 250 × 175 mm. Calculate the ground area in square metres if the scale is
(a) 1:2000
(b) 1:500

Solution
(a) RF = 1:2000

i.e., = 1 mm (plan) = 2000 mm (ground)

therefore
1 mm^2 (plan) = 2000 × 2000 mm^2 (ground)
Plan area = 175 × 250 mm^2

therefore
Ground area = 175 × 250 × 2000
$$\times 2000 \text{ mm}^2$$

$$= 175 \times 250 \times \frac{2000}{1000} \times \frac{2000}{1000} \text{ sq. m.}$$

$$= 175\,000 \text{ m}^2$$

(b) *By formula:*
Plan area $\quad = \text{Ground area} \times (RF)^2$

therefore

Ground area $\quad = (175 \times 250) \times 500^2 \times \text{mm}^2$

$$= 175 \times 250 \times \frac{500}{1000} \times \frac{500}{1000} \text{ m}^2$$

$$= 10\,937.5 \text{ m}^2$$

Check on Solution Scale (a) is four times smaller than (b); therefore there will be sixteen times the area for the same size plan.

$$10\,937.5 \times 16 = 175\,000 \text{ m}^2$$

Example 6 A plot of land was surveyed and found to have an area of 2000 m^2. If it is plotted on a plan, scale 1:500, what will be the plan area in mm^2?

Solution

RF = 1:500

i.e., 1 mm (plan) = 500 mm (ground)

therefore 1 mm^2 (plan) = 500 × 500 mm^2 (ground)

$$= \frac{500 \times 500}{1000 \times 1000} \text{ m}^2$$

$$= 0.25 \text{ m}^2$$

Ground area = 200 m^2

therefore Plan area $= \frac{2000}{0.25}$ mm^2 = 8000 mm^2

Alternatively: Plan area = Ground area × (RF)^2

$$= 2000 \times \frac{1}{500} \times \frac{1}{500} \text{ m}^2$$

$$= 2000 \times \frac{1000}{500} \times \frac{1000}{500} \text{ mm}^2$$

$$= \underline{8000 \text{ mm}^2}$$

Example 7 A plot of land is in the form of a rectangle in which the length is twice the breadth. When surveyed, it was found to have an area of 16 722.54 m^2. Calculate the length of the sides as drawn on a plan whose scale is 1:10 560

(S.A.N.C.D.)

Solution

Plan area = Ground area × RF^2

$$= 16\,722.54 \times \frac{1}{10\,560^2} \text{ m}^2$$

$$= \frac{16\,722.54 \times 1000 \times 1000 \text{ mm}^2}{10\,560 \times 10\,560}$$

$$= 149.96 \text{ mm}^2$$

Let dimensions be $2x$ by x

Then Area $= 2x^2$
therefore

$$2x^2 = 149.96 \text{ mm}^2$$
$$x^2 = 75.00 \text{ mm}^2$$

therefore $x = 8.66$ mm
therefore Plot measures $\underline{8.66 \text{ mm} \times 17.32 \text{ mm}}$

Test questions

1 Describe three ways in which the scale of a map or plan may be expressed. Outline their advantages and disadvantages.

2 An area of ground lying between a straight road and a boundary fence has an actual area of 6250 square metres.
(a) The area, as measured on an old plan whose scale has been obliterated, scales 1000 mm^2. What is the scale of the old plan?
(b) The main survey line runs along the edge of the road. If the distances from the ends of the survey line to the boundary are 36.90 and 25.60 m respectively, measured at right angles to the survey line, calculate the length of the main line.

3 A small building site has a polygonal shape as shown in Fig. 2.3. It is plotted on a 1:500 plan but the scale has been mislaid. Calculate the area in square metres after measuring the sides by ordinary rule in millimetres.

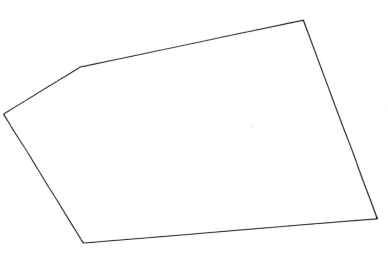

Figure 2.3

4 In a town centre redevelopment, an existing war memorial is to be replaced by a new concrete one. A model of the concrete block scale 1:5 has dimensions of 1 metre high by 500 mm wide by 500 mm long. If the concrete model weighs 500 kg calculate the dimensions and weight of the real memorial.

5 The scale of a plan is 1:200. The distance between two points X and Y on the plan measures 435 mm. Calculate the actual ground distance between the points.

6 The dimensions of a room on a 1:50 plan are 60 mm × 85 mm. Calculate the area of the room in m^2.

7 A building measures 10 m × 6 m. It is to be plotted on an A4 sheet of drawing paper to scale 1:50.

Calculate the margins at the bottom and left-hand side of the sheet if the building is to be plotted centrally on the sheet.

(The dimensions of an A4 sheet are 296 mm × 210 mm.)

8 Figure 2.4 shows an irregular parcel of ground overlain by a sheet of transparent graph paper. Each square on the graph paper is 1 cm². Estimate the number of squares covering the parcel of ground and calculate the area of ground in m² given that the plan scale is 1:2500.

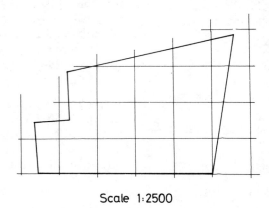

Scale 1:2500

Figure 2.4

3. Ordnance Survey maps and plans

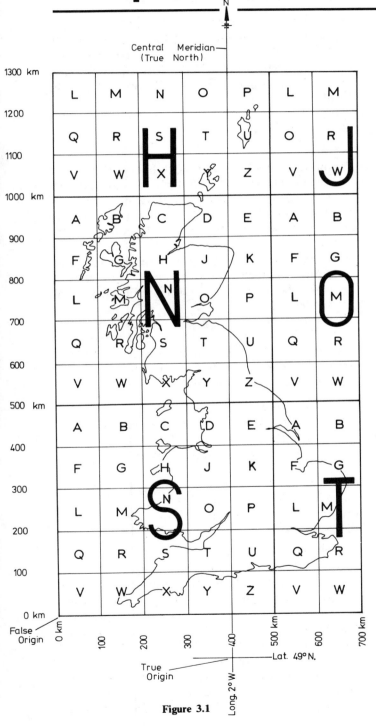

Figure 3.1

Many different maps and plans are published by the Ordnance Survey, from the very small-scale maps covering the whole country to the very large-scale maps covering only major urban areas.

In 1938, a major resurvey of Great Britain was commenced by the OS and as a result of the survey all ordnance maps have the National Grid superimposed on them. The National Grid is merely a symmetrical network of parallel lines crossing at right angles, like a sheet of graph paper. The axes of the Grid are the 2° west line of longitude and 49° north line of latitude. The intersection of these two lines forms the true origin of the National Grid (Fig. 3.1).

On the globe of the earth, all lines of latitude are parallel and run truly east to west. All lines of longitude, however, meet at the North and South poles. They all point to true north but obviously are not parallel.

Since all east–west lines of the National Grid are parallel to the 49° north line of latitude they point truly east–west. The north–south lines of the Grid are parallel to the 2° west line of longitude and cannot therefore point to true north. The direction in which they do point is called Grid North.

The 49° north line of latitude was chosen as the axis for the National Grid because all points in Great Britain lie to the north of it, while the 2° west line of longitude was chosen because it runs roughly centrally through the country. As a result it is called the Central Meridian.

The use of these two lines as axes means that grid coordinates of points west of the Central Meridian would be negative while points on the mainland of the extreme north of Scotland would have north coordinates in excess of 1000 kilometres. In order to keep all east–west coordinates positive and all north coordinates less than 1000 kilometres, the origin of the National Grid was moved northwards by 100 km and westwards by 400 km to a point south–west of the Scilly Isles. This point is called the false origin (Fig. 3.1). Any position in Great Britain is therefore known by its eastings followed by its northings which are respectively the distances east and north of the false origin.

Commencing at the false origin, a rectangular grid of 100 km square covers the country as shown in Fig. 3.1.

Each 500 km block has a prefix letter H, J, N, O, S, T (anagram ST JOHN) and comprises 25 squares of 100 km side.

Each 100 km square has a separate letter of the alphabet (I being excepted) which refers to the south–west corner of the square. The National Grid coordinates of the centre of Trafalgar Square, London, are 530030 m East and 180420 m North, therefore the point falls within the 100 km square TQ. In order to quote the coordinates of Trafalgar Square to the nearest metre, the 100 km square is quoted first followed by the remainder of the east coordinate and finally the remainder of the north coordinate, that is

$$TQ\ 3003080420$$

Example 1 The National Grid coordinates of Ben Nevis are 216745E, 771270N while those of Cardiff Castle are 318100E and 176610N.

Quote (i) the NG 1 metre and (ii) the NG 100 metre reference of each point.

Solution

Ben Nevis   (i) NN1674571270 (ii) NN167712
Cardiff Castle (i) ST1810076610 (ii) ST181766

Scales of OS maps

Ordnance Survey maps can be conveniently divided into two classes.

1. *Small-scale maps:*
   RF   1:625 000
                 1:250 000
                 1:50 000
                 1:25 000

             $2\frac{1}{2}$ inches to 1 mile (approx.)

2. *Large-scale maps (plans):*
   1:10 000   6 inches to 1 mile (approx.)
   1:2500   25 inches to 1 mile (approx.)
   1:1250   50 inches to 1 mile (approx.)

The 1:25 000 map can be considered as being the smallest map of value to the builder. It is a convenient map to use as a base and is shown in Fig. 3.2(b).

It is important to notice the difference between a map and a plan. A plan will accurately define widths of roads, sizes of buildings, etc.; in other words, every feature is exactly true to scale. A map, on the other hand, is a representation, no matter how accurately it may be drawn.

As an example, a winding country road about the width of a car, measures almost 1 millimetre on a 1:50 000 map. This represents 50 metres, far in excess of the actual width of the road.

Of the Ordnance Survey productions only the 1:1250 and 1:2500 can be considered to be plans in the strict sense of the word.

Map references

It has already been shown in Fig. 3.1 that the major blocks of the National Grid have sides of 100 km. Thus block NS is bounded on the North and South by the 700 km and 600 km Northing lines and on the West and East by the 200 km and 300 km Easting lines.

Figure 3.2(a) shows this block subdivided into one hundred blocks, 10 km square.

1:25 000 scale map
Taking that 10 km square block bounded by Grid lines 670 km North, 660 km North, 250 km East, and 260 km East (Fig. 3.2(b)), and drawing it to scale 1:25 000, produces a square of 400 mm side. This is the typical format for the small-scale 1:25 000 maps.

Points of detail are shown by conventional signs, variations in height by contours at 5 m vertical intervals, and the National Grid is superimposed at 1 km intervals.

Each map has a unique reference referred to its south-west corner. Since the side of the map represents 10 km, the reference is given to the 10 km figure. Thus the reference of the map shown in Fig. 3.2(b) is derived as follows:

     South-west corner
        250 000 m East 660 000 m North
   = 250 km East   660 km North
   10 km Grid reference = NS 56

The 1:25 000 map is the smallest scale map of value in the construction industry. It is the base map on which the various series of larger scale maps and plans are built.

1:10 000 scale maps
Figure 3.2(c) shows a ground square of 5 km side drawn to a scale of 1:10 000 to produce the format for the 1:10 000 scale series of maps. Certain Town and Country Planning matters and Development proposals are shown on this scale. The map really represents one-quarter of the area of the 1:25 000 scale map shown in Fig. 3.2(b).

Details on these maps are shown true to scale and only the widths of narrow streets are exaggerated. Surface relief is shown by contour lines at 10 metre

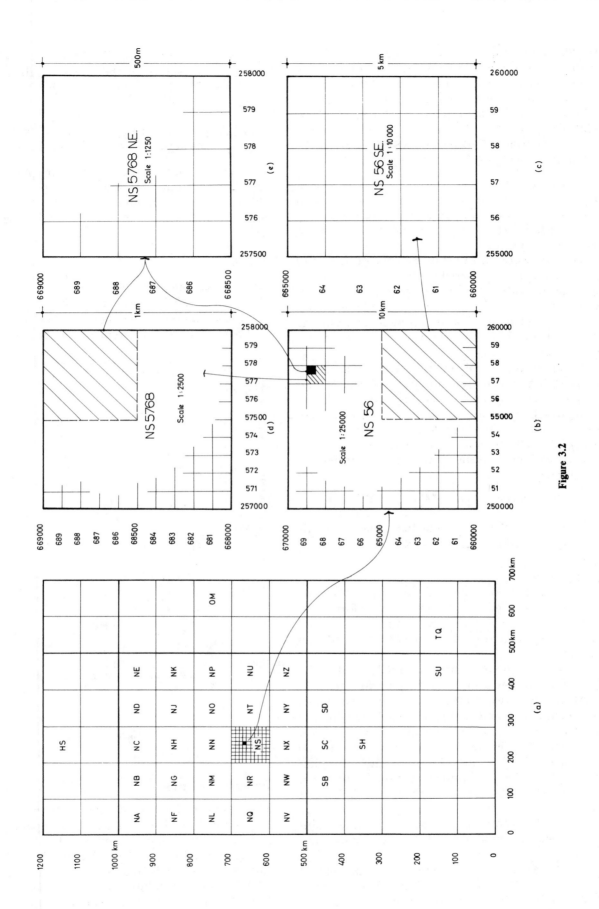

Figure 3.2

vertical intervals in mountainous areas and 5 metre vertical intervals in the rest of the country. the National Grid is superimposed at 1 km intervals.

The map is referenced as a quadrant of the 1:25 000 map of which it is a part. Thus the 1:10 000 scale map of Fig. 3.2(c), being the south-east quadrant of the 1:25 000 map NS 56, is given the reference NS 56 SE.

1:2500 scale plan

Figure 3.2(d) shows a 1 km square taken from 1:25 000 map NS 56 and enlarged to scale 1:2500 to produce the format for the 1:2500 National Grid series of plans.

The whole of Great Britain, except moorland and mountain areas, is covered by this series. All details are true to scale, and areas of parcels of land are given in acres and hectares. Surface relief is shown by means of bench marks and spot heights, and the National Grid is superimposed at 100 metre intervals.

This plan is probably the mnost commonly used plan in the construction industry. Most site and location plans are shown to this scale. Facsimile examples 1:2500 plans are given in this chapter— Figs 3.3 and 3.7. The plan reference is again made to the south-west corner and is given to 1 km. Thus the reference of the map shown in Fig. 3.2(d) is derived as follows:

South-west corner = 257 000 m   668 000 m
                                    East   North
                   = 257 km East 668 km North
    1 km Grid reference = NS 5768.

1:1250 scale map

Figure 3.2(e) shows a ground square of 500 m side drawn to a scale of 1:1250 to produce the format for the 1:1250 National Grid series of plans. The plan represents one-quarter of the area of the 1:1250 plan of Fig. 3.2(d).

These maps are the largest scale published by the Ordnance Survey and cover only urban areas.

All details are true to scale. Surface relief is shown by bench marks and spot heights and the National Grid is carried at 100 m intervals.

The plan is referenced as a quadrant of the 1:2500 plan of which it is a part. Thus the 1:2500 scale plan of Fig. 3.2(e), being the north-east quadrant of the 1:2500 plan NS 5768, is given the reference NS 5768 NE.

It has been found that it is more convenient and economical to produce the 1:25 000 maps and 1:2500 plans in pairs.

The large-scale 1:2500 plan covers an area of 2 kilometres east–west by one kilometre north–south. The grid line forming the western edge of the sheet

is always an even number. Thus plan NS 5768 (Fig. 3.2(d)) would be the eastern half of sheet NS 5668–5768. This reference may be shortened to read NS 56/5768.

Example 2 Figure 3.3 shows part of a 1:2500 O.S. sheet measuring 2 km by 1 km. Quote:

(a) The reference numbers of the 1 km plans which together constitute this sheet.
(b) The reference numbers of the 1:1250, 1:25 000, and 1:10 000 scale maps which show the same area.

Solution

(a) NS 5267 and NS 5367.
(b) Scale 1:1250—reference number NS 5267 SW
     Scale 1:25 000—reference number NS 56
     Scale 1:10 000—reference number NS 56 NW

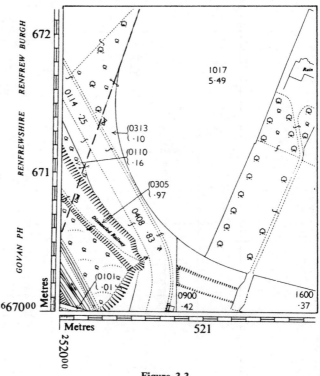

Figure 3.3

Types of OS maps

The maps produced by the Ordnance Survey Department are best described as topographical maps. As the name implies, they show the topography of the ground, that is, the hills and valleys, rivers, roads and man-made features.

The large-scale plans are often referred to as Cadastral maps though this is not strictly true. The word Cadastral means that the plans show property boundaries, when in fact they show the physical boundaries of property, for example, fence lines, boundary walls, etc. The physical boundary is not necessarily the legal boundary, and to this extent the name is misapplied.

In order to read and understand ordnance maps, it is necessary to understand scales, grid references, and directions—all of which have been mentioned. The methods of showing the rise and fall of the ground (or surface relief) and the conventional signs for features on the map must also be mastered.

Depiction of surface relief

On the large-scale and small-scale maps, there are three methods of showing surface relief.

1. By contour lines

A contour line is a line joining all points of equal height. The 1:10 000 scale map is the largest scale which shows contours, the interval being 10 m, 5 m or the metric equivalent of 50 ft depending upon the nature of the terrain and available survey information.

On other small-scale maps the interval varies with the scale and series of the map, e.g., on some of the second series 1:50 000 scale maps the contour interval is 10 m, while on others the interval is the metric equivalent of 50 ft.

2. By spot levels

As the name would imply, a spot level is simply a dot or spot on the map, the level of which is printed alongside, and this is the method of showing variations in elevation on the large-scale maps. These levels are shown to the nearest 100 mm, that is, one decimal place.

3. Bench marks

Bench marks (Fig. 3.4) are permanent marks chiselled on buildings showing the height of that particular point above Ordnance Datum. Sometimes the bench mark is a bronze plaque let into the wall on which an arrow records the height.

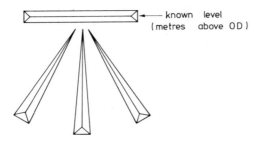

known level (metres above OD)

Figure 3.4

Ordnance Datum is the point of reference to which all bench marks, spot levels, and contour lines are referred, and is the mean level of the sea as recorded over many years at Newlyn Harbour, Cornwall.

The values of these bench marks are shown to the nearest centimetre on the 1:2500 and 1:1250 sheets. The latest series of these sheets show only the positions and not the values of the bench marks. In order to find the value of a particular bench mark, it is necessary to scale its coordinates from a plan and consult a bench mark list published by the OS where the location and value of the bench mark may be read off against its coordinates.

Conventional signs

Since it is impossible truly to represent each individual feature on any map, a code of signs has been devised for use with each scale of OS map. As far as possible, the signs are facsimiles of the actual object, e.g., conifers and deciduous trees, cliffs, and quarries; but very often recourse is made to simply printing the initials of the feature, e.g., T for telephone kiosk, and PH for public house.

A selection of signs and abbreviations used on the 1:2500 and 1:1250 sheets is shown in figures 3.5 and 3.6.

Parcel numbers and areas on 1:2500 sheets

In the context of Ordnance Survey maps, a parcel is an area of ground which may be a single feature, e.g., a field, or it may consist of several small areas grouped together, e.g., some small islands in a river or a combination of a field, ditch and wood.

Each parcel is given a four-figure reference number representing the National Grid 10 metre reference of the centre of the parcel. Thus in Fig. 3.7 the wooded area has a NG reference 1084 derived from the coordinates 5410E 3984N.

Where a sheet edge cuts across any particular parcel, part of the reference number is the mid-point of that sheet edge intercepted by the parcel. Thus in Fig. 3.7 the area shown cross-hatched has a reference number 0082 derived from coordinates 5400E 3982N.

Any parcel lying in the corner of a sheet is given an arbitrary reference number from 0001 to 0006. This system is necessary since obviously two sheet edges are intercepted by the parcel and the mid-points of both sheet edges would not necessarily lie within the parcel. Thus the reference number of the corner parcel in Fig. 3.7 is 0002, whereas the number derived from the mid-points of the parcel sheet

Scale 1:2500

Beer House	BH	Guide Post	GP	Signal Bridge	S Br
Bench Mark	BM	Letter Box	LB	Signal Post	SP
Boundary Post	BP	Level Crossing	LC	Spring	Spr
Boundary Stone	BS	Mile, or Mooring Post	MP	Stone	S
Capstan, Crane	Cn, C	Mile Stone	MS	Sundial	SD
Cart Track	CT	Pillar, Pole, Post or Pylon	P	Tank or Track	Tk
Drinking Fountain	D Fn	Police Call Box	PCB	Telephone Call Box	TCB
Electricity Pillar, Post or Pylon	El P	Police Telephone Pillar	PTP	Traverse Station	ts
Fire Alarm Pillar	FAP	Post Office	PO	Water Point	Wr Pt
Flagstaff	FS	Public House	PH	Water Tap	Wr T
Foot Bridge	FB	Pump	Pp	Weighbridge	WB
Footpath	FP	Revision Point	rp	Well	W
Fundamental Bench Mark	FBM	Signal Box	SB	Wind Pump	Wd Pp
High or Low Water Mark of Medium Tides	H or LWMMT			High or Low Water Mark of Ordinary Spring Tides	H or LWMOST

Figure 3.5

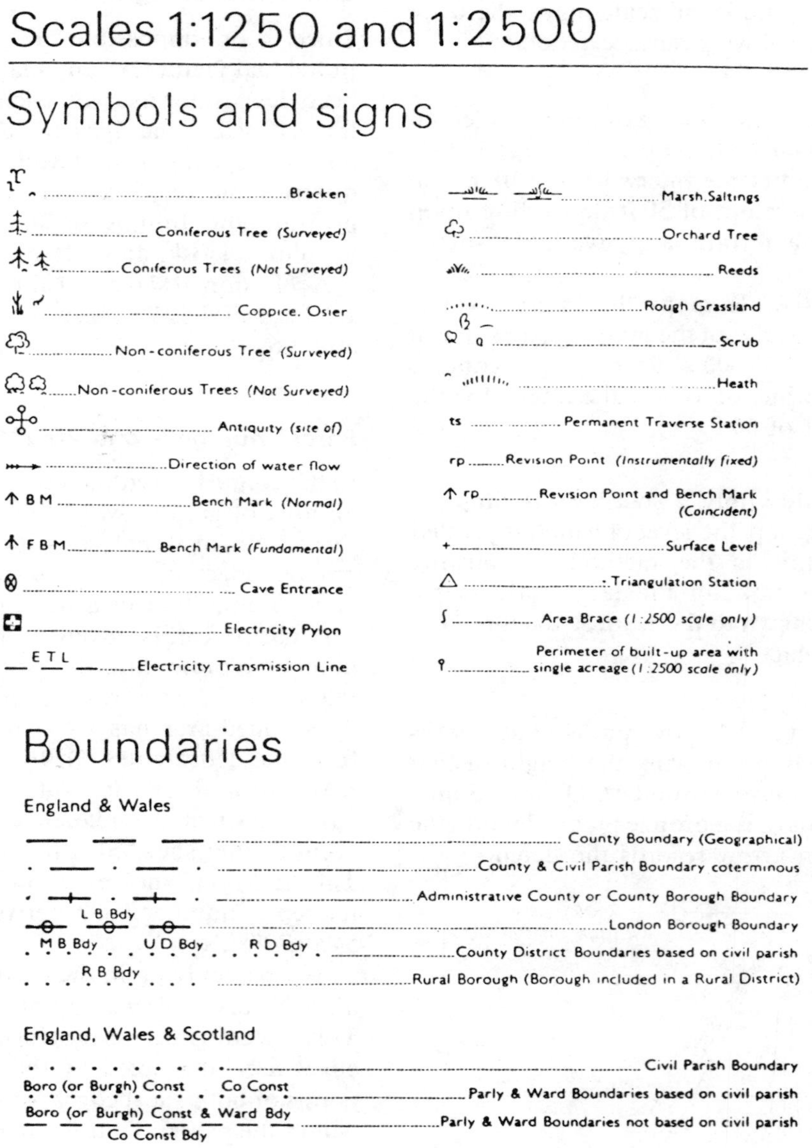

Scales 1:1250 and 1:2500

Symbols and signs

Bracken	Marsh, Saltings	
Coniferous Tree (Surveyed)	Orchard Tree	
Coniferous Trees (Not Surveyed)	Reeds	
Coppice, Osier	Rough Grassland	
Non-coniferous Tree (Surveyed)	Scrub	
Non-coniferous Trees (Not Surveyed)	Heath	
Antiquity (site of)	ts — Permanent Traverse Station	
Direction of water flow	rp — Revision Point (Instrumentally fixed)	
↑ BM — Bench Mark (Normal)	↑ rp — Revision Point and Bench Mark (Coincident)	
↑ FBM — Bench Mark (Fundamental)	Surface Level	
Cave Entrance	△ — Triangulation Station	
Electricity Pylon	∫ — Area Brace (1:2500 scale only)	
ETL — Electricity Transmission Line	Perimeter of built-up area with single acreage (1:2500 scale only)	

Boundaries

England & Wales

	County Boundary (Geographical)
	County & Civil Parish Boundary coterminous
L B Bdy	Administrative County or County Borough Boundary
	London Borough Boundary
M B Bdy U D Bdy R D Bdy	County District Boundaries based on civil parish
R B Bdy	Rural Borough (Borough included in a Rural District)

England, Wales & Scotland

	Civil Parish Boundary
Boro (or Burgh) Const Co Const	Parly & Ward Boundaries based on civil parish
Boro (or Burgh) Const & Ward Bdy / Co Const Bdy	Parly & Ward Boundaries not based on civil parish

Figure 3.6

(continued)

Scotland

(Not with parish)	(Coincident with parish)	
Co Cnl Bdy	Co Cnl Bdy	....County Boundary (Geographical)
Co of City Bdy	Co of City Bdy	....County Council Boundary
Burgh Bdy	Burgh Bdy	....County of the City Boundary
Dist Bdy	Dist Bdy	....Burgh Boundary
		....District Council Boundary

Examples of Boundary Mereings

FF RH	..................................Symbol marking point where boundary mereing changes
Und	..................................Undefined boundary
Def	..................................Original boundary feature destroyed or defaced

C B	Centre of Bank	E K	Edge of Kerb
C C	Centre of Canal. etc.	F F	Face of Fence
C D	Centre of Ditch. etc.	F W	Face of Wall
C R	Centre of Road. etc.	S R	Side of River. etc.
C S	Centre of Stream. etc.	T B	Top of Bank
C O C S	Centre of Old Course of Stream	Tk H	Track of Hedge
C C S	Centre of Covered Stream	Tk S	Track of Stream
4ft R H	4 feet from Root of Hedge		

Figure 3.6 (continued)

SPECIMEN SHOWING METHODS OF BRACING AND NUMBERING OF AREAS

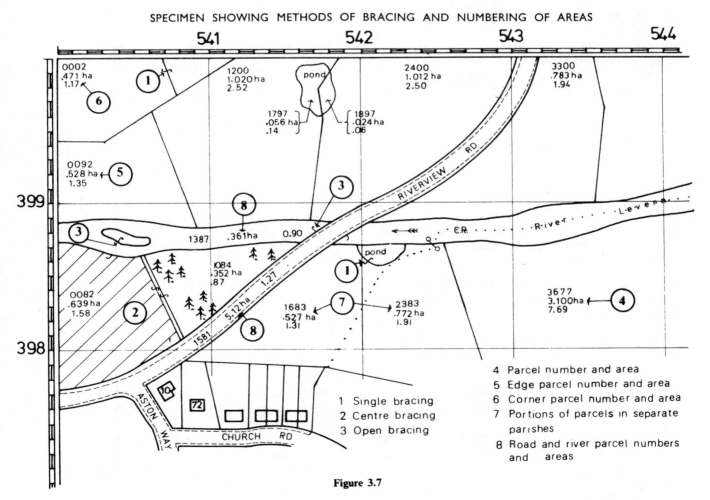

4 Parcel number and area
5 Edge parcel number and area
6 Corner parcel number and area
7 Portions of parcels in separate parishes
8 Road and river parcel numbers and areas

1 Single bracing
2 Centre bracing
3 Open bracing

Figure 3.7

edges would be 0696. When plotted this point is only just within the area.

On the latest series of 1:2500 sheets the area of each parcel is given in hectares and in acres. Both of these figures are printed immediately below the reference number. The wooded area is therefore parcel number 1084 of area 0.352 hectares or 0.87 acres.

Whenever possible, parcel boundaries are natural features such as hedges or streams. Thus the wooded area 1084 is bounded by the river, hedge and road. However, wherever an administrative boundary, e.g., parish or district boundary, divides an enclosure, each part is treated as a separate parcel. Parcel numbers 1683 and 2383 are parts of the same field, divided by a parish boundary into areas of 0.527 hectares and 0.772 hectares respectively.

Whenever a parcel is composed of two or more separate areas of land, they are joined on the sheet by an elongated symbol called a bracing. Figure 3.7 shows several types of bracing, each of which is self-explanatory.

Example 3 Using Fig. 3.7 determine:
(a) The area of the northernmost pond.
(b) The parcel number in which lies the island in the river Levern.
(c) The coordinates of the eastern extremity of parcel number 3300.
(d) The area of the field with coordinates 5420 3982 as approximate centre.

Solution
(a) 0.08 ha (b) 0082 (c) 5436 3996 (d) 1.299ha

Test Questions

1 The reference number of an OS map scale 1/2500 is SK5265. List the reference numbers of the plans immediately adjoining it.

2 Give the reference number of the 1:1250 plan containing the point 533545E and 180202N (Tower Bridge, London).

3 Make a diagram to show the relationship between the True North and Grid North meridians at a point whose NG reference is SN5461.

4 Refer to Fig. 3.7. Calculate the areas of the plots of ground at 10 Aston Way and 72 Church Road.

5 Figure 3.8 shows diagrammatically the 1.2500 scale Ordnance Survey plan SK 5265. List the reference numbers of the eight plans immediately adjacent to it.

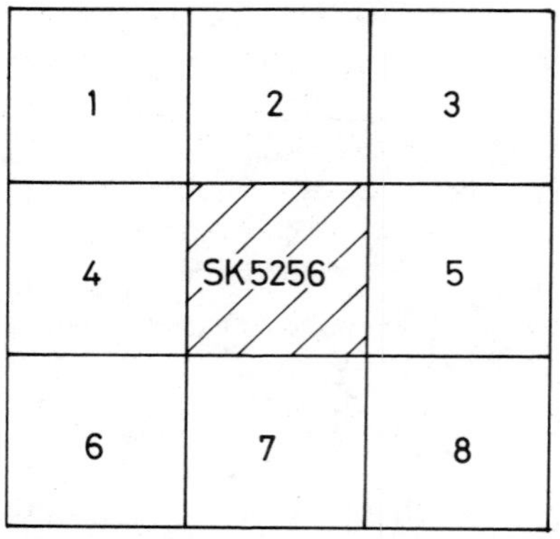

Figure 3.8

6 Figure 3.9 shows diagrammatically the 1.1250 scale Ordnance Survey plan SK 3657SE. List the reference numbers of the fifteen plans adjacent to it.

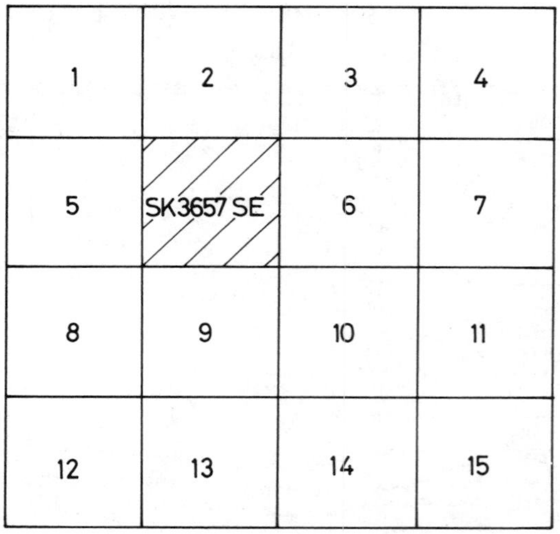

Figure 3.9

7 Figure 3.10 is a facsimile 1.2500 scale OS plan.
(a) Calculate the area of the pond in hectares and acres.
(b) State the boundaries of parcel number 1565.
(c) Describe the parish boundary running diagonally across the plan.
(d) State the parcel number which includes the island in the river Eden.
(e) List the conventional signs and abbreviations which appear on the plan numbered 1 to 10.

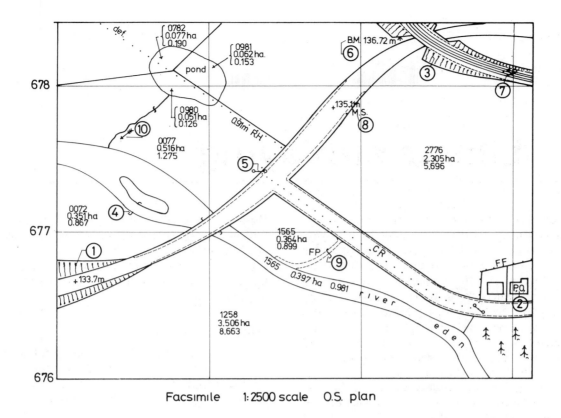

Facsimile   1:2500 scale   O.S. plan

Figure 3.10

4. Linear surveying

Linear surveying, or chain surveying, is probaby the most elementary method of surveying, relying purely on the accurate determination of length. No attempt is made to obtain angular measurements.

The principle of *trilateration* is employed on linear surveys. The area to be surveyed is divided into a series of well-shaped triangles all of whose sides are measured. As a triangle is the simplest geometrical figure to plot, a plan can be readily and speedily constructed from a trilateration survey.

Before studying the procedure used in linear surveying, it is necessary to understand fully, the new metric measuring instruments, the errors which can arise from their use, and the techniques required to obtain an accurate measurement over rough undulating ground often in adverse climatic conditions.

Linear measuring equipment

All linear measuring equipment should conform to the British Standard Specification 4484:1969. Such instruments will then have clear legible graduations and unambiguous figuring. The instruments required on a linear survey vary from the simple folding 1-metre rule to the heavy 20-metre steel wire land chain.

Table 4.1 shows the lengths, graduations and method of figuring of the principal linear measuring instruments.

Folding boxwood rule

The 1-metre folding rule is made from high quality boxwood with one central brass joint and two folding hinges. The rule is made in such a way that the fine graduations are always protected (Fig.4.1).

Steel pocket rule

The rule is contained in a strong lightweight case, which is 50 mm wide at the base to enable inside measurements to be taken easily. The rule is graduated on one side only with fine graduations on the bottom edge. The top edge shows 5 mm coarse graduations. The 1 metre graduation mark is shown in contrasting colour and a quick reading repeater figure, 1 m, occurs at every 100 mm to facilitate reading, without scanning the instrument (Fig. 4.2).

Measuring instrument	Length (metres)	Graduations			Method of figuring
		Major	Inter	Fine	
Folding rule	1	10 mm	5 mm	1 mm	10 mm intervals using 3-digit numbers
Folding and multi-folding rods	1, 1.5, and 2	10 mm	5 mm	1 mm	Ditto
Steel pocket rule	2	10 mm	5 mm	1 mm	Ditto
Steel tapes	10, 20 30	100 mm	10 mm	5 mm	100 mm intervals in decimals of a metre
		First and last metres further subdivided into divisions of			
Synthetic tapes	10, 20, 30	100 mm	50 mm	10 mm	50 mm marks are denoted by arrows
Land chains	20	1 m	—	200 mm	Yellow tallies at 1 metre intervals. 5 and 10 m shown by red tallies

Table 4.1

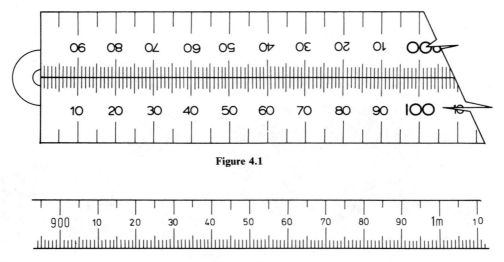

Figure 4.1

Figure 4.2

Steel tapes

These tapes are made from hardened tempered steel painted with a durable white enamel. The tapes are 10 mm wide with black and red figures and black graduations. Alternatively, the tape may be sheathed in a heavy-duty nylon or plastic, making the tape very tough and the graduations almost indestructible. The tapes are housed in open-frame plastic winders fitted with a quick rewind handle. The tapes are subdivided in millimetres throughout, figured at every 10 mm with quick-reading metre figures in red at every 100 mm and whole metre figures are in red (Fig. 4.3).

Synthetic tapes

These tapes are manufactured from multiple strands of fibreglass and coated with PVC. Fibron is impervious to water and can be wound back into the case without damage even when wet. The tapes are graduated throughout in metres and decimals, the finest graduation being 10 mm (Fig. 4.4).

Land chains

Land chains are manufactured from hardened and tempered black enamelled steel wire or galvanized wire. The chains may be 20 m or 30 m long and consist of a series of wire links connected by three small oval links. They are fitted with heavy brass swivel handles which are included in the measurement as shown in Fig. 4.6.

The chain is not a common measuring tool and is used mainly in agricultural and forestry surveys.

Use of the tape

Figure 4.7(a) shows a survey line AB marked on the ground by two pegs. The distance AB is shorter than one length of tape (or chain). The measurement of the line AB is obtained by unreeling the tape and straightening it along the line between the pegs. The zero point of the tape (usually the end of the handle) is held against station A by the rear tape man (called the follower). The forward end of the tape is read

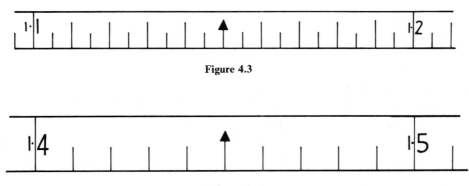

Figure 4.3

Figure 4.4

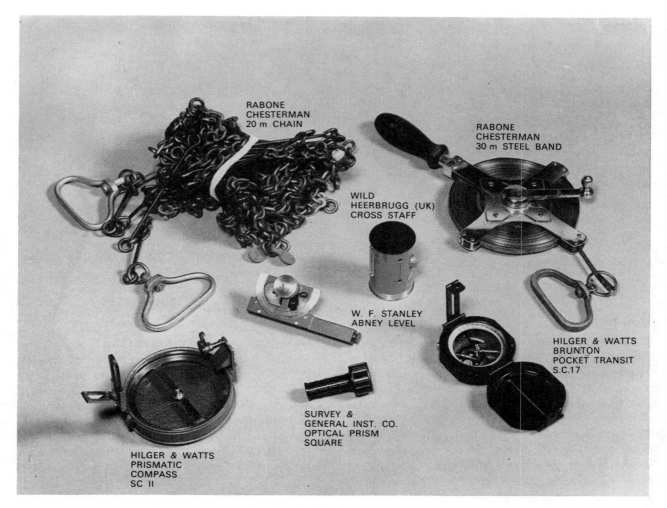

RABONE
CHESTERMAN
20 m CHAIN

WILD
HEERBRUGG (UK)
CROSS STAFF

RABONE
CHESTERMAN
30 m STEEL BAND

W. F. STANLEY
ABNEY LEVEL

HILGER & WATTS
BRUNTON
POCKET TRANSIT
S.C.17

SURVEY &
GENERAL INST. CO.
OPTICAL PRISM
SQUARE

HILGER & WATTS
PRISMATIC
COMPASS
SC II

Figure 4.5

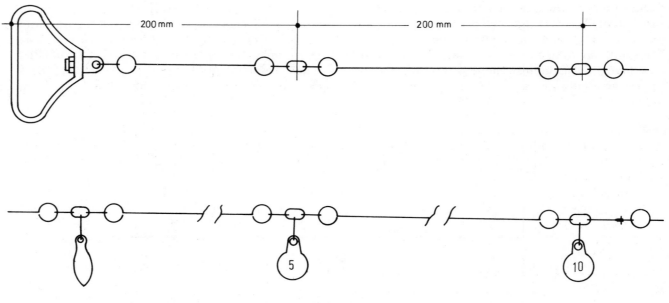

200 mm

200 mm

5

10

Figure 4.6

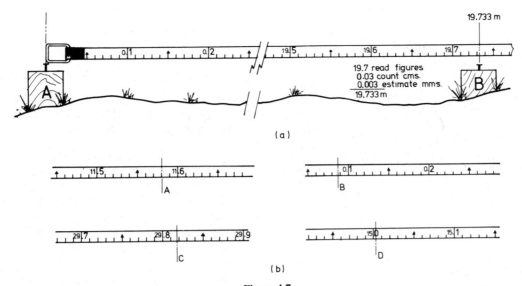

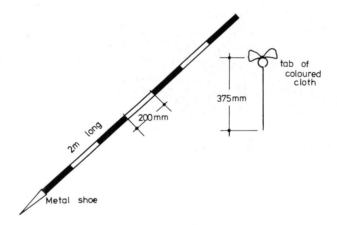

Figure 4.7

against station B by the forward tape man (called the leader) after he has carefully tightened it.

Example 1 Figure 4.7(b) shows tape readings A, B, C, and D. Read the metres and decimals from the tape, count the centimetres and estimate the millimetres to give the correct readings at each point.

Solution
A—11.580 m   B—0.088 m   C—29.818 m
D—15.003 m

On surveys most of the lines will be considerably longer than one tape or chain length and a sound operational technique is required. Two ancillary pieces of equipment are necessary, namely, ranging rods and marking arrows.

Ranging rods are 2-metre-long, round, wooden poles, graduated into 200 mm divisions and painted alternately red and white. They have a pointed metal shoe for penetration into the earth.

Marking arrows are made from steel wire, 375 mm long, pointed at one end and having a 30 mm loop at the other, to which is tied a piece of brightly coloured cloth. They are made up in sets of ten. Both instruments are shown in Fig. 4.8.

Two surveyors are required to measure a long line. The leader's job is to pull the tape in the required direction and mark each tape length. He carries with him a known number of arrows and a ranging rod. The follower's job is to align the tape and count the tape lengths.

In Fig. 4.9 a line AB is to be measured across a gently sloping grassy field.

Figure 4.8

The follower holds the zero end of the tape against station A and the leader pulls the tape towards station B. When the tape has been laid out the leader holds his ranging rod vertically approximately on the line. The follower signals to him to move it until it is exactly on the line AB. The tape is tightened between the newly erected rod and station A and an arrow is pushed into the ground at the 20 mark of the tape.

The follower moves forward to this new point and the whole procedure is repeated for the remainder of the line until station B is reached. The follower gathers the marking arrows and the number of tape lengths measured is the number of arrows carried by the follower. The portion of tape between the last arrow and station B is then measured and added to the number of complete tape lengths to produce the total length of the line.

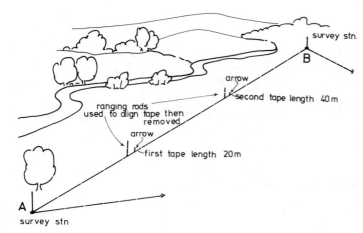

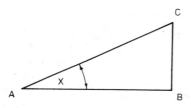

Figure 4.9

Inclined measurements

When any measured distance is to be shown on a plan, the horizontal distance is required and any inclined distance must be converted to its horizontal equivalent before plotting.

Figure 4.10 shows a survey line measured between two stations A and C. The line is not horizontal. Trigonometrically, the inclined distance is the hypotenuse of a right-angled triangle ABC.

Figure 4.10

In $\triangle ABC$

$$AB/AC = \cos x$$

therefore $AB = AC \times \cos x$

i.e., plan length = slope length $\times$ cos inclination.

The angle of inclination is measured in the field using some form of clinometer, the most common instrument being the Abney level.

Abney level

This instrument consists of a square-section sighting tube, 127 mm long, fitted with a draw tube extension, which extends to 178 mm. The draw tube is fitted with a pin-hole sight, and a horizontal cross wire, at the viewing end, completes the sighting arrangements.

Screwed to the rectangular sighting tube is a semi-circular arc graduated in degrees and read by a vernier. A spindle running through the arc carries a bracket holding a spirit level. A highly polished mirror is fitted within the sighting tube at an angle of 45° to the line of sight, allowing the observer to view simultaneously the spirit level via the mirror and some distant target against the crosshair.

To measure an angle of inclination the Abney level is placed to the eye so that the bubble is apparent in the mirror. The sighting tube is tilted to observe the forward station and the slow motion screw controlling the spirit level is activated until the view shown in Fig. 4.11 is obtained. The spirit level will now be in the centre of its run and the vernier will have been moved by the slow motion screw over the graduations giving the angle of inclination to 10 minutes of arc.

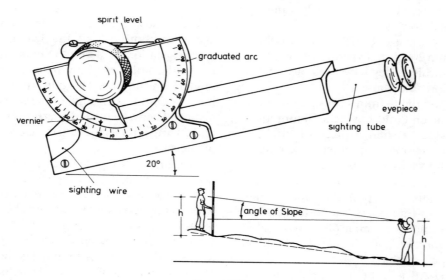

Figure 4.11

Step taping

An alternative method of obtaining horizontal measurements, without using angle-measuring instruments, is that known as step-taping. This is a field method where the horizontal distances are obtained directly.

Three men are required, one leader, one follower, and one observer. The follower's duties are exactly as before; namely aligning the leader, holding the end of the tape on the marks left by the leader, and collecting the arrows.

Once aligned by the follower, the leader holds the tape horizontally. Considerable tension is required to straighten the tape and avoid sagging. The horizontal position is estimated by the observer, who signals to the follower and leader when this position has been attained. On receiving his signal, the leader drops a drop-arrow (a marker arrow with a lead weight attached) from the handle of the chain, thus transferring the horizontal distance to the ground (Fig. 4.12). Alternatively, a plumb-bob or plumbed ranging rod may be used for the transference.

The length of steps which can be adopted is limited by the gradient. At no time should the tape be above the leader's eye level, because plumbing becomes very difficult. As the gradient increases the length of step must therefore decrease. The maximum length of unsupported chain should not exceed 8 metres.

When the observer has noted the length of the first step in his book, the second and third steps are measured and the procedure repeated until the whole line has been measured. The summation of the steps will produce the required horizontal distance.

Step chaining can also be performed when measuring uphill. In this case, the follower has to hold the handle of the tape above the mark left by the leader. For this purpose he uses a plumb-bob.

Simultaneously therefore, the follower is applying tension to the tape horizontally and plumbing the handle above a ground mark. The observer may assist in the plumbing, but generally it is more difficult to step-chain uphill than down.

Errors in measurement

All measurements made with tapes are subject to some form of error, no matter how carefully any line is measured. The error may be due to the surveyor's carelessness or inexperience; it may be due to physical site conditions; or it may be inherent in the instrument being used.

Errors are divided into three classes.

1. Gross errors

Gross errors are blunders. They arise from inexperience, carelessness or lack of concentration on the part of the observer. Examples of gross errors are:

(a) Misreading the tape graduations. This is the most common error in survey operations, e.g., six metres and forty millimetres is 16.040 m not 16.400 m.

(b) Miscounting the number of tape lengths. During the measurement of long lines the surveyor may lose count of the number of tape lengths measured. A careful count of the number of arrows used on a long line should be made before and after the measurement of the line in an attempt to minimize the occurrence of this particular error.

(c) Booking errors. The booker may simply write down the wrong measurement particularly in windy wet weather when it is difficult to hear and write.

All of these errors may be detected by measuring every line twice.

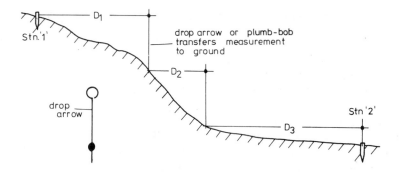

Figure 4.12

2. Constant errors

Constant errors are those which occur no matter how often a line is measured and checked. The error will always be of the same sign for any one tape or for any given set of circumstances. Examples of constant errors are:

(a) *Misalignment of the tape* This error is perhaps the simplest which demonstrates this class of error. If any line is to be measured between two points it must be measured in a straight line. Suppose, however, that a tree stump is in the direct line of measurement, and the distance is simply measured around it, producing a deviation, *e*, from the line, at the obstruction. The length of line measured will be too long no matter how often the line is measured and regardless of the number of chains or tapes used.

By using the theorem of Pythagoras and a binomial extension the error can be shown to be

$$e^2/L$$

where *e* is the amount of deviation of the tape and *L* is the total measured length.

Example 2 A survey line PQ is measured round a large tree stump causing a deviation from the straight line of 1.0 m. The measured length PQ is 110.65 m. Calculate the correct length of the straight line PQ.

Solution

Measured length	=	110.65 m
Deviation $e^2/L = 1.0^2/110.65$	=	0.01 m
therefore Correct length	=	110.64 m

From the example it is evident that the error is not serious even for such a large deviation. In a linear survey, however, offsets have to be measured from the line and complications arise when the tape is not straight.

(b) *Standardization* It is very important, before commencing a survey, that the measuring instrument being used is exactly the right length. It must be compared with some standard length, probably a new tape kept solely for that purpose. If the tape or chain is not the standard length it will give a wrong measurement. Chains especially are prone to this error because of stretching of the small oval links.

Suppose a tape is used to measure a room and the width is found to be 10 m exactly. If a knot is tied in the tape, obviously making it shorter, and the room is again measured, the width will appear to be greater than 10 m. This shows that a short tape, when used to measure distances, will produce

lengths which are greater than the correct length, and conversely a long tape will produce lengths which are less than the correct length. The correction to any measured length is found from the formula:

$$c = (L - l) \text{ per tape/chain length}$$

where
c = correction
L = actual length of tape
l = nominal length of tape.

Example 3 A line AB is measured using a tape of nominal length 20 m and is found to be 65.32 m long. When checked against a standard, the tape was found to be 50 mm too long. Calculate:
(a) the correction to the length AB;
(b) the correct length AB.

Solution

Nominal length (l) of tape $= 20.00$ m

No. of tape lengths in line AB $= \dfrac{65.32}{20.00} = 3.266$

Actual length (L) of tape $= 20.05$ m

Correction c $= (20.05 - 20.00) \times 3.266$

$= +0.16$ m

Correct length AB $= (65.32 + 0.16)$ m

$= \underline{65.48 \text{ m}}$

Example 4 A survey line XY is measured using a 30 m tape which owing to previous damage is actually 40 mm short. The line measured 147.36 m. Calculate:
(a) the correction to the measured length XY;
(b) the corrected length of line XY.

Solution

Nominal length (l) of tape $= 30.00$ m

Actual length (L) of tape $= 29.96$ m

No. of tape lengths $= 147.36/30.0 = 4.912$

Correction c $= (29.96 - 30.00) \times 4.912$

$= -0.20$ m

therefore

Correct length XY $= 147.36 - 0.20 = \underline{147.16 \text{ m}}$

(c) *Slope* All distances measured on a survey are slope lengths and must be converted into plan lengths before plotting. Slope lengths are longer than plan lengths hence a constant error will be made by ignoring the inclination.

The formula (derived previously) for converting slope to plan lengths is as follows:

Plan length = Slope length × cos angle of inclination

Example 5 A survey line AB was measured along a 6° gradient. The slope distance was 49.75 m. Calculate the plan length of the line.

Solution

Plan AB = Slope AB × cos 6°

= 49.75 × cos 6°

= 49.48 m

Some survey lines will have several changes of gradient along their lengths. In such cases each inclined section is treated separately and its plan length is calculated as in Example 5. All of the plan lengths are then added to produce the total horizontal length of the line.

Example 6 Figure 4.13 shows a straight line ABCD which has three distinct changes of gradient along its length. Each gradient was obtained by Abney level. Calculate the plan length of the line AD given the following field results:

Line	Section	Slope length (m)	Inclination
AD	AB	84.40	−5.0°
	BC	47.21	+2.0°
	CD	39.47	+6.5°

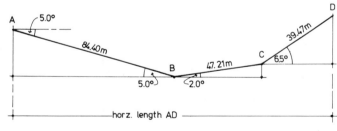

Figure 4.13

Solution

Plan length AB = 84.40 × cos 5° = 84.08
  BC = 47.21 × cos 2° = 47.18
  CD = 39.47 × cos 6.5° = 39.22

Total length AD = 170.48 m

In practice it is likely that a combination of errors will be present and separate corrections will have to be made for standardisation and slope.

Example 7 A survey line XY is measured along a steep gradient of 8° by two survey teams A and B.

Team A used a 30 m fibron tape which when checked was found to have shrunk by 50 mm. The measured length was 85.24 m.

Team B used a 20 m steel chain which was actually 100 mm too long. Their measured length was 84.60 m.

Calculate the most probable horizontal corrected length XY.

Solution

Team A
Actual tape length (L)	= 29.95	
Nominal tape length (l)	= 30.00	
($L - l$)	= −0.05	
No. of tape lengths	= 85.24/30 = 2.841	
Correction	= −0.05 × 2.841	
	= −0.14 m	
Corrected length	= 85.24 − 0.14	
	= 85.10 m	

Team B
Actual chain length (L)	= 20.10	
Nominal length (l)	= 20.00	
($L - l$)	= +0.10	
No. of chain lengths	= 84.60/20 = 4.230	
Correction	= +0.10 × 4.230	
	= +0.42	
Corrected length	= 84.60 + 0.42	
	= 85.02 m	
Average slope length	= ½ (85.10 + 85.02)	
	= 85.06 m	
Plan length XY	= 85.06 × cos 8°	
	= 84.23 m	

3. Human errors

This class of error arises from defects of human sight and touch, when marking the various tape or chain lengths and when estimating the readings on the tape when they do not quite coincide with a graduation mark. The chances are that two people will mark or estimate slightly differently. It is also reasonable to assume that no one person will overestimate or underestimate every reading, nor is he likely to mark every tape length too far forward. These small errors tend to be compensatory and have relatively little significance at this level of surveying.

Conclusions

In every measured length there will be errors which fall into one of three classes.

1. *Gross errors*   These are mistakes which should not occur in practice provided each line is measured at least twice.

2. *Constant errors*   All of these errors are cumulative and have a very great effect on the accuracy of any length. Every tape should be checked to ensure that it is the correct length and each tape length must be accurately aligned. The gradient must be determined along each measured length. In linear surveying, temperature and tension corrections are *NOT* necessary. They are applied in theodolite traversing (Chapter 13) where accuracy is essential.

3. *Human errors*   These are small residual errors of human sight and touch and tend to be compensating.

Procedure in linear surveying

Where comparatively small areas have to be surveyed a linear survey might be used. As already mentioned, the principle of linear surveying is to divide the area into a number of triangles, all the sides of which are measured. The errors which can arise when measuring have already been discussed. It is obvious therefore that great care must be exercised for every measurement.

Reconnaissance survey

On arrival at the site, the survey team's first task is to make a reconnaissance survey of the area, that is, the team simply walks over the area with a view to establishing the best sites for survey stations. The sites must be chosen with care and are in fact governed by a considerable number of factors.

1. *Working from the whole to the part*
This is the fundamental rule of all survey operations. It can be illustrated by a simple analogy. When a rectangular jigsaw puzzle is being manufactured, the whole picture is painted then broken down into several hundred pieces. The alternative would be to manufacture one piece then add the others to form the picture. Clearly the latter method would give a very inaccurate picture and certainly would not make a rectangle when fitted together.

The area to be surveyed is treated as a whole and is then broken down into several triangles rather than the reverse.

2. *Formation of well-conditioned triangles*
The triangles into which the area is broken should be well-conditioned, that is, they should have no angle less than 30° nor greater than 120°. These are minimum conditions. The ideal figure is an equilateral triangle and every effort should be made to have triangles whose angles are all around 45° to 75°. The reason is simply that it is much easier and much more accurate to plot such a triangle as opposed to plotting a badly conditioned triangle.

3. *Good measuring conditions*
All of the lines of the survey must be accurately measured and it is sound practice to select lines which are going to be physically easy to measure. Roads and paths are usually constructed along even gradients and present good measuring conditions. Lines which change gradient frequently are best avoided. Heavy undergrowth, for example, or railway embankments and cuttings, present difficulties and may make an otherwise good line completely immeasurable.

4. *Permanency of the stations*
The survey stations may have to be used at some future date when setting-out operations take place. They may, therefore, have to be of a permanent nature. Examples of permanent marks are shown in Fig. 4.14.

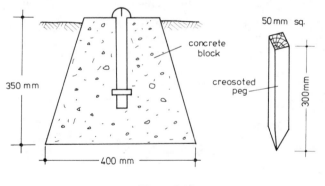

Figure 4.14

The marks must be sited in places which do not inconvenience anyone: for example, concrete blocks can do considerable damage to ploughs, etc., and cannot be placed in the middle of a field.

5. *Referencing the stations*
When the stations have to be used again it is necessary to be able to find them easily. If possible they should be placed near some permanent objects, fence posts, gates, bus stops, lamp standards, etc., from which measurements may be taken to the survey stations. This is known as 'referencing the station'. Each station should be referenced in

case any check measurements are to be taken in cases of survey errors. A typical situation for a station is shown referenced in Fig. 4.15.

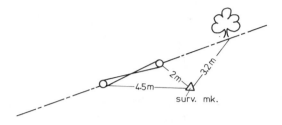

Figure 4.15

6. *Obstacles to measuring*

When siting any station with consideration for the factors outlined above it may be found that the survey line will cross a pond, river, or railway cutting, which will present a considerable problem to measurement. While such problems can be overcome they should be avoided if at all possible.

7. *Intervisibility of stations*

Strictly speaking it is necessary to be able to see only from any one station to the other two stations of any triangle. Check measurements have to be made, however, and wherever possible an attempt should be made to see as many stations as possible from any one station.

8. *Check measurements or tie lines*

Before the survey is complete, check measurements must be made. In Fig. 4.16(a) a check line AF has been measured.

On completion of the plotting, the scaled distance of this line must agree with its actual measured length. If it does, the survey is satisfactory, but if not, there is an error in one or more lines and each must therefore be remeasured until the error is found; hence the necessity for referencing each station. This procedure could take a considerable time and a better method is to check each triangle in turn by providing it with its own check line. When plotted triangle by triangle, each must check and if it does not it is then necessary to measure only the sides of the faulty triangle. Figure 4.16(b) shows the two methods of checking a triangle.

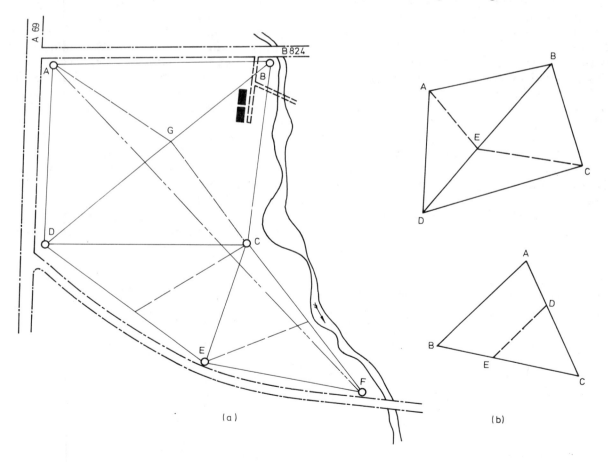

(a)

(b)

Figure 4.16

Conducting a survey

Once the stations have been selected, the various lines are measured, using the methods previously described. It must be borne in mind continuously that the plan length is required and all gradients must be carefully observed and measured, or step-taping must be employed.

The three sides and check line of each triangle should be measured before another triangle is attempted. In Fig. 4.16(a) the order of measuring the sides is BA, AD, DB, and check line GA. In measuring line AG a ranging pole should be left at station G. It will then serve as a check point for triangle BCD, producing a check line GC. Only the lines BC and CD remain to be measured to complete the second triangle. There is therefore economy of movement of the survey party in using this technique, resulting in a considerable saving of time.

In Fig. 4.16(a) the survey lines are sited as close as possible to the details which have to be surveyed. These details include the roadways, river, fences, and buildings.

Short distances, called offsets, are measured to these points of detail from positions along the main survey lines. These latter points are called chainages. Thus, any point of detail must have at least two measurements to fix its position, namely a chainage and offset.

Wherever possible, the offsets are measured at right angles to the survey lines, the right angle being judged by eye. The maximum length of offset is determined by the scale to which the plan is to be plotted.

On average, the naked eye can detect a distance of 0.25 mm, on paper. Thus when a survey is plotted to a scale of 1:1000, the eye will be able to detect (0.25×1000) mm actual ground size. Thus any error of over 0.25 metre will be detectable in the plotting. The average person is able to judge a right angle to $\pm 3°$.

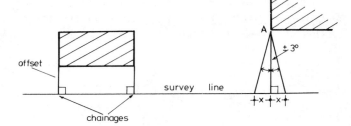

Figure 4.17

Therefore, in Fig. 4.17 the detail point A will be fixed to within $\pm x$ metres. If x exceeds 0.25 metre, it will be plottable, therefore, the maximum offset at which this error will be indiscernible will be

$$(0.25 \text{ cotangent } 3°) \text{ metres}$$

$$= 0.25 \times 19.08 \text{ metres}$$

$$= 5.00 \text{ metres approx.}$$

At this scale (1:1000) no offset should be allowed to exceed 5 metres and as the plotting scale decreases so the maximum length of offset increases.

It may be necessary or desirable to fix certain points of detail by more than one offset, whereupon the estimation of right angles ceases. In Fig. 4.18(a) certain details are fixed from two chainage points by 'oblique' offsets. This is a more accurate method and is used to fix important details, like the corners of a house. It is necessary to use this method when the maximum allowable right-angled offset is to be exceeded.

The principle of well-conditioned triangles should again be adhered to, and the distance between chainage points should be approximately equal to the oblique offsets.

When a building or wall, etc., lies at an angle to a survey line, it may be desirable to use 'in line' offsets. Such offsets are very similar to 'oblique' offsets but

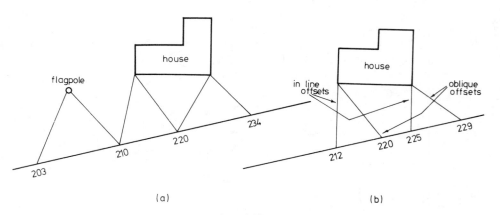

Figure 4.18

have the advantage that they are measured on the line of the detail feature and when plotting is being done a 'bonus' is thereby obtained.

Recording the survey

One of the most important and probably most underrated points in a linear survey is recording the information. There is little point in measuring accurately if the survey notes are not clear and intelligible. At the outset, a clear legible style of 'booking' must be adopted, if confusion is to be avoided at a later date.

The survey field book on which the notes are recorded, measures 200 mm × 110 mm. It is normally rainproof and consists of about 100 leaves, ruled with two red centre lines, about 15 mm apart, running up and down both sides of each leaf. The double red centre lines represent the chain or tape as it lies on the ground.

1. *Referencing the survey*
The first task in booking, is to make a reference sketch of the survey as a whole. The sketch shows the main framework with the stations in their correct relationship (Fig. 4.19).

As each line is measured, its length is written alongside the line, together with any gradient values. The measurements are always written in the direction of travel and arrows, indicating gradients, always point downhill. From this reference sketch, the basic famework can be plotted. Very often, the survey will not be plotted by the surveyor who made the survey, but by a draughtsman. Consequently, the sketch must be absolutely clear.

2. *Booking the details*
In Fig. 4.16, the details to be surveyed from the main survey lines, include the roads, buildings, and stream. These are surveyed by taking offsets at

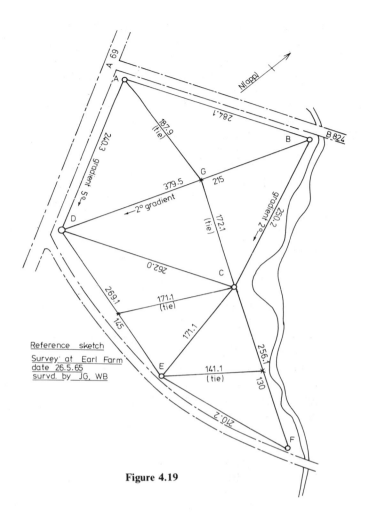

Figure 4.19

selected chainages, along the main survey lines. Figure 4.20 shows line BC, being measured in the direction of B towards C. The field book is opened at the back and station B is positioned on the chain line (double red centre lines) at the bottom of the

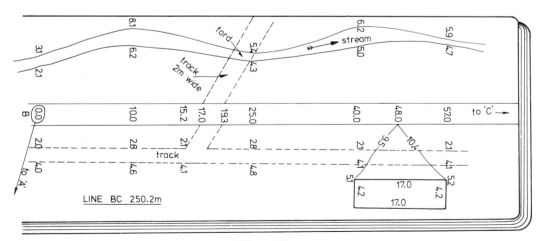

Figure 4.20

page. The booking proceeds up the page. Should it be necessary to use more than one page for any line, the page is simply flicked over and the booking continues on the reverse side. In this way, continuity of the line is preserved, in the field book.

This method of booking is called the 'double line' method and all chainages are noted within the red lines. The various features to which offsets are taken, are drawn in their relative positions and the offsets to right or left of the chain line are noted graphically. Virtually no attempt is made to draw to scale. The emphasis is placed rather on relative positioning of the features and on clear legible figuring.

A second method of booking, the 'single line' method, finds favour among a large number of surveyors. A blank field book is used, whereon a single pencil line is drawn to represent the chain line. Chainages are noted alongside the chain line and offsets shown graphically to right and left, as before. Figure 4.21 shows the same information as in Fig. 4.20 but as booked in 'single line' style.

The choice of method is largely personal, there being no overwhelming advantages in either. The double line method does have the advantage that the chainages are contained within the red lines and there is therefore no possibility of their being confused with offsets. On the other hand, where features of any width, such as streams, roadways, etc., cross the chain line, their shape is distorted. In Fig. 4.20 a farm track crosses the chain line at an angle, at chainages 17 and 19.3 metres. It appears to have a distinct change of direction. In Fig. 4.21 the same track crosses the chain line and is undistorted because the chain line has no width.

Example 8 Figure 4.22 shows a small area of some 2½ hectares, bounded on the north by a stream and on the other sides by roadways. The area is to be developed as a housing estate and an accurate plan is required.
(1) Devise a suitable framework for the survey.
(2) What should be the maximum length of right-angled offset?
(3) Using a 1:2500 scale as a chain, show the field book notes of a section of a chain line, passing in close proximity to the buildings.

Solution
(1) See page 33.
(2)
Maximum discernible distance on plan 0.25 mm
  Scale 1:2500, i.e., 1 mm = 2.5 m

Therefore 0.25 mm = 0.625 m

Maximum length of offset = $(0.625 \times \text{cotangent } 3°)$ m
  = (0.625×19.08) m
  = 11.9 m
  = 12 m approx.

(3) See page 34.

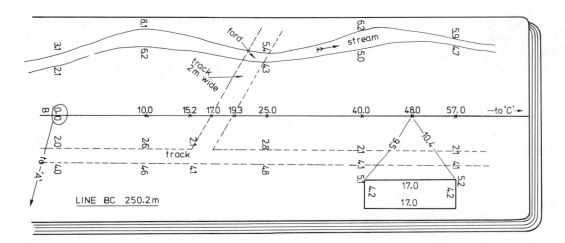

Figure 4.21

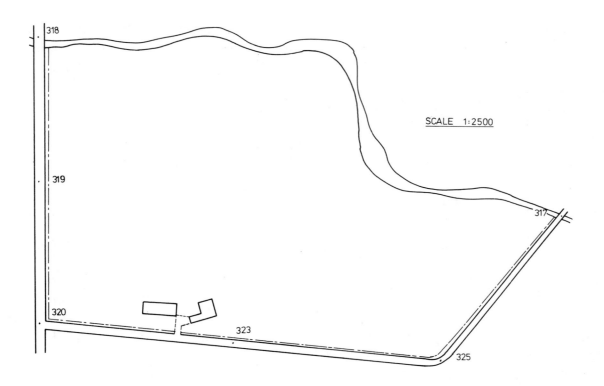

Figure 4.22

Solution to Example 8

(1)

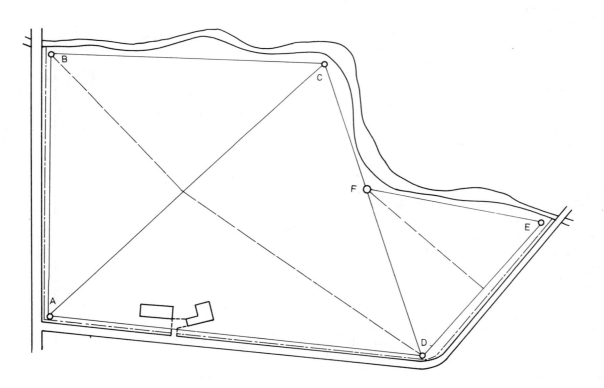

Figure 4.23a

(3)

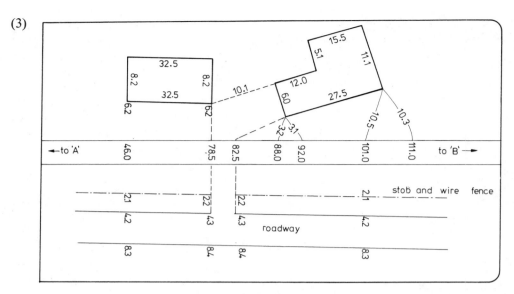

Figure 4.23b

Negotiating obstructions

Despite the precautions taken during reconnaissance to site the survey stations in suitable positions, it might be impossible to avoid some obstacles to measuring. Such obstacles might well be in the form of a wood, small hill, change of gradient, river, railway cutting, or even a building. They can generally be overcome by using 'field geometry' methods.

In some of the methods it is essential to be able to construct a right angle accurately either by linear means or by using some form of small hand instrument.

Constructing a right angle

1. *By linear means*

In Fig. 4.24, B is a point on the main survey line AC from which a right angle is to be erected. Equal distances BX and BY are laid off to right and left on the point B. From X and Y equal arcs XZ and YZ are described and the point Z defined. BZ will then be at right angles to the survey line AC. A right-angled triangle can also be established using the principle of Pythagoras. The basic relationship of the three sides is

$$(2n + 1):2n(n + 1):2n(n + 1) + 1$$

If $n = 1$ then the relationship becomes 3:4:5.

In Fig. 4.25, AB is the survey line and B the point at which the right angle is to be established. BC is measured out for a distance of 6 metres and C is accurately aligned on the survey line. The chain handle is held at B and the 18 metre marker held at C. The 8 metre marker is then pulled laterally from

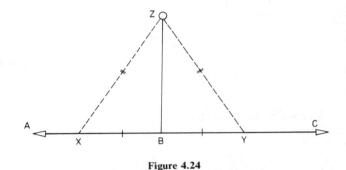

Figure 4.24

B forming a triangle BCD. The length BD is 8 metres and CD is 10 metres, while BC has already been measured as 6 metres. Thus, a triangle has been formed whose sides are in the ratio 3:4:5 and a right angle has thereby been established at B.

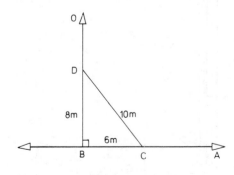

Figure 4.25

Sometimes, a right angle has to be established from a point, to a survey line. In Fig. 4.26, X is an external point and AB is a survey line. If X is less than one chain length from the line, a right angle can be established quite easily by holding the chain handle at X and swinging the chain in an arc, cutting the survey line at points C and D. The required angle to X will occur at E, the mid-point of CD.

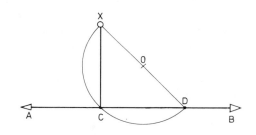

Figure 4.26

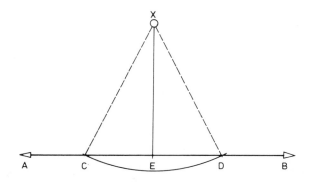

Figure 4.27

In Fig. 4.27 X is more than one chain from the survey line. The point C on the survey line, at which the right angle will occur, is roughly estimated and the distance CX paced or estimated. The same distance is then stepped out to point D. Length XD is accurately measured and bisected at O. If an arc of a circle is described with radius OX, it will cut the survey line at points D and C. The angle XCD is a right angle since it stands on a diameter of XD of a circle.

2. *By using hand instruments*
There are several small hand instruments to assist the surveyor in setting out right angles.

The cross staff Probably, the simplest instrument is the cross staff which is essentially a metal cylinder in which four slots are cut at 90° intervals around the cylinder.

The instrument is held on top of a ranging pole, or a special rod and a sight taken along the survey line. A right angle can then be set out by sighting through the cross slits, Fig. 4.28.

The optical square The instrument is contained in a metal case 60 mm diameter. The eye is placed to a pinhole aperture through which the direct and reflected objects are seen to coincide at 90°. The instrument makes use of the law of reflection which states that when a ray of light strikes a mirror it is reflected at the same angle in the same plane but on opposite sides of the normal, Fig. 4.29.

Figure 4.30 shows an interior view of the instrument. Mirror A, fully silvered, is placed at 45° to mirror B which is silvered on its upper half only. Any ray of light from aperture X is reflected by the mirrors A and B to the pinhole sight at Z. This ray of

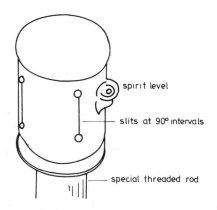

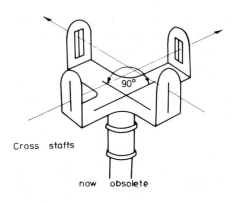

Figure 4.28

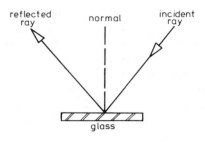

Figure 4.29

light is at right angles to the ray of light from aperture Y by the law of reflection.

The observer holds the instrument over the selected point and sights to target Y through the unsilvered half of the mirror B. An assistant holding

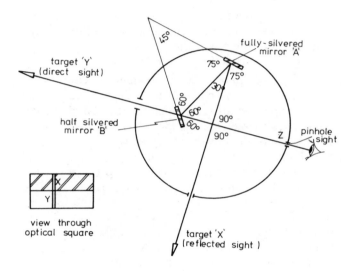

Figure 4.30

a ranging pole is directed at X until the pole is seen to coincide exactly with target Y in the upper silvered half of mirror B, whereupon a right angle will have been formed.

Like most other surveying instruments, the optical square will go out of adjustment from time to time and it must be properly adjusted if errors are to be avoided. To test the instrument three poles P, Q, R are set out in a straight line about 50 metres apart (Fig. 4.31). The instrument is set up at station Q and a perpendicular QS$_1$ set out from line PQ. The instrument is now inverted and a second perpendicular QS$_2$ set out from line QR. If the instrument is in adjustment S$_1$ and S$_2$ will be coincident. If they do not coincide a pole is inserted midway between S$_1$ and S$_2$ at point S. Angles SQP and RQS will be right angles. The adjustable mirror is

now turned by its key until pole S is seen to coincide with pole R. The instrument is re-inverted and a sight taken to station P. The pole at S should coincide with station P. If not the test is repeated until this condition is satisfied.

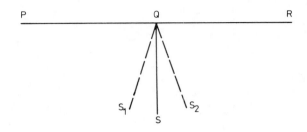

Figure 4.31

The prism square In this simple instrument, use is made of a pentagonal glass prism PQRST (Fig. 4.32). The faces PT and RS, if produced, contain 45°. The observer holds the prism at eye level and looks over the instrument to view directly target 1. The incident rays carrying the image of target 2 are reflected to the observer's eye. When target 1 and the reflected image of target 2 coincide a right angle will have been formed.

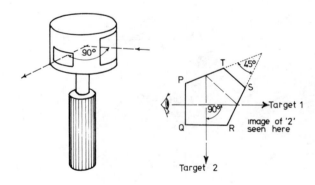

Figure 4.32

Obstacles

An obstacle or obstruction is any object which obstructs the direct measurement or ranging of any line. They can be divided into several groups.

1. *Obstructions to ranging only*

Where a survey line has been established and both ends are not intervisible, a straight line must still be established between the points before the measurement can be conducted. In Fig. 4.33 survey stations A and D are not intervisible because of the intervening high ground.

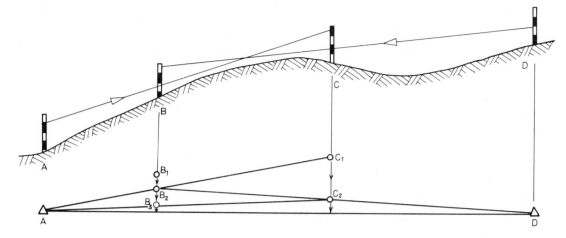

Figure 4.33

Interpoles B_1 and C_1 are placed between A and D, with no attempt being made to align them. The only condition governing their positioning is, that C_1 must be visible from A and B_1 visible from D.

On the line C_1A, pole B_1 is ranged to its new position B_2, forming a straight line AB_2C_1. On the line B_2D, pole C_1 is ranged to C_2. B_2 is then ranged into new position B_3, on the line AC_2, and the procedure is continued until A, B, C, and D form a straight line.

2. Obstructions to measuring only

(a) *Obstruction can be measured around* Figure 4.34 shows a pond which lies directly on the line of survey stations X and Y, making measurement of that part of the survey line over the pond impossible. At a point A near the pond, a right angle is set out to point B by any of the previously described methods, and the distance A–B is measured. The distance from point B to point C, on the survey line at the far end of the pond, is measured. By the theorem of Pythagoras, the inaccessible distance A–C is calculated.

$$AC = \sqrt{(BC^2 - AB^2)}$$

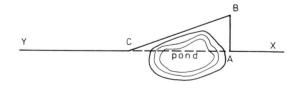

Figure 4.34

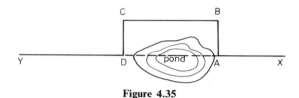

Figure 4.35

Figure 4.35 shows an alternative method of obtaining the distance across the pond. At points A, B, C, and D, right angles are set out and distances AB, BC, and CD measured. If all setting out is accurate AB will equal CD and the inaccessible distance AD will be equal to CB.

Many other methods are available for obtaining this distance but one, in particular (Fig. 4.36), is worthy of closer study, because it is a method used for setting out parallel lines.

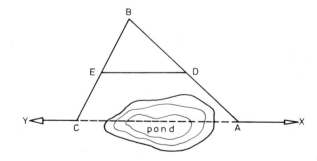

Figure 4.36

At point A, any random line is set out to point B. The distances AB and BC are measured accurately and bisected at D and E respectively. Triangles BED and BCA are similar and DE is parallel to CA. The sides of the triangles have been laid out in proportion of 1 to 2 and CA is therefore equal to twice ED.

$$\frac{DE}{CA} = \frac{BD}{BA} = \frac{1}{2}$$

therefore CA = 2DE

(b) *Obstruction cannot be measured around*   The obstructions envisaged under this heading are rivers and railway cuttings, exceeding one chain in width. Again, many variations are possible. Figure 4.37 shows a survey line XY crossing a railway cutting; at point A, a right angle is set out to B and the distance AB measured and bisected at C. At B another right angle is set out towards D and the point established where this line intersects the line EC produced. Lines BD and AE are parallel and triangles AEC and BCD are congruent. The inaccessible distance AE therefore equals BD.

An alternative method of determining the unknown distance is shown in Fig. 4.38.

A right angle is again set out at point A, any point B established and the distance AB measured. At B an optical square is used to sight C, across the railway cutting. A right angle is set out from B and the point D established on the line XY such that CBD is the right angle. Lines BD and AD are measured.

In the figure, triangles BAD and CBD are similar, since both have a right angle and the angle D is common to both. All three angles in both triangles are therefore equal: a condition for similarity.

$$\text{Therefore } \frac{CD}{BD} = \frac{BD}{AD}$$

$$\text{therefore } CD = \frac{BD^2}{AD}$$

$$\text{But } CD = CA + AD$$

$$\text{therefore } (CA + AD) = \frac{BD^2}{AD}$$

$$\text{therefore } CA = \left(\frac{BD^2}{AD} - AD \right)$$

3. *Obstructions to measuring and ranging*
Despite taking every precaution to avoid obstacles, the occasion arises where a building or wood lies on the survey line and the line can neither be measured nor ranged. Figure 4.39 shows a typical situation, where survey line XY is interrupted by a copse.

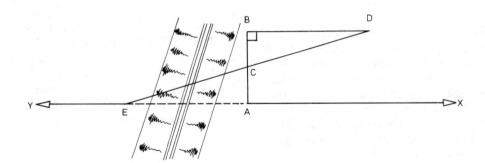

Figure 4.37

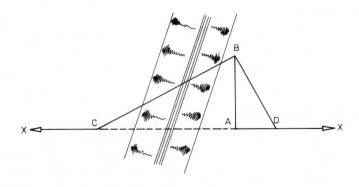

Figure 4.38

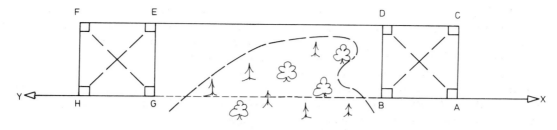

Figure 4.39

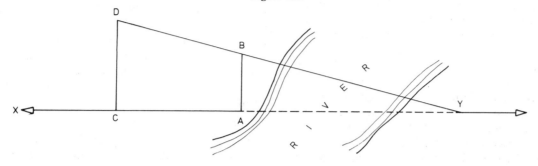

Figure 4.40

At points A and B, right angles are set out to the points C and D respectively and AC made equal to BD. The length CD when checked must equal the length AB. Line CD is then ranged forward to points E and F and the lengths DE and EF measured. Further right angles are set out at E and F, the length EG and FH measured equal to AC and BD, and points G and H established. On checking GH, it should be found equal to EF. Points G and H are then on the line XAB and when the line GH is ranged forward it should point to Y. The inaccessible distance GB will equal ED.

Example 9 A survey line XY is interrupted by a wide river and the distance across the river is required. On the near bank a right angle is set out at point A for a distance of 20 metres and a point B established. At point C, 40 metres back from A on the line AX, a second right angle is set out to D, on the same side of AX as point B. The point D is aligned with Y on the far bank and with B, already established. The line CD is found to measure 31 metres. Calculate the length of the inaccessible portion AY.

Solution (Fig. 4.40)
Triangles YAB and YCD are similar

$$\text{therefore } \frac{YA}{AB} = \frac{YA + AC}{CD}$$

$$\text{i.e., } \frac{YA}{20} = \frac{YA + 40}{31}$$

therefore $31\ YA = 20\ YA + 800$

$$\text{therefore } YA = \frac{800}{11}$$

$$= 72.72 \text{ metres}$$

Plotting the survey

Undoubtedly the surveyor's main objective is to achieve accuracy in his field operations. It can safely be stated, however, that unless his results can be depicted accurately, legibly and pleasingly on paper, his proficiency in the field is robbed of much of its value.

While it can be said that a natural flair for drawing can greatly assist the surveyor in the presentation of his results, it is none the less true that a systematic approach to plotting will produce a perfectly acceptable plan of the results of the field work.

Results of field work

The specimen booking in Figs 4.41 and 4.42 shows a small partially developed area in which is situated a small factory building and car park.

Equipment required for plotting

1. Paper. The survey when plotted may have to be referred to frequently, over a number of years, and it is essential therefore that the material on which it is plotted should be stable. The paper must be well

seasoned, and should be mounted on brown holland backing. Modern 'plastic' drawing materials such as Permatrace are excellent.

2. The introduction of the metric system has simplified the problem of buying suitable scales, since almost all scales used nowadays are direct multiples of each other. The British Standard 1347 scale rule contains no fewer than eight scales—1:1, 1:5, 1:20, 1:50, 1:100, 1:200, 1:1250, and 1:2500. Other common plotting scales, 1:500 and 1:1000, can be derived simply by multiplying the scale units of the 1:50 and 1:100 scales, by the factor 10. Best plotting scales are made of boxwood or ivory, though once again some excellent scales are produced in plastics.

3. Beam compasses. Since the lines of chain surveys can be hundreds of metres long, large radius compasses are necessary. Ordinary compasses are unwieldy at large radius and beam compasses are used in plotting work. Figure 4.43 illustrates a set of beam compasses.

4. Other equipment includes two set squares, 45° and 60°, varying grades of pencils, 4H to H, sharpened to a fine point, paper weights, curves, inking pens, pricker pencil, rubbers, steel straight edge, and springbow compasses.

Procedure in plotting

Orientation

Most maps and plans are drawn and interpreted looking north towards the top of the paper. It is customary on chain surveys to find north by some means, for example, by compass or more roughly by the sun and a watch. The plotting material should always be oriented, to north such that the top and bottom of the paper are respectively north and south.

Rough sketch

If a sketch of the survey is roughly drawn to scale it will greatly facilitate centralizing the survey on the drawing paper and will result in a much more balanced appearance.

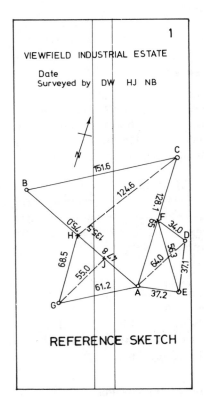
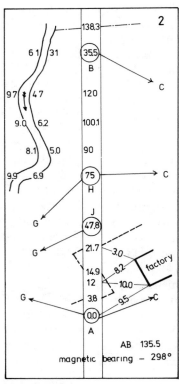
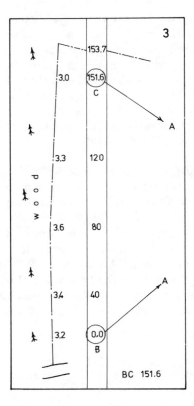

Figure 4.41

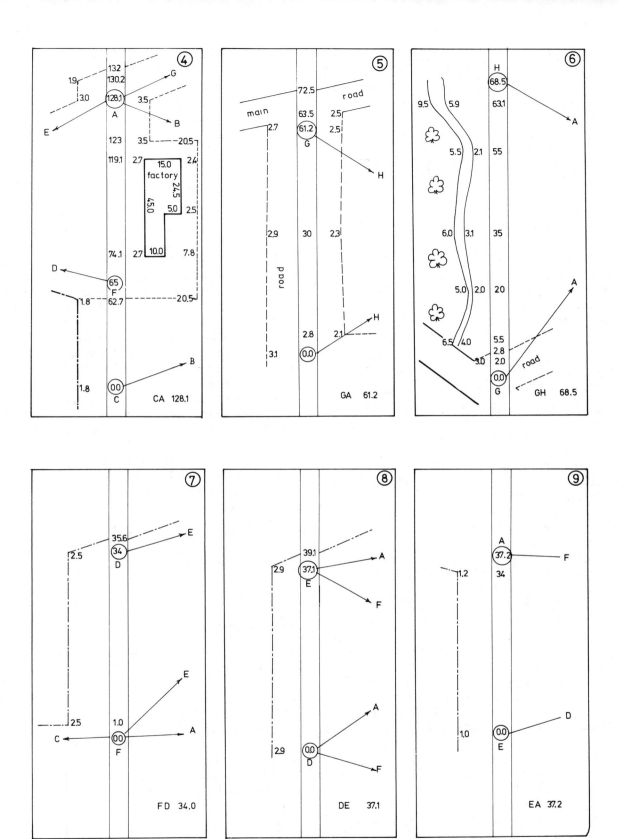

Figure 4.42

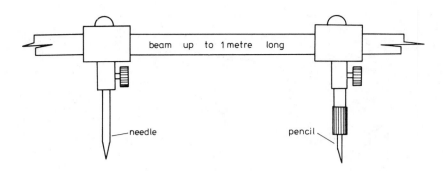

beam up to 1 metre long

needle

pencil

Figure 4.43

Scale

A line scale, filled or open divided, is next drawn along the bottom. This scale drawn on the paper is necessary to detect possible shrinkage or expansion of the drawing material.

Plotting the framework

The details in Figs 4.41 and 4.42 are to be plotted to a scale of 1:1250. Examination of the reference sketch, Fig. 4.41, shows that north is approximately along the line AC and the plotting is therefore oriented such that AC points towards the top of the drawing paper. If the reference sketch is also treated as being the rough sketch previously referred to, it would indicate that point 'A' is almost the centre of the figure, in an east–west direction, while 'G' is the southernmost station.

Station 'A' can therefore be fixed as the commencing point for plotting.

From A, the line AC is drawn and scaled accurately. An arc of a circle of length AB is constructed using the beam compasses and a second arc of length CB is drawn to intersect the first arc in B. The station H, along the line AB, is marked and the distance HC is scaled. The scaled length should agree closely with the actual measured length of HC. Should it fail to do so, it indicates that either the plotting of one or more lines of the triangle ABC is in error or the field measurement of one or more of the lines is wrong. This is precisely the purpose of the tie line HC.

Assuming that the scaled length HC agrees with the field length, the triangle AHG is then plotted from the line AH in the same way as was the triangle ABC, and the length of its check line is compared with the actual measured length of 55.0 metres.

The triangles AFD and AFE are then plotted in like manner thereby completing the construction of the framework. The various survey lines are then lightly drawn in ink and each survey station marked by a small circle.

Details It now remains to plot the details along each chain line.

On the line BC, for example, the various chainages of 0 m, 40 m, 80 m, and 120 m are marked off. Short right-angled lines lying to the left of the chain line are marked off and the various offsets scaled along them. At 40 m, an offset of 3.4 m is scaled to the left and at 80 m a distance of 3.6 m is scaled. When these points are themselves joined by a straight line they form the line of the fence lying to the left of survey line BC.

Along line AB, there are several oblique offsets to the corners of the factory building. These offsets are arced by springbow compasses, the intersection of a 9.5 m arc from A and a 10.0 m arc from chainage 12 m forming a corner of the factory building.

Along each chain line the plotting of the detail is carried out with great care until the whole survey has been plotted.

The various details are then drawn in ink and a suitable title, north point etc., are added.

Figure 4.44 shows the survey completely plotted to a scale of 1:1250.

Test questions

1 (a) A plan is to be constructed from a chain survey such that distances of 0.5 metre can be read from it. What is the minimum scale to which the plan should be made and what should be the maximum length of offset during the survey?

(b) Show how a right angle can be constructed to a survey line from a point 25 metres to one side using a 20 metre chain.

(c) Show how the measurement of a survey line can be continued over a river which is wider than the chain length.

2 Linear measurements are open to errors of three classes:

(a) Gross
(b) Constant
(c) Accidental

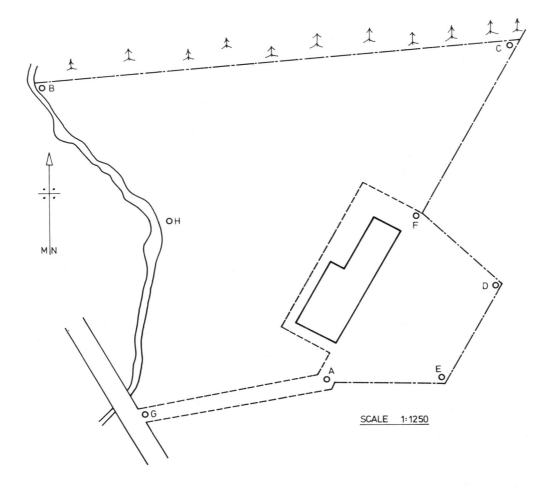

Figure 4.44

Write a short account (about one page) on the major errors in each class indicating how they are minimized.

3 Figure 4.45 shows Riverview farm Hazelburgh drawn to a scale of 1:1000.
(a) Draw on the plan a system of lines which would enable the area to be surveyed by linear means only. Stations A and B shown on the plan must be incorporated in the survey.
(b) Using a scale as a chain, draw up a page from a surveyor's field book showing clearly the offsets required to survey the barn, house, and driveway from line AB.

4 A survey line PQ was measured along a regular 5° slope using a 20 m tape. Given that the tape was 10 mm long and that the measured length was 48.46 m, calculate the plan length of the line.

5 Figure 4.46 shows the station points and lines relating to the survey of a plot of land. Utilizing the measured data shown in the figure, plot to a scale of 1.500 the main survey lines including all the detailed information on lines BC and CD.

(City and Guilds of London Institute)

6 (a) Describe the principle of trilateration.
(b) Define (i) offsets, (ii) tie lines, (iii) station points.
(c) List four possible causes of error in taking linear measurements.
(d) Describe how to measure a line up a steep slope.

(City and Guilds of London Institute)

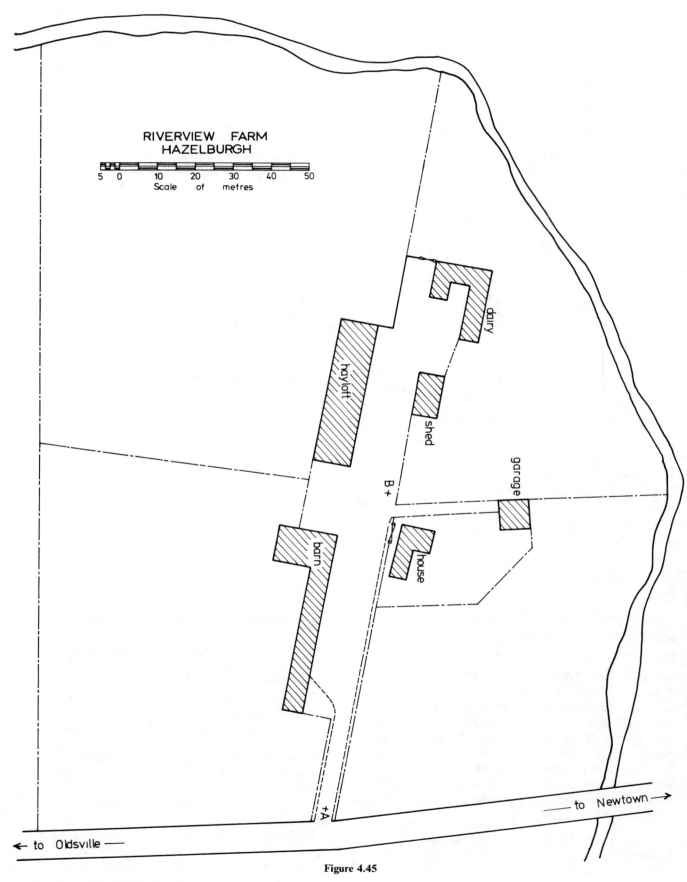

RIVERVIEW FARM
HAZELBURGH

5 0   10   20   30   40   50
Scale of metres

dairy

hayloft

shed

garage

B+

barn

house

+A

to Newtown →

← to Oldsville →

Figure 4.45

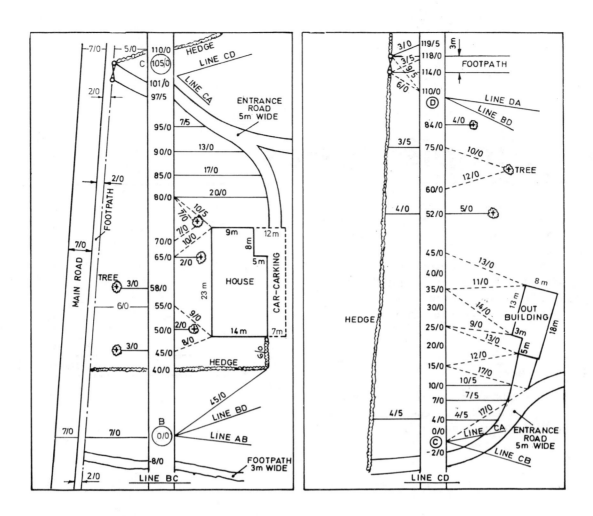

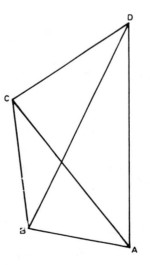

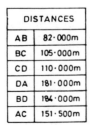

DISTANCES	
AB	82·000m
BC	105·000m
CD	110·000m
DA	181·000m
BD	184·000m
AC	151·500m

Figure 4.46

5. Simple optics

In Chapter 4, several simple surveying instruments were considered, some of which made use of the law of reflection.

In the succeeding chapters many types of surveying instruments, for example, levels and theodolites, will be examined, and a knowledge of simple optics is an essential prerequisite to an understanding of them.

The law of reflection

If a ray of light strikes a mirror or a piece of plane glass as in Fig. 5.1 it will be reflected in such a way that angle PP_1I equals angle PP_1R.

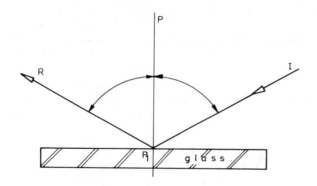

Figure 5.1

The angle of incidence PP_1I equals the angle of reflection PP_1R.

The law of refraction

As the lamp at I moves nearer to the perpendicular PP_1 the incident ray of light IP_1 will cease to be reflected and will instead, pass through the glass to emerge on the other side of the glass as a ray parallel to the incident ray. The emergent ray will, however, have been bent or 'refracted' along the path IP_1E_1E (Fig. 5.2). Refraction will take place when angle PP_1I is less than 41°.

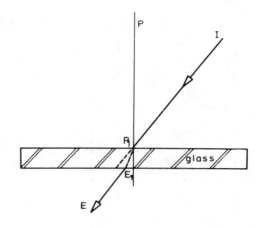

Figure 5.2

It will be seen from the figure that the ray is bent towards the perpendicular when passing through the glass and away from the perpendicular when emerging into the air. Since glass is denser than air it follows that a ray of light will be bent towards the 'normal' (perpendicular) when passing from one substance into another denser substance and will be bent away from the normal when passing from one substance into a less dense substance.

When the lamp at I is moved closer to P until finally, I coincides with P, the ray of light will pass through the glass unrefracted.

Lenses

If a curved surface is now considered it will be seen in Fig. 5.3 that the ray of light will behave in the same manner as above. It will be bent towards the normal when entering the glass and away from the normal on leaving the glass. The normal is defined as the line perpendicular to the tangent to the curved surface at any point along its length. The lens illustrated is called a double convex lens, because of the distinctive outward facing curves of the faces of the lens.

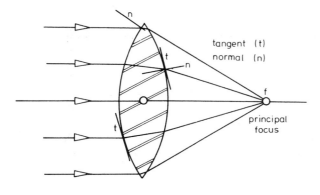

Figure 5.3

If the lens curves were inward facing, they would be called concave, thus forming a double concave lens (Fig. 5.4(a)).

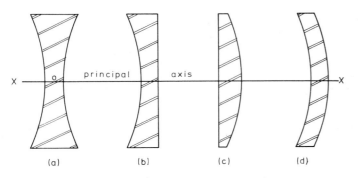

Figure 5.4

If only one face of the lens is curved and the other is plain, the lens is plano-concave (b) or plano-convex (c). Should one face be concave and the other convex, a concavo-convex lens would result. Figure 5.5 shows rays of light passing through firstly a double concave lens and secondly two plano-convex lenses.

The double concave lens has the effect of spreading rays of light while the effect of having two plano-convex lenses facing each other is to form a simple magnifying glass.

In Figs 5.4 and 5.5 the lines XX form the principal axis of each lens and pass through the optical centre O. Any ray of light passing through the optical centre continues in its same direction unrefracted.

Modern surveying telescopes are in their simplest form tubes with a double convex lens forming an object glass at the sighting end and two plano-convex lenses forming an eyepiece at the viewing end.

In Fig. 5.6, line XX is the principal axis and O is the optical centre of the lens. Rays AA_1 and BB_1 pass through the optical centre and are captured on a glass screen AB called the diaphragm or reticule, thus forming an image. Rays A_1C and B_1D entering the lens parallel to the principal axis are refracted such that they converge at a point f, called the principal focus, and thereafter meet the reticule at A and B respectively. The length Of is the focal length of the lens. The image AB formed by the rays is real and inverted. This image will be in focus, however, for only one position of the object sighted. If the object A_1B_1 is moved further away from the telescope, the image formed at the reticule AB, will be out of focus as in Fig. 5.7.

When an internal double concave lens is introduced into the telescope between the object lens and the principal focus f, the rays of light are spread by movement of the lens along the telescope until B_1OB and B_1DB converge at the reticule thereby bringing point B into focus. Point A will, of course, be brought to focus simultaneously in the same way (Fig. 5.8).

In effect the principal focus f is being moved along the principal axis with movement of the internal lens. The lens is moved along the telescope on a rack and pinion arrangement by turning the focusing screw on the side of the telescope.

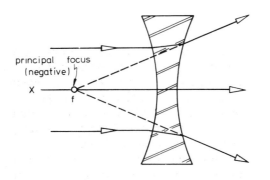

Figure 5.5

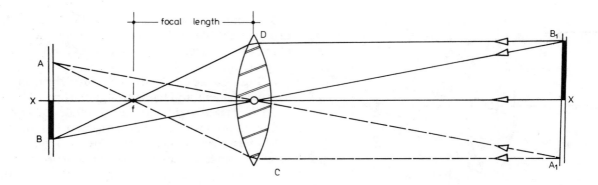

Figure 5.6

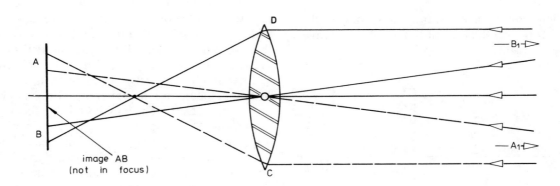

Figure 5.7

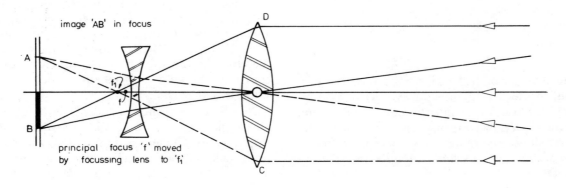

Figure 5.8

Reticule and eyepiece

The reticule on which the image is focused is a circle of plane glass etched with a system of lines called crosshairs. Various forms are shown in Fig. 5.9.

Figure 5.9

The image formed on the reticule is very small and has to be enlarged by the eyepiece. The eyepiece used in modern levels consists of two plano-convex lenses which are at a distance apart of $\frac{2}{3}$ of the focal length of any one of them. It is called a Ramsden eyepiece. The paths of the rays of light through this eyepiece are clearly shown in Fig. 5.10. The net result is an enlargement of the real image AB, seen by the eye as an apparent or virtual image at $A_{11}B_{11}$.

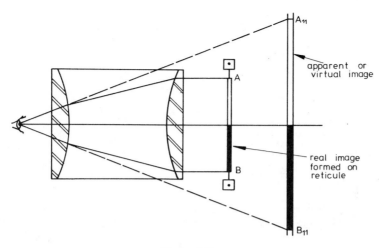

Figure 5.10

Parallax

It should be abundantly clear from the foregoing descriptions that the image must be clearly formed on the reticule by proper focusing of the telescope. It is possible, however, for the eye to discern the image even when it is not properly focused on the reticule. In Fig. 5.11 the image has been formed slightly in front of the reticule. (It could equally well have been formed slightly behind it.) The eye can still see the crosshairs which are situated some short distance behind the image.

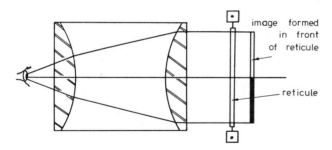

Figure 5.11

It follows therefore, that if the observer's head is moved up or down or from side to side, the crosshairs will appear to move with respect to the image making accurate sighting impossible. This phenomenon is known as parallax and must be eliminated by the following procedure:

1. Observe the sky through the telescope and screw the eyepiece out or in until the crosshairs appear black and sharply defined. The eyepiece is thereby correctly focused on the reticule.
2. Sight some distant object and focus the telescope until the image is clearly defined.

3. The eyepiece is now clearly focused on the reticule, as also is the image, If the eye is moved, there should be no apparent movement of the crosshairs and parallax has been eliminated.

The surveying telescope

The function of each lens has been explained above and it now remains to assemble the various lenses in a telescope and trace the paths of light through them in order to see clearly the virtual image of the target sighted by the observer. Figure 5.12(b) shows the assembled telescope and various light paths through it from the target to the eye, while Fig. 5.12(a) is an external view of the telescope.

This type of telescope, with certain refinements, is common to most surveying instruments.

While not part of the optics, it is perhaps convenient at this stage to consider another feature common to most instruments, namely the spirit level.

Spirit level

A spirit level can be simply considered to be a miniature glass version of a beer barrel. Its shape should therefore be readily visualized. The internal surface is ground to the radius required; the larger the radius, the more sensitive the bubble. It is almost completely filled with ether or alcohol, such liquids having a low freezing point. The air in the remaining space will always find its way to the highest point and when the spirit level is properly levelled the air bubble will take up a central position as in Fig. 5.13. If the spirit level were moved along the circumference of a circle formed by

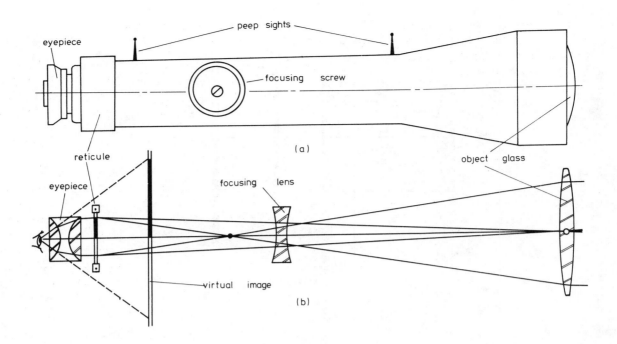

Figure 5.12

rotating the radius of curvature, until the bubble moved over one division, say 2 mm long, etched on the glass surface it might have moved through an angle of, say, 20 seconds.

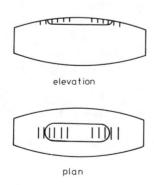

elevation

plan

Figure 5.13

The value of 1 division = 20 seconds is the sensitivity of the bubble and from the sensitivity, the radius of curvature of the bubble tube can be found.

In any circle the circumference can be found from the formula

$$\text{Circumference} = 2\pi r$$

Radius is unknown

Angle at centre $\qquad$ = 20 seconds

since $\qquad 360° = 2\pi$ radians

$$1° = \frac{2\pi}{360} \text{ radians}$$

and $\qquad 1'' = \frac{2\pi}{360 \times 60 \times 60}$

$$= \frac{1}{206\,265} \text{ radians}$$

therefore $20'' = \frac{20}{206\,265}$ radians

In the example

$$\text{Radius} = \frac{\text{Circumference}}{\text{Angle in radians}}$$

$$= (2 \div 20/206\,265) \text{ mm}$$

$$= \left(\frac{20 \times 206\,265}{20} \right) \text{ mm}$$

$$= \left(\frac{20 \times 206\,265}{20 \times 1000} \right) \text{ m}$$

$$= 20.63 \text{ m}$$

The spirit level is always attached to the instrument in such a fashion that it can be adjusted relative to the instrument if required. Figure 5.14 shows the spirit level attached to the telescope with one end free to pivot while the other end can be raised or lowered by means of the capstan screws.

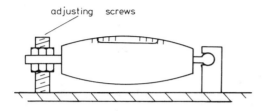

Figure 5.14

Coincidence bubble reader

In surveying instruments the bubble of the spirit level is set by eye to the centre of the bubble tube graduations. Many modern instruments however incorporate an optical system whereby the images of both ends of the bubble are viewed side by side in the same field of view using a coincidence bubble reader.

A system of 45° prisms reflects the images of the ends of the bubble as in Fig. 5.15. The observer sees both ends of the bubble through the viewing eyepiece of a level or theodolite. As the spirit level is made to tilt in these surveying instruments, the ends of the bubble are seen to move relative to each other until they coincide in the field of view whereupon the bubble is accurately centred. A large increase in accuracy in setting the bubble is achieved by this means.

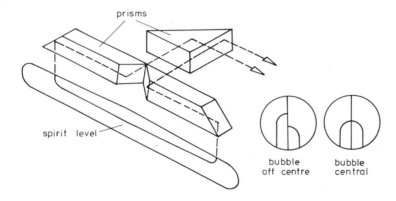

Figure 5.15

6. Levelling

Chapter 4 dealt with the simple principles of representing the earth's features in two dimensions on a plan. To be of any practical value however the third dimension, namely the height of the feature, must be shown by some means on the plan. In surveying, these heights are found by levelling. The heights of the points are referred to some horizontal plane of reference called the datum.

Figure 6.1 shows a table and chair standing on a level floor. The difference in height between them could be found using a rule in a variety of ways:

1. By measuring upwards from a horizontal plane, namely the floor.
2. By measuring downwards from a horizontal plane, namely the ceiling.

3. By measuring downwards from an imaginary horizontal plane established by a spirit level as in Fig. 6.1.

If the spirit level were held at eye level, say 1.500 m, and the horizontal plane extended over the table and chair, the plane would cut the rule held on the table at 0.750 m and on the chair at 1.050 m.

In practical levelling, the simple rule is replaced by a levelling staff while the spirit level is replaced by a surveying instrument called a 'level'. The level is, in fact, only a spirit level attached to a telescope which is mounted on a tripod. If the table and chair are replaced by two points on the surface of the earth, the simple illustration in Fig. 6.1 becomes an actual levelling exercise in Fig. 6.2.

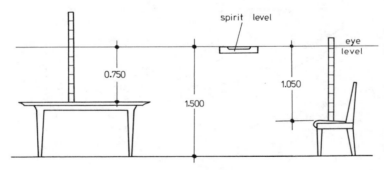

Figure 6.1

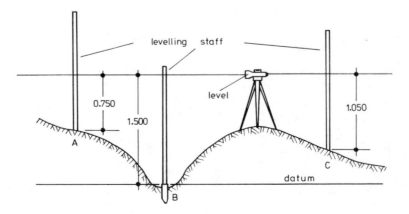

Figure 6.2

(a) Watt's SL15 tilting level showing ball and socket joint

(b) Kern GK1A automatic level showing jointed head tripod

(c) Sokkisha TTL6 tilting level showing coincidence bubble reader and three footscrews levelling arrangement

Figure 6.3

It will be seen that the height of point A above datum is 1.500 − 0.750 = 0.750 m while the height of C is 1.500 − 1.050 = 0.450 m above datum.

The datum, in this case, is an imaginary horizontal plane through the top of peg B.

If any Ordnance Survey map is examined, the heights of several points on the map will be seen to be shown by spot levels. These heights are measured above a datum called Ordnance Datum (OD) which is actually the mean level of the sea recorded at Newlyn Harbour, Cornwall, over the period 1915 to 1921. From this datum, levellings have been conducted throughout the country and the levels of numerous points permanently established by bench marks (Fig. 3.4).

Any levelling done on a building site, e.g., in Fig. 6.2, could therefore be referred to OD by simply taking a reading to a bench mark instead of the peg B. On many sites, however, this will not be necessary and a peg concreted into the ground would serve as a bench mark.

Levelling instruments

Levels are categorized into three groups: (1) dumpy levels; (2) tilting levels; and (3) automatic levels. Some are shown in Fig. 6.3.

1. Dumpy levels

Figure 6.4 shows a dumpy level reduced to its simplest form.

It comprises the following parts:

(a) *Trivet stage*. The open threaded flat base plate which attaches the instrument to the tripod.

(b) *Levelling arrangement*. Conventional three-screw type where the feet of the screws fit into the trivet stage.

(c) *Tribrach*. The main 'platform' which sits on top of the levelling screws and carries the remainder of the instrument.

It should be noted at this juncture that the trivet stage remains fixed in position, its sole function being to hold the instrument firmly on the tripod. The tribrach, however, can be tilted by movement of the levelling screws. The three parts are collectively known as the levelling head.

(d) *Telescope*. Mounted on a vertical spindle which is free to rotate within the tribrach. The optical arrangement of the telescope has already been discussed. The principal axis is known as the line of sight or line of collimation.

(e) *Spirit level*. Mounted on the telescope in the manner previously described.

Setting up the instrument (*temporary adjustments*)
The temporary adjustments are performed every time the instrument is set up. Three distinct operations are involved.

(i) *Setting the tripod*. This is probably the most underrated aspect in setting up any surveying instrument, yet a few seconds spent at this point will save much time and effort later on.

Two legs of the tripod should be pushed firmly into the ground. If the tripod is being set on sloping ground, the two legs should be downhill. The third leg is manoeuvred until the tripod head is approximately level and is then pushed firmly into the ground.

(ii) *Levelling the instrument*. The levelling screws are positioned as in Fig. 6.5. The telescope is turned until it lies along the line of any two screws, say B and C, and the position of the spirit level bubble is

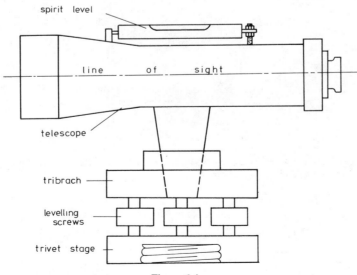

Figure 6.4

observed. The levelling screws are held by the thumb and forefinger of each hand and the screws

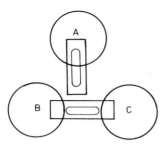

Figure 6.5

turned in opposite directions. The bubble will be seen to move along the bubble tube in the direction of movement of the left thumb; hence the 'left thumb rule'. The movement is continued until the bubble is centralized. If the telescope is now turned through 90° it will lie over screw A. Using screw A only and moving it by the left thumb, the bubble is centralized once more. Theoretically a horizontal plane through the telescope is established by these two levelling operations but in practice, both operations will have to be repeated two or three times before the bubble remains central for both positions.

(iii) *Parallax elimination*. Parallax must be removed before any observations can be made to a levelling staff. The method of removing parallax has been described in Chapter 5 dealing with optical principles.

The dumpy level is now ready to observe and a sight can be taken to a staff held in any position, whereupon, the bubble of the spirit level should be central. In practice this might well not be the case for a variety of reasons, namely:

(a) Imperfect adjustment of the instrument.
(b) Wind pressure.
(c) The observer's movements around the tripod.
(d) Soft ground causing the instrument to sink.
(e) Unequal expansion of the various parts of the instrument by the sun.

Before the staff is read, however, the bubble must be perfectly centred, therefore continual small adjustments of the spirit level have to be made by the only means available, namely the footscrews. Each movement of the screws, however, will alter the height of the line of sight causing errors. These errors will be small and of practically no significance but the whole operation of relevelling several times is very time-consuming and annoying.

This disadvantage is overcome by using a tilting level.

2. Tilting levels

A typical tilting level is shown in Fig. 6.6.
It comprises the following parts:

(a) *The levelling head.* The levelling head is made ↓ of the same three components as the dumpy level, namely the trivet stage, the levelling arrangement, and the tribrach.

The levelling arrangement is either a conventional three screw system as on a dumpy level or some form of ball and socket joint (Fig. 6.6(a)) which allows much faster levelling of the instrument to take place.

It is used in conjunction with a small relatively insensitive circular spirit level mounted on the tribrach. The tribrach can thus be levelled completely independently of the telescope and main spirit level. The instrument can be rotated about the vertical axis and the bubble of the circular spirit level will remain in the centre indicating that the tribrach is approximately level.

(b) *The telescope.* On a tilting level, the telescope is not rigidly attached to the tribrach but is supported by a central pivot. It is therefore capable of a small amount of vertical movement. This movement gives the instrument a tremendous advantage over a dumpy level. The vertical movement is imparted to the telescope by a tilting screw passing through the tribrach at the eyepiece end of the telescope. A spring loaded return, mounted on the tribrach at the objective end of the telescope works in sympathy with the tilting screw to elevate or depress the telescope (Fig. 6.6(a)).

(c) *Spirit level.* The main spirit level is mounted on top or on the side of the telescope and has already been described in Chapter 5.

Setting up the tilting level

1. The tripod is set up with the levelling head approximately level and secured to the instrument by the ring clamp (Fig. 6.6(a)) or fastening screw (Fig. 6.6(c)).
2. The bubble of the circular spirit level is centralized by one of the three methods, (a), (b) or (c) below, thus making the vertical axis of the instrument approximately vertical and the line of sight approximately horizontal.
   (a) If the instrument has three footscrews (Fig. 6.6(b)) the bubble is moved in the *x* direction using footscrews 1 and 2 in the same manner as with a dumpy level. Without rotating the instrument, the bubble is moved in the *y* direction using footscrew 3 until it is centred.
   (b) Using a ball and socket joint (Fig. 6.6(a)), the ring clamp is released and the bubble is centralized by hand and the clamp tightened.

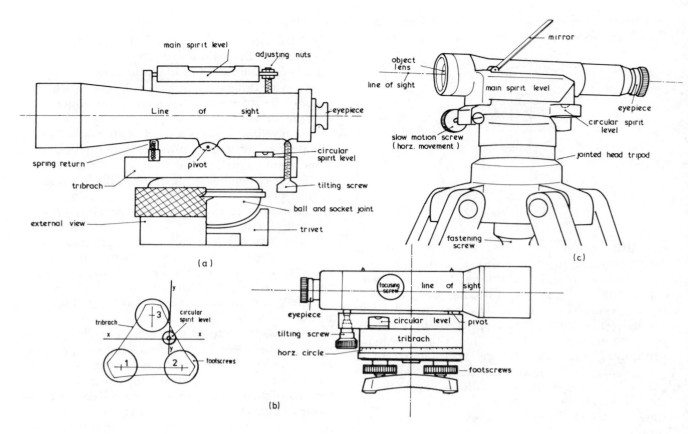

Figure 6.6

(c) Using the jointed head principle (Fig. 6.6(c)), the fastening screw is released and the instrument is shifted over the spherical surface of the tripod head by hand until the bubble is centralized. The fastening screw is then tightened.

3. Parallax is now eliminated by focusing the eyepiece onto the diaphragm until the cross lines appear sharp and clearly defined.

4. The staff is sighted using the slow motion screw if necessary and brought into sharp focus using the focusing screw on the side of the telescope.

5. The main spirit level is accurately centralized using the tilting screw and a reading taken on the staff.

When the staff is removed to another station the bubble of the main spirit level will move off-centre. However, a small rotation of the tilting screw will quickly bring it back to centre and another sight can be taken. In contrast to the dumpy level this relevelling will not alter the height of the plane of collimation since the telescope is pivoted centrally.

Example 1 In surveying examinations the following question is frequently asked:

'What is the *essential* difference between a dumpy and tilting level?'

The essential difference between a dumpy and a tilting level is that the vertical axis of the dumpy level is made perfectly vertical by the levelling operation. The telescope being fixed at right angles to the vertical axis will revolve around it sweeping out a horizontal plane. The telescope of the tilting level possesses a limited amount of vertical movement and can be made horizontal even though the vertical axis of the instrument is not truly vertical.

Coincidence bubble reader
In both the dumpy and tilting levels, the bubble of the spirit level is set by eye, to the centre of the bubble tube graduations. Many modern tilting levels, however, incorporate a coincidence bubble reader, the principles of which have been explained in Chapter 5 (Fig. 5.15). Figure 6.7 shows a WILD tilting level utilizing such a system.

3. Automatic levels
In conventional tilting levels the line of sight is, or should be, parallel to the axis of the telescope. It is only horizontal when the bubble of the properly adjusted spirit level is central.

In the automatic levels the line of sight is levelled automatically (within certain limits) by means of an

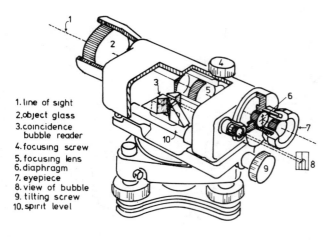

1. line of sight
2. object glass
3. coincidence
   bubble reader
4. focusing screw
5. focusing lens
6. diaphragm
7. eyepiece
8. view of bubble
9. tilting screw
10. spirit level

Figure 6.7

optical compensator inserted into the path of the rays through the telescope.

There are currently over fifty different automatic levels produced by various manufacturers and most of these instruments utilize a different compensator. Fundamentally, the compensator is a form of prism suspended like a pendulum and acting as a mirror

to direct the incoming horizontal ray from the staff through the centre of the diaphragm.

Basic principle of the compensator
Figure 6.8(a) shows a telescope in which two mirrors have been placed at 45° to the telescope axis.

The horizontal ray of light entering the objective glass through the optical centre will be reflected at 90° from mirror A onto mirror B where it will once again be reflected at 90° to pass through the centre of the diaphragm C.

In Fig. 6.8(b) the telescope has been tilted through a small angle of 1°. Relative to the horizontal plane the mirrors A and B therefore lie at an angle of 44°.

The horizontal ray of light (solid line) entering through the optical centre of the object glass strikes mirror A, is reflected to strike mirror B, and is reflected from it at an angle of 44°, that is, diverging from the original ray (shown dashed) by 1°. It no longer passes through the centre of the diaphragm.

If mirror A could be maintained in a position at 45° to the horizontal, however, the incoming horizontal ray would be reflected vertically from the surface towards B. It will strike mirror B at 46° and be reflected from it at the same angle thereby converg-

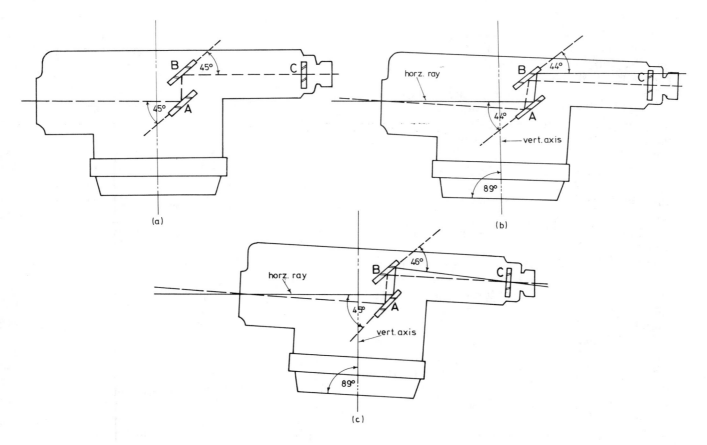

Figure 6.8

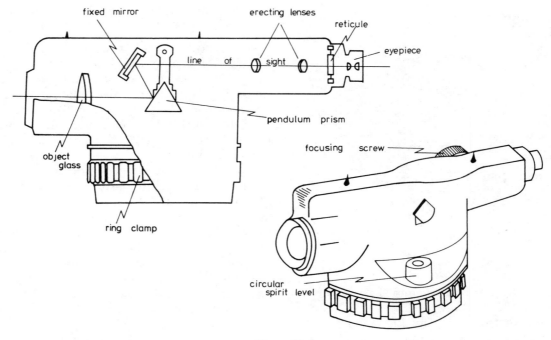

fixed mirror   erecting lenses

reticule

eyepiece

line of sight

pendulum prism

object glass

ring clamp

focusing screw

circular spirit level

Figure 6.9

ing on the original ray at 1° to pass through the centre of the diaphragm C (Fig. 6.8(c)).

Using this system the compensator (mirror A) must be placed exactly mid-way between the object glass and diaphragm.

It must be emphasized at this point that the deviation of 1° in the example could not be tolerated in practice. This figure was used for illustrative purposes only. The maximum deviation is of the order of ± 15 minutes of arc.

The system, as previously described, is applied almost exactly in the Nikon AP level. It has an automatic compensation prism suspended by a special metal plate mounted on ball bearings to maintain the horizontal line of sight automatically (Fig. 6.9).

There are several variations of this system but in all of them there is fundamentally some form of prism suspended like a pendulum which directs the horizontal ray through the centre of the diaphragm.

In the Hilger & Watts Autoset Levels the compensator consists of two suspended prisms and one fixed prism (Fig. 6.10). When the telescope is perfectly horizontal the ray of light follows the path shown. If the telescope is tilted by an angle x each of the suspended prisms will deflect the ray by $2x$ and cause it to pass through the centre of the diaphragm.

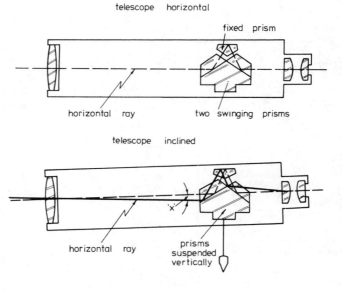

telescope horizontal

fixed prism

horizontal ray   two swinging prisms

telescope inclined

horizontal ray

prisms suspended vertically

Figure 6.10

Figure 6.11 shows a WILD automatic engineer's level. The compensator consists of two fixed prisms and one suspended prism. The pendulum and damping devices have been omitted in the figure for clarity.

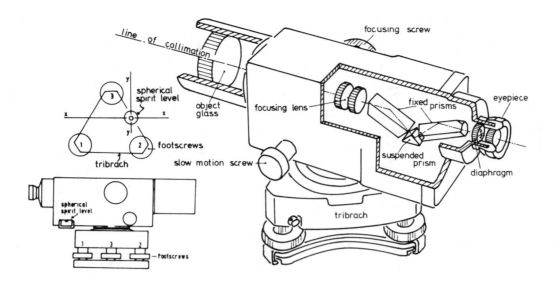

Figure 6.11

Damping systems

Most pendulum devices are suspended on wires (Sokkisha B2, Wild NAK). A few are suspended in the force field of a permanent magnet (Kern GKIA). Naturally the pendulum device oscillates and these oscillations must be speedily damped before a reading can be taken on a staff. The damping is achieved by attaching the compensator to a piston which is free to move backwards and forwards within a short cylinder (Fig. 6.12). The compensator's oscillations move the piston causing air to be expelled from the cylinder. This creates a vacuum which opposes the motion of the piston and brings the compensator to rest in a fraction of a second.

Alternatively, the damping of the compensator is achieved magnetically. A small magnet is built into the telescope housing. It produces a north–south magnetic field. The pendulum compensator is positioned within this magnetic field and the oscillations of the pendulum produce a braking force which is directly proportional to the movement of the pendulum.

Setting up the Automatic Level:

1. The instrument is set up with the levelling head approximately level and the instrument securely attached using the fastening screw.
2. On all automatic levels there is a small circular spirit level which is centred in exactly the same

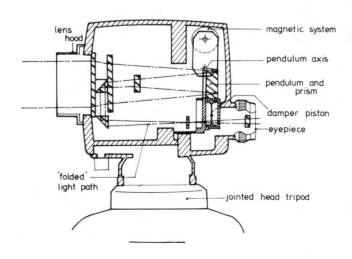

Figure 6.12

manner as a tilting level via a three-screw arrangement, a ball and socket joint or a jointed head system.

3. When the spirit level has been centred the vertical axis of rotation of the instrument is approximately vertical. The compensator automatically levels the line of sight for every subsequent pointing of the instrument.

4. Parallax is eliminated as before, the staff is sighted and brought into focus and the staff reading noted.

Levelling staff

The levelling staff should conform with British Standard Specification 4484 (1969). A portion of such a staff is shown in Fig. 6.13. The length of the instrument is 3 m, 4 m, or 5 m, while the width of reading face must not be less than 38 mm. Different colours must be used to show the graduation marks in alternate metres, the most common colours being black and red, on a white ground. Major graduations occur at 100 mm intervals, the figures denoting metres and decimal parts. Minor graduations are at 10 mm intervals, the lower three graduation marks of each 100 mm division being connected by a vertical band to form a letter E. Thus, the 'E' band covers 50 mm and its distinctive shape is a valuable aid in reading the staff.

Minor graduations of 1 mm may be estimated.

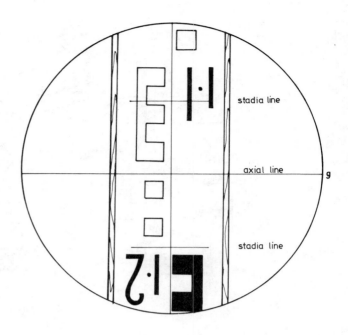

Figure 6.14

In Figure 6.13 various staff readings are shown to illustrate the method of reading. Owing to the optical principles of the telescope, staff readings will be inverted with many instruments. Figure 6.14 shows such an inverted view through the telescope of a levelling staff set up about eight metres away.

Example 2 Make a list of the staff readings 'a' to 'f' in Fig. 6.13 and reading 'g' in Fig. 6.14.

Solution

a — 1.960 m   b — 2.033 m   c — 1.915 m
d — 1.978 m   e — 2.050 m   f — 2.002 m
g — 1.156 m

Taking a reading

The method of taking a reading differs slightly with the type of levelling instrument being used. In general, however, the instrument is set up and levelled accurately, then parallax is eliminated.

The staffholder holds the staff on the mark and ensures that it is held perfectly vertically facing towards the instrument.

The observer directs the telescope towards the staff and using the focusing screw on the side of the telescope brings the staff clearly into focus. If parallax has been properly eliminated, the diaphragm should be sharply outlined against the staff and there should be no apparent movement of the crosshairs over the staff as the head is moved up and down.

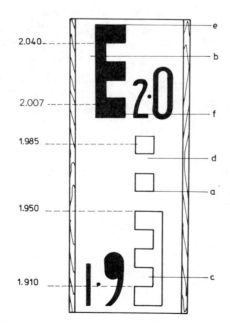

Figure 6.13

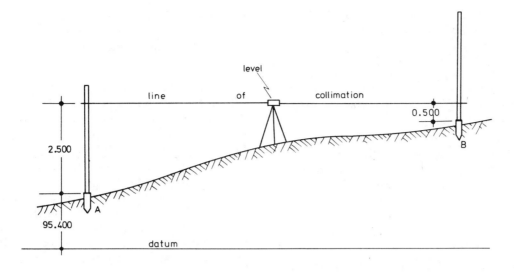

Figure 6.15

If a tilting level is being used, the main spirit level is now accurately centred. The dumpy and automatic levels will already be levelled accurately. The observer must now remove his hands from the tripod and instrument. If he leans on the tripod at this stage the line of collimation will be tilted upwards and a wrong reading will be taken.

Of the two sets of figures on either side of the axial line, the lower is noted. The 10 mm graduations are counted and the final millimetre placing is estimated. The complete reading is then booked, the spirit level checked, the staff reading taken once more and checked against the booked figures.

The staffholder is then directed to the next staff station where the reading procedure is repeated.

Levelling between two points

In order to find the difference in level between any two points on a building site, a level tripod and staff are used. In Fig. 6.15, A and B are two points approximately 60 m apart. Their difference in level is required.

The level is set up approximately mid-way between the points, accurately levelled, and a first reading is taken to a staff held vertically on 'A'. Let the reading be 2.500 m. The staff is then transferred to 'B', held vertically on the station and a second reading taken. Let this reading be 0.500 m.

From the sketch it is obvious that point B is higher than 'A' by 2.500 − 0.500 = 2.000 m. In other words the ground rises from A to B by 2.000 m.

In this example, if the height of the ground at 'A' is 95.400 m above datum, the height of the ground at B above datum can be readily deduced. Since the ground rises by 2.000 from A to B, the point B must lie 95.400 + 2.000 = 97.400 m above datum.

This is the basis for all levelling work no matter how long or complex the particular levelling may be. It is essential therefore to understand fully the above principle.

In general terms where the height above datum of any point is required a staff reading is taken to it and compared with a staff reading taken to a point of known height above datum. A comparison of the readings indicates a rise or fall of the ground between them. The unknown height is then determined by adding the rise to, or subtracting the fall from, the known height.

Booking and reducing the readings

Rise and fall system

All levellings must be properly recorded in a field book ruled as in Table 6.1.

BS	IS	FS	Rise	Fall	Reduced level	Distance	Remarks
2.500					95.400		A. Ground level
		0.500	2.000		97.400		B. Ground level

Table 6.1

At any instrument setting the first sight taken is called a 'backsight' (BS). In Fig.6.15 the sight to 'A' is therefore a backsight and the reading of 2.500 is entered in the BS column. The description of the observed point is entered in the 'remarks' column.

The final sight taken is called the 'foresight' (FS). In the example the foresight is the sight to 'B' and the reading 0.500 is entered in the FS column.

The rise or fall of the ground is calculated by *always* subtracting the second reading from the first. If a positive result is obtained there is a rise between the points. Similarly a negative result indicates a fall of the ground.

In the example: BS 'A'   =   2.500
            FS 'B'   =   0.500
          Difference (A − B) = +2.000 (Rise)

In levelling the term 'reduced level' is used to denote the height of any point above datum. Since 'A' lies 95.400 m above datum, this figure is entered in the reduced level column alongside 'A'.

The reduced level of 'B' is the algebraic sum of the reduced level A and the rise or fall from A to B.

       Reduced level A   =   95.400
       Rise A to B   =   +2.000
Therefore Reduced level B   = +97.400 m

Example 3   The data shown in Table 6.2 relate to three different levellings. Calculate the reduced level of the second point in each case by completing the table.

Flying levelling

Where the two points A and B lie a considerable distance apart or are at largely different elevations, more than one instrument setting is necessary. In Fig. 6.16, points A and B lie about 250 metres apart. The reduced level of A is 23.900 m and the reduced level 'B' is required.

The instrument is set about 40 m from A (position 1) and a backsight of 4.200 m taken. The staff is removed to a point about 40 m beyond the instrument and a foresight (0.700 m) taken. The reduced level of X is therefore calculated, as before:

      Backsight to A   =   4.200
      Foresight to X   =   0.700
      Difference (A − X)   = +3.500 m (Rise)

      Reduced level A   =   23.900
      Rise A to X   =   +3.500
      Reduced level X   27.400 m

Table 6.3 shows the field booking and reduction as it would be done on lines 1 and 2 of the table.

	BS	IS	FS	Rise	Fall	Reduced level	Distance	Remarks
(a)	3.250					135.260		Bench mark 'A'
			1.130					Kerb 'B'
(b)	0.752					73.270		Temp. bench mark
			2.896					Peg A
(c)	2.111					55.210		O.S. bench mark
			2.397					Ground level K

Table 6.2

Solution

(a) 3.250 − 1.130 = (Rise 2.120) + 135.260   = 137.380 m (Red. lev. B).

(b) 0.752 − 2.896 = (Fall 2.144) + 73.270   = 71.126 m (Red. lev. Peg A).

(c) 2.111 − 2.397 = (Fall   0.286) + 55.210   = 54.924 m (Ground lev. K).

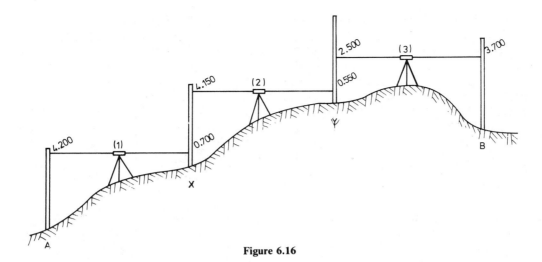

Figure 6.16

No further staff readings can be taken beyond X because the line of sight will run into the ground. The level is therefore removed to position (2), the staff again held on X facing towards the instrument and a sight is taken. Since this is the first sight at the new setting, it is by definition a backsight and the reading 4.150 is entered in the BS column. The reading must be entered on line 2 because this is the line referring to point X (Table 6.4). The staff is then removed to Y and a foresight taken. The reading 0.550 is entered in the field book on line 3 in the FS column. The reduced level of Y is calculated thus:

Backsight to X	=	4.150
Foresight to Y	=	0.550
Difference (X − Y)	=	+3.600 (Rise)

Reduced level of X	=	27.400
Rise X to Y	=	3.600
Reduced level of Y	=	31.000 m

It should be noticed that the booking and reduction of the second instrument setting is exactly the same as for the first setting and indeed all levelling reductions are performed in exactly this manner.

Since the line of sight would again run into the ground, if continued beyond Y the level is removed to position (3). The staff is held on Y and turned to face towards the new instrument position. A backsight is taken resulting in a reading of 2.500. A foresight is then taken to 'B' where the reading is 3.700. Table 6.5 shows the readings entered in their appropriate columns on lines 3 and 4 respectively. The reduced level of 'B' is calculated thus:

Backsight to Y	=	2.500
Foresight to B	=	3.700
Difference (Y − B)	=	−1.200 (Fall)

Reduced level Y	=	31.000
Fall Y to B	=	−1.200
Reduced level B	=	29.800

	BS	IS	FS	Rise	Fall	Reduced level	Distance	Remarks
Line 1	4.200					23.900		A. Ground level
Line 2			0.700	3.500		27.400		X. Ground level CP
Line 3								
Line 4								

Table 6.3

	BS	IS	FS	Rise	Fall	Reduced level	Distance	Remarks
Line 1	4.200					23.900		A. Ground level
Line 2	4.150		0.700	3.500		27.400		X. Ground level CP
Line 3			0.550	3.600		31.000		Y. Ground level CP

Table 6.4

	BS	IS	FS	Rise	Fall	Reduced level	Distance	Remarks
Line 1	4.200					23.900		A. Ground level
Line 2	4.150		0.700	3.500		27.400		X. Ground level CP
Line 3	2.500		0.550	3.600		31.000		Y. Ground level CP
Line 4			3.700		1.200	29.800		B. Ground level
Line 5	10.850		4.950	7.100	1.200	29.800		
Line 6	−4.950			−1.200		−23.900		
Line 7	= 5.900			= 5.900		= 5.900		

Table 6.5

The points X and Y are points where both foresights and backsights are taken as already seen, the instrument position is changed between the foresight and backsight and the points are called 'change points'. The letters CP are often inserted in the remarks column but are not absolutely necessary.

The complete reduction is shown in Table 6.5. As in all surveying operations, a check should be provided on the arithmetic. Lines 5, 6, and 7 provide such a check.

A moment's thought will show that the last reduced level is calculated as follows:

Last reduced level = first reduced level + all rises − all falls

therefore last reduced level = first reduced level + sum rises − sum falls

Each rise or fall, however, is the difference between its respective backsight and foresight, therefore, the sum of the rises minus the sum of the falls must equal the sum of the backsights minus the sum of the foresights.

The complete check is therefore:

(Last reduced level − first reduced level) = (sum rises − sum falls) = (sum BS − sum FS)

that is (29.800 − 23.900) = (7.100 − 1.200) = (10.850 − 4.950) = 5.900

This particular levelling is an example of flying levelling. The shortest route between the points A and B is chosen and as few instrument settings as possible are used.

Example 4 The data shown in Table 6.6 relate to a flying levelling from an Ordnance Survey Bench Mark (OSBM) to a proposed temporary bench mark (TBM). Calculate the reduced level of the TBM.

Series levelling

When the reduced levels of many points are required, the method known as series levelling is used.

In Fig. 6.17 the reduced levels of six points A to F are required.

The instrument is set up accurately and a sight taken to A. Since this is the first sight it is of course a backsight. Its value is entered in the BS column.

BS	IS	FS	Rise	Fall	Reduced level	Remarks
2.57					85.36	OSBM
1.03		2.16				CP1
2.03		1.52				CP2
		2.16				TBM

Table 6.6

Solution

2.57 − 2.16 = (Rise 0.41) + 85.36 = 85.77 (Red. lev. CP 1).

1.03 − 1.52 = (Fall 0.49) + 85.77 = 85.28 (Red. lev. CP 2).

2.03 − 2.16 = (Fall 0.13) + 85.28 = 85.15 (Red. lev. TBM).

Check (BS col. − FS col.) = (Rise col. − Fall col.) = (Red. lev. TBM − (Red. lev. OSBM) = −0.21.

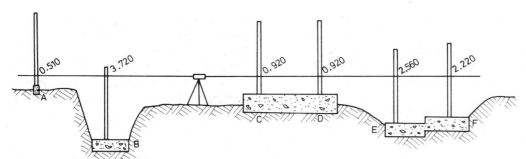

Figure 6.17

BS	IS	FS	Rise	Fall	Reduced level	Distance	Remarks
0.510					107.520		A. Ground level
	3.720			3.210	104.310		B. Foundation level 1
	0.920		2.800		107.110		C. Foundation level 2
	0.920				107.110		D. Foundation level 2
	2.560			1.640	105.470		E. Foundation level 3
		2.220	0.340		105.810		F. Foundation level 4
0.510		2.220	3.140	4.850	105.810		
2.220			−4.850		−107.520		
−1.710			−1.710		−1.710		

Table 6.7

The points B, C, D, and E are sighted in turn and finally point F is observed. Point F is the last sight taken and is therefore, by definition, a foresight. The readings taken at B, C, D, and E are intermediate between the first and last sights and are in fact called intermediate sights. The readings are entered in the IS column as in Table 6.7.

The rise or fall is required between successive points A to B, B to C, C to D, etc.

Backsight A	=	0.510
Intermediate sight B	=	3.720
Difference		−3.210 Fall

Intermediate sight B	=	3.720
Intermediate sight C	=	0.920
Difference		+2.800 Rise

Intermediate sight C	=	0.920
Intermediate sight D	=	0.920
Difference		0.000

Intermediate sight D	=	0.920
Intermediate sight E	=	2.560
Difference		−1.640

Intermediate sight E	=	2.560
Foresight F	=	2.220
Difference		+0.340 Rise

Check Sum of backsights
(one only) 0.510
Sum of foresights
(one only) 2.220
Difference −1.710 Fall

Sum of rises
(2.800 + 0.340) = 3.140
Sum of falls
(3.210 + 1.640) = 4.850
Difference −1.710 Fall

The arithmetic checks at this stage and it is therefore safe to proceed to the reduced level column. As in the previous example the reduced levels are obtained by algebraically adding the rises or falls to the first reduced level in succession.

Reduced level A	=	107.520
− fall (3.210) A to B	=	104.310
+ rise (2.800) B to C	=	107.110
No rise or fall C to D	=	107.110
− fall (1.640) D to E	=	105.470
+ rise (0.340) E to F	=	105.810
Check Last reduced level	=	105.810
− first reduced level	=	−107.520
Difference	=	−1.710 Fall

The arithmetical check provided in all levellings checks solely that the observed readings entered in the field book have been correctly computed. It does not prove that the reduced level of any point is correct. If verification of each level is required, the field work must be repeated from a different instrument set up point as in Example 5.

Example 5 The levelling shown in Fig. 6.17 was re-observed from a second instrument station. The observations were as follows:

A—0.240 B—3.450 C—0.655 D—0.650
E—2.290 F—1.955

Enter these results in a levelling table, then calculate and check the reduced levels of the stations from the temporary bench mark A (reduced level 107.520 m).

The relevelling shows that there is a 5 mm difference in the reduced levels of points C and F, which is acceptable.

Figure 6.18 shows a small site where reduced levels are required around the perimeter at various identifiable features. One benchmark is available.

Whenever possible the levelling instrument should be set in the centre of the site and observations made to all points sequentially as in Example 5. Frequently this will not be possible for the following reasons:

1. Buildings obstruct the lines of sight.
2. The lengths of some of the sights are too great.
3. The point being observed may be at a higher level than the instrument. The horizontal line of sight through the instrument therefore strikes the

Solution (Table 6.8)

BS	IS	FS	Rise	Fall	Reduced level	Remarks
0.240					107.520	A (TBM)
	3.450			3.210	104.310	B
	0.655		2.795		107.105	C
	0.650		0.005		107.110	D
	2.290			1.640	105.470	E
		1.955	0.335		105.805	F
0.240		1.955	3.135	4.850	105.805	
−1.955			−4.850		−107.520	
−1.715			−1.715		−1.715	

Table 6.8

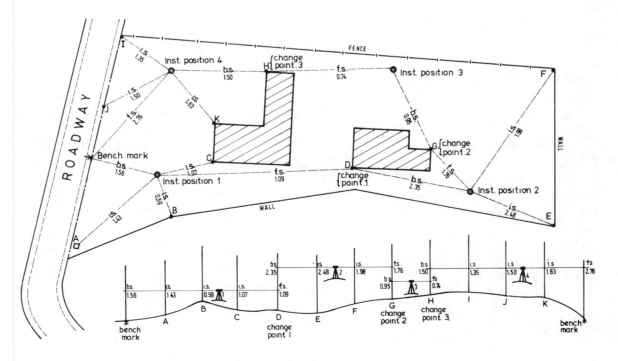

Figure 6.18

ground before reaching the point.

4. The point being observed may be at a much lower level than the instrument in which case the line of sight through the instrument passes over the staff held on the point.

In Fig. 6.18 some of these difficulties are present, so a set-up position has to be chosen from which the bench mark is visible. The instrument is set up at position 1 and a backsight is taken to the BM, (RL 35.27 m) followed by sights to as many other points as possible or desirable. In this case only the points A, B, C, and D are visible from the instrument.

Intermediate sights are taken to points A, B, and C, and a foresight taken to the final point D. The readings are entered in the field book (Table 6.9 section ①) and point D is noted as a change point.

It is now possible *though not usual practice* to reduce the levels in exactly the same manner as before. The relevant calculations are detailed below and entered in Table 6.9, section ②.

(a) Calculation of rise or fall between successive points BM to A, A to B, B to C, and C to D.

Backsight to BM,   1.56
− Intermediate sight to A   − 1.43
        = + 0.13 Rise

Intermediate sight to A   1.43
− Intermediate sight to B   −0.59
        = + 0.84 Rise

Intermediate sight to B   0.59
− Intermediate sight to C   − 1.07
        = − 0.48 Fall

Intermediate sight to C   1.07
− Intermediate sight to D   − 1.09
        = − 0.02 Fall

(b) Calculation of reduced levels. These are obtained by algebraically adding the rises or falls to the first reduced level in succession.

Reduced level BM,   = 35.27
+ rise (0.13) BM to A = 35.40 (RL A)
+ rise (0.84)   A to B = 36.24 (RL B)
− fall (0.48)   B to C = 35.76 (RL C)
− fall (0.02)   C to D = 35.74 (R,L, D)

The field work and associated calculations finished on point D which is a change point. The second series of readings must therefore begin on point D which in effect has become a temporary bench mark.

The instrument is removed to instrument set-up point 2 and a backsight taken to the point D. It is entered in the field book, Table 6.9, section ③, in

B.S.	I.S.	F.S.	RISE	FALL	REDUCED LEVEL	DISTANCE	REMARKS
1.56					35.27	not	Bench mark
①	1.43		0.13	②	35.40	required	A manhole
	0.59		0.84		36.24		B fence
	1.07			0.48	35.76		C corner of building
2.35		1.09		0.02	35.74		D corner of building change point 1
③	2.48			0.13	35.61		E fence
	1.98		0.50	④	36.11		F fence
0.95	⑤	1.76	0.22		36.33		G corner of building change point 2
1.50		0.74	0.21	⑥	36.54		H corner of building change point 3
	1.35		0.15		36.69		I fence
⑦	1.50		⑧	0.15	36.54		J fence
	1.63			0.13	36.41		K corner of building
		2.76		1.13	35.28		Bench mark
6.36		6.35	2.05	2.04	35.28		
−6.35			−2.04		−35.27		
=0.01			=0.01		=0.01		

Table 6.9

the BS column on line D. Thus there are two field readings on the same line since they were taken to the same station, i.e., point D.

The levelling is continued with another series of readings taken to any further accessible points. These are intermediate sights to E and F and a foresight to G which becomes the second change point.

(a) These results are calculated and entered in the field book:
Table 6.9, section ④.

Backsight to D	2.35
− Intermediate sight to E	− 2.48
=	− 0.13 Fall

Intermediate sight to E	2.48
− Intermediate sight to F	− 1.98
=	+ 0.50 Rise

Intermediate sight to F	1.98
− Foresight to G	− 1.76
=	+ 0.22 Rise

(b) The reduced levels E, F, and G are calculated from change point D.

Reduced level   = 35.74
− fall (0.13) to E = 35.61 (RL E)
+ rise (0.50) to F = 36.11 (RL F)
+ rise (0.22) to G = 36.33 (RL G)

The instrument is removed to set up point 3 and the third series of levels is taken. From this position only the points G and H can be seen. Point G is a backsight. The reading is entered in the field book, Table 6.9, section ⑤ in the BS column, on line G. Point H is a foresight and is the next change point.

The relevant calculations to obtain the reduced level of H, shown in Table 6.9, section ⑥, are as follows:

(a) Backsight to G	0.95
− Foresight to H	− 0.74
=	+ 0.21 Rise

(b) Reduced level G = 36.33
+ rise (0.21) to H = 36.54 (RL H)

From instrument position 4, the fourth series of levels is observed beginning with a backsight to station H and finishing with a foresight to the BM. They are calculated in the manner of the foregoing paragraphs. The observations and relevant results are shown in Table 6.9, sections ⑦ and ⑧.

(a) Backsight H	1.50
− Intermediate sight I	− 1.35
=	+ 0.15 Rise

Intermediate sight I	1.35
− Intermediate sight J	− 1.50
=	− 0.15 Fall

Intermediate sight J	1.50
− Intermediate sight K	− 1.63
=	− 0.13 Fall

Intermediate sight K	1.63
− Foresight L	− 2.76
=	− 1.13 Fall

(b) Reduced level H   = 36.54
+ rise (0.15) to I = 36.69 (RL I)
− fall (0.15) to J = 36.54 (RL J)
− fall (0.13) to K = 36.41 (RL K)
− fall (1.13) to L = 35.28 (RL)

The arithmetical check shows that the levels have been correctly calculated and also shows that an error of 10 mm has been made in the field work. The normal acceptable limit of error is ± 20 mm for this levelling.

It was stated earlier that it is unusual to calculate the results in parallel with the field work for the simple reason that if any observation is wrongly read the calculated results must also be wrong. It is therefore wise to complete the field work and verify that the levelling closes before calculating any results.

The logical order in any levelling calculation is therefore:
1. Complete the field work.
2. Check that the sum BS column = sum FS column within the acceptable limit of error. If the error exceeds these limits there is no point in continuing the calculation.
3. Calculate rises and falls.
4. Check that (sum Rise column − sum Fall column) = (sum BS column − sum FS column).
5. Calculate reduced levels.
6. Check that (Last RL − First RL) = (sum Rise column − sum Fall column).

Example 6 (a) Figure 6.9(a) shows the positions of a level and staff set up on the line of a proposed drainage installation.

Draw up a page of a level book and reduce the readings, applying the necessary checks.

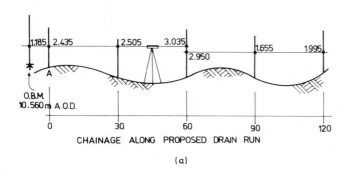

CHAINAGE ALONG PROPOSED DRAIN RUN

(a)

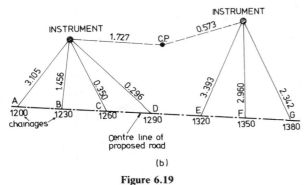

(b)

Figure 6.19

(b) Figure 6.19(b) shows levelling information observed as part of a proposed road. The readings 1.727 and 0.573 were on a change point while the remainder of the readings at points A, B, C, D, E, F,

and G were taken at 30 m intervals along the line of the proposed road.

Given that the reduced level of point A is 13.273 m above datum, calculate the reduced levels of all points along the road.

(City and Guilds of London Institute Surveying and Levelling Examination)

Solution   See Tables 6.10 and 6.11

Collimation system of reduction

An alternative method of reducing the levels is that commonly known as the collimation system of reduction.

It should be remembered that the line of collimation is the line joining the optical centre of the object glass to the axial line of the reticule. When the telescope is rotated it will sweep out a horizontal plane known as the plane of collimation, Fig. 6.20(a).

In Fig. 6.20(a) the plane of collimation cuts a levelling staff held on a point A where the reduced level is 205.500 m. The staff reading is 2.400 m.

In this method of reduction the height of the plane of collimation above datum is required for every instrument setting. Clearly, in the figure the height of the plane of collimation (HPC) above datum is the reduced level of A (205.500 m) plus the staff reading 2.400 m at that point.

BS	IS	FS	Rise	Fall	Reduced level	Distance	Remarks
1.185					10.560		OBM
	2.435			1.250	9.310	0	
	2.505			0.070	9.240	30	
2.950		3.035		0.530	8.710	60	
	1.655		1.295		10.005	90	
		1.995		0.340	9.665	120	
4.135		5.030	1.295		9.665		
−5.030			−2.190		−10.560		
−0.895			−0.895		− 0.895		

Table 6.10

BS	IS	FS	Rise	Fall	Reduced level	Distance	Remarks
3.105					13.273	1200	TBM A
	1.456		1.649		14.922	1230	B
	0.350		1.106		16.028	1260	C
	0.296		0.054		16.082	1290	D
0.573		1.727		1.431	14.651		Change point
	3.393			2.820	11.831	1320	E
	2.960		0.433		12.264	1350	F
		2.342	0.618		12.882	1380	G
3.678		4.069	3.860	4.251	12.882		
−4.069			−4.251		−13.273		
−0.391			−0.391		− 0.391		

Table 6.11

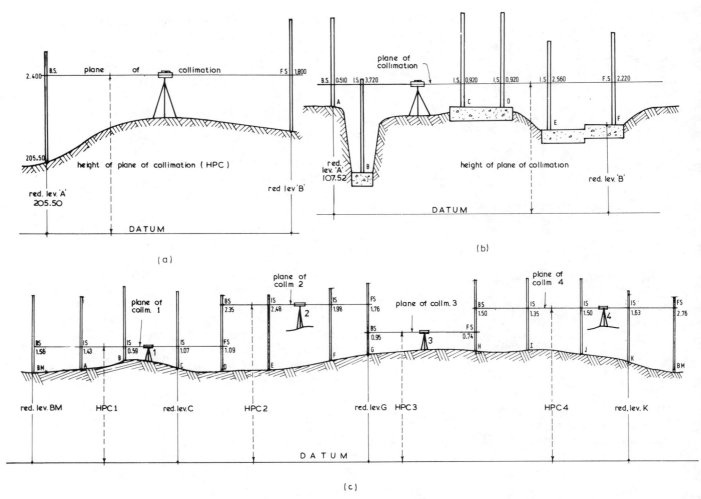

Figure 6.20

That is, HPC = Reduced level A + staff reading
= 205.500 + 2.400
= 207.900 m

If another reading of 1.800 m is observed at B, the reduced level of the point B can be readily found. The plane of collimation is still at a height of 207.900 when pointing to B and since ground level B lies 1.800 m below the HPC the reduced level of B is HPC minus staff reading B.

That is, Reduced level B = HPC − staff reading
= 207.900 − 1.800
= 206.100 m

Generally, then, the height of the plane of collimation is the reduced level of any point plus the staff reading *AT THAT POINT* and the reduced level of any other point is the HPC minus the staff reading at the point.

The field book is ruled differently, of course, and Table 6.12 shows the book ruled for the collimation system of reduction.

Since the actual observations are in no way altered, the sighting to A is a backsight while B is a foresight. The HPC 207.900 is entered on line 1. The HPC is written only once, opposite the backsight, and is understood to refer to the whole of that instrument setting. The reduced level of B, 206.100, is entered on line 2 in its appropriate column.

	BS	IS	FS	HPC	Reduced level	Distance	Remarks
Line 1	2.400			207.900	205.500		A. Ground level
Line 2			1.800		206.100		B. Ground level

Table 6.12

In Fig. 6.20(b) the reduced levels of six points are required. (The figure and the readings used are those of Fig. 6.17 where the levels were calculated by the Rise and Fall Method.)

(a) The height of collimation is found as before.

$$\begin{aligned} \text{HPC} &= \text{Reduced level A} + \text{staff reading (BS)} \\ &= 107.52 + 0.51 \\ &= \underline{108.03 \text{ m}} \end{aligned}$$

The HPC is entered in the field book in line 1. It is then held to apply to the whole of that instrument setting and need not be repeated down the whole HPC column.

(b) The reduced level of every other point is found thus:

$$\text{RL (any point)} = \text{HPC} - \text{staff reading at that point}$$

therefore
$$\begin{aligned} \text{RLB} &= 108.03 - 3.72 = 104.31 \\ \text{RLC} &= 108.03 - 0.92 = 107.11 \\ \text{RLD} &= 108.03 - 0.92 = 107.11 \\ \text{RLE} &= 108.03 - 2.56 = 105.47 \\ \text{RLF} &= 108.03 - 2.22 = 105.81 \end{aligned}$$

The reduced levels are entered in the RL column (Table 6.13) opposite their respective staff readings. They agree with the levels calculated in Table 6.7 by the rise and fall method.

Example 7 The levelling shown in Fig. 6.20(b) was checked from a second instrument station. The observations were as follows:

A—0.240 B—3.450 C—0.655 D—0.660
E—2.290 F—1.955

Enter these results in a levelling table then calculate the reduced levels of the stations from the temporary bench mark A (RL 107.520 m).

For comparison of levels, see Table 6.8.

Solution See Table 6.14

BS	IS	FS	HPC	Reduced level	Distance	Remarks	
0.51			108.03	107.52		A	Ground level
	3.72		108.03	104.31		B	Foundation level 1
	0.92		108.03	107.11		C	Foundation level 2
	0.92		108.03	107.11		D	Foundation level 2
	2.56		108.03	105.47		E	Foundation level 3
		2.22	108.03	105.81		F	Foundation level 4
0.51		2.22		105.81			
−2.22				−107.52		Check	
−1.71				− 1.71			

Table 6.13

BS	IS	FS	HPC	Reduced level	Distance	Remarks	
0.240			107.760	107.520		A	(TBM)
	3.450		107.760	104.310		B	Foundation level 1
	0.655		107.760	107.105		C	Foundation level 2
	0.650		107.760	107.110		D	Foundation level 2
	2.290		107.760	105.470		E	Foundation level 3
		1.955	107.760	105.805		F	Foundation level 4
0.240		1.955		105.805			
−1.955				−107.520		Check	
−1.715				− 1.715			

Table 6.14

Series levelling—multiple settings

Figure 6.20(c) shows a whole series of points where the reduced levels are required. (The figure and the readings are those of Fig. 6.18 where the levels were reduced by the rise and fall method.)

(a) The height of collimation of the first instrument setting is found by adding the reduced level of the bench mark and the backsight reading observed to it.

$$\begin{aligned} \text{HPC (1)} &= \text{Reduced level BM} \\ &\quad + \text{staff reading (BS)} \\ &= 35.27 + 1.56 \\ &= \underline{36.83 \text{ m}} \end{aligned}$$

The HPC is entered in the field book (Table 6.15, line 1).

(b) The reduced levels of stations A, B, C, and D, observed from instrument position 1, are obtained by subtracting their staff readings from the HPC, thus:

$$\begin{aligned} \text{reduced level (any point)} &= \text{HPC} \\ &\quad - \text{staff reading} \end{aligned}$$

therefore

reduced level A = 36.83 − 1.43 = 35.40 m
reduced level B = 36.83 − 0.59 = 36.24 m
reduced level C = 36.83 − 1.07 = 35.76 m
reduced level D = 36.83 − 1.09 = 35.74 m

These levels are entered in the field book (Table 6.15, lines 2–5) opposite their respective staff readings.

(c) The field work and calculations finished with a foresight to point D which is a change point. The instrument is removed to set-up point 2 and a backsight taken to station D. The instrument therefore has a new height of collimation which is found as before:

$$\begin{aligned} \text{HPC (2)} &= \text{Reduced level (change point)} \\ &\quad + \text{staff reading (BS)} \\ &= 35.74 + 2.35 \\ &= \underline{38.09 \text{ m}} \end{aligned}$$

The HPC (2) is entered in the field book (Table 6.15, line 5) opposite the BS, to which it refers.

(d) The reduced levels of the second set of points are obtained as before by subtracting their staff readings from the HPC, thus:

$$\begin{aligned} \text{reduced level (any point)} &= \text{HPC (2)} - \text{staff} \\ &\quad \text{reading} \end{aligned}$$

therefore

reduced level E = 38.09 − 2.48 = 35.61 m
reduced level F = 38.09 − 1.98 = 36.11 m
reduced level G = 38.09 − 1.76 = 36.33 m

These levels are entered in the field book (Table 6.15, lines 6–8) opposite their respective staff readings.

(e) Point G is a change point. The instrument is removed to set-up point 3 and a backsight observed to change point G immediately followed by a foresight to H.

$$\begin{aligned} \text{HPC (3)} &= \text{Reduced level G + BS} \\ &= 36.33 + 0.95 \\ &= \underline{37.28 \text{ m}} \text{ (Table 6.15, line 8)} \end{aligned}$$

$$\begin{aligned} \text{Reduced level H} &= \text{HPC (3)} - \text{foresight} \\ &= 37.28 - 0.74 \\ &= 36.54 \text{ m (Table 6.15, line 9)} \end{aligned}$$

(f) The instrument is set up at station 4 and the levelling is closed from change point H to the bench mark.

$$\begin{aligned} \text{HPC (4)} &= \text{Reduced level H + backsight} \\ &= 36.54 + 1.50 \\ &= \underline{38.04 \text{ m}} \text{ (Table 6.15, line 9)} \end{aligned}$$

$$\begin{aligned} \text{Reduced level (any point)} &= \text{HPC (4)} \\ &\quad - \text{staff readings} \end{aligned}$$

Reduced level I = 38.04 − 1.35 = 36.69 m
Reduced level J = 38.04 − 1.50 = 36.54 m
Reduced level K = 38.04 − 1.63 = 36.41 m
Reduced level BM = 38.04 − 2.76 = 35.28 m

The levels are entered in the field book (Table 6.15, lines 10–13).

Check: The commonly applied check is shown in lines 14, 15 and 16. It is the same as the check in the rise and fall system in that the difference between the first and last reduced levels equals the difference between the sum of the BS column and the sum of the FS column, namely 0.01.

While this check is very widely used it does *not* completely check the levelling and many serious errors have arisen on site because this fact is not appreciated. Suppose for the moment, that the RL of point B had been found to be 37.24 m instead of 36.24 m. The check shown above would still work and indeed if the reduced level of *every* intermediate sight were wrong, the check would continue to be successful. The reason for this is simply that a reduced level does not depend on the value of the preceding level as it does in the rise and fall system.

Line	BS	IS	FS	HPC	Reduced level	Remarks
1	1.56			36.83	35.27	BM
2		1.43			35.40	A
3		0.59			36.24	B
4		1.07			35.76	C
5	2.35		1.09	38.09	35.74	D
6		2.48			35.61	E
7		1.98			36.11	F
8	0.95		1.76	37.28	36.33	G
9	1.50		0.74	38.04	36.54	H
10		1.35			36.69	I
11		1.50			36.54	J
12		1.63			36.41	K
13			2.76		35.28	BM
14	6.36	12.03	6.35		35.28	
15	−6.35				−35.27	
16	0.01				0.01	

Table 6.15

The complete arithmetical check shown below is complex and is almost certainly the reason for the adoption of the simple check shown above.
Complete check:

Sum of reduced
levels (except first) = sum (each height of collimation
× number of IS and FS
observed from each)
− sum (IS column
+ FS column)

In Table 6.15:

Sum of RLs except first = 432.65 m
Sum of HPC × No. of IS and FS =

$$
\begin{aligned}
36.83 \times 4 &= 147.32 \\
+\ 38.09 \times 3 &= +114.27 \\
+\ 37.28 \times 1 &= +\ 37.28 \\
+\ 38.04 \times 4 &= +152.16 \\
\hline
 &\ 451.03
\end{aligned}
$$

Sum of IS column 12.03
Sum of FS column 6.35 = 18.38

(451.03 − 18.38)   = 432.65 m

Example 8 The level booking sheet (Table 6.16) shows the ground levels through which it is proposed to run a drain, rising from Manhole 1 to Manhole 5, at a gradient of 1:40. Calculate the

(SCOTVEC—Ordinary National Diploma in Building)

Backsight	Intermediate sight	Foresight	Height of collimation	Reduced level	Distance	Remarks
1.579				100.000	0.0	BM 1 on cover plate Manhole No. 1
	1.295				20.0	
	1.873				40.0	
	2.018				60.0	
	1.884				80.0	Manhole No. 2 Cover level
	1.625				100.0	
2.441		1.000			105.0	Start of steps
	1.807				118.5	Top of steps
	1.495				122.3	Manhole No. 3 Cover level
		1.020		102.000	135.0	BM 2 on cover plate Manhole No. 5

Table 6.16

Solution

Backsight	Intermediate sight	Foresight	Height of collimation	Reduced level	Distance	Remarks
1.579			101.579	100.000	0.0	BM 1 on cover plate Manhole No. 1
	1.295			100.284	20.0	
	1.873			99.706	40.0	
	2.018			99.561	60.0	
	1.884			99.695	80.0	Manhole No. 2 Cover level
	1.625			99.954	100.0	
2.441		1.000	103.020	100.579	105.0	Start of steps
	1.807			101.213	118.5	Top of steps
	1.495			101.525	122.3	Manhole No. 3 Cover level
	1.807			101.213	129.6	Manhole No. 4 Cover level
		1.020		102.000	135.0	BM 2 on cover plate Manhole No. 5
4.020 −2.020 = 2.000	13.804	2.020		102.000 −100.000 = 2.000		

Table 6.17

Check:

Sum of RL (except 1st) = 1005.73
$$= (101.579 \times 6)$$
$$+ (103.02 \times 4)$$
$$- (13.804 + 2.020)$$
$$= \underline{1005.73}$$

reduced levels of each staff position, and apply a full arithmetic check.

Comparison of methods of reduction

The rise and fall system provides a complete check on the total working with ease, whereas the collimation system of checking is very tedious.

The rise and fall system takes longer to work out but takes much less time to check than does the collimation system. The overall time is much the same for both systems.

The rise and fall system should be used where the levelling involves a great number of intermediate sights.

Undoubtedly, the collimation system is best for setting out levels and since a great deal of setting out has to be done on building sites, this is probably the reason for the popularity of the method amongst building engineers.

Inverted staff readings

In all of the previous examples on levelling, the points observed all lay below the line of sight. Frequently on building sites, the reduced levels of points above the height of the instrument are required, e.g., the soffit level of a bridge or underpass, the underside of a canopy, the level of roofs, eaves, etc., of buildings. Figure 6.21 illustrates a typical case.

The reduced levels of points A, B, C, and D on the frame of a multi-storey building require checking. The staff is simply held upside down on the points A and C and booked with a *minus sign* in front of the reading, e.g., − 1.520. Alternatively, the reading may be put in brackets, e.g., (1.520) or an asterisk may be put alongside the figure, e.g., 1.520*. Such staff readings are called inverted staff readings.

The readings in Table 6.18 refer to the levelling in Fig. 6.21.

Reduction by rise and fall method

The rise or fall is required between successive points as before. The second reading is subtracted from the first as follows:

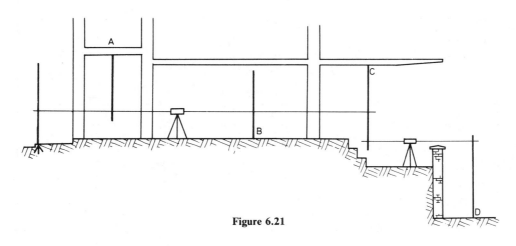

Figure 6.21

BS to Bench Mark	=	(1.750)
IS to frame 'A'	=	−(−3.100)
Difference to BM to A	=	+4.850 Rise
IS to frame A	=	(−3.100)
IS to floor level B	=	−(1.490)
Difference A to B	=	−4.590 Fall
IS to floor level B	=	1.490
FS to Canopy C	=	−(−2.560)
Difference B to C	=	+4.050 Rise
BS to Canopy C	=	(−4.210)
FS to Kerb D	=	−(4.200)
Difference C to D	=	−8.410 Fall

The reduced levels are obtained by the algebraic addition of the rises and falls as before.

The arithmetical check is applied in the usual manner. The BS and FS columns are added algebraically, the inverted staff readings being regarded as negative.

(Last RL − first RL) = (sum rises − sum falls)

= (sum BS − sum FS)

= 68.200 − 72.300 = 8.900 − 13.000

= −2.460 − 1.640 = −4.100 m.

Example 9 Calculate the reduced levels of points A, B, C, and D in Fig. 6.21 using the height of collimation method.

BS	IS	FS	Rise	Fall	Reduced level	Distance	Remarks
1.750					72.300		Bench mark
	−3.100		4.850		77.150		A Frame (lift shaft)
	1.490			4.590	72.560		B Floor level
−4.210		−2.560	4.050		76.610		C Canopy
		4.200		8.410	68.200		D Kerb
−2.460		+1.640	8.900	13.000	68.200		
−1.640			−13.000		−72.300		
−4.100			−4.100		−4.100		

Table 6.18

BS	IS	FS	HPC	Reduced level	Remarks
1.750			74.050	72.300	Bench mark
	−3.100			77.150	A Frame lift shaft
	1.490			72.560	B Floor level
−4.210		−2.560	72.400	76.610	C Canopy
		4.200		68.200	D Kerb
−2.460	−1.610	1.640		68.200	
−1.640				−72.300	
−4.100				− 4.100	

Table 6.19

Check:

Sum of RLs except 1st = 294.52

$$= (74.050 \times 3) + 72.400 - (-1.61 + 1.64)$$
$$= 222.15 + 72.40 - 0.03$$
$$= \underline{294.52}$$

Errors in levelling

As in all surveying operations, the sources and effects of errors must be recognized and steps taken to eliminate or minimize them. Errors in levelling can be classified under several headings.

1. *Gross errors*
Mistakes arising in the mind of the observer. They may be due to carelessness, inexperience, or fatigue.

(a) *Wrong staff readings*   This is probably the most common error of all in levelling. Examples of wrong staff readings are: misplacing the decimal point, reading the wrong metre value and reading the staff wrong way up.

(b) *Using the wrong crosshair*   Instead of reading the staff against the axial line, the observer reads against one of the stadia lines. This error is common in poor visibility.

(c) *Wrong booking*   The reading is noted with the figures interchanged, e.g., 3.020 instead of 3.002.

(d) *Omission or wrong entry*   A staff reading can easily be written in the wrong column or even omitted entirely.

(e) *Spirit level not centred*   The staff is read without centring the bubble.

All of these errors can be small or very large and every effort must be made to eliminate them. The only way to eliminate gross errors is to make a double levelling, that is, to level from A to B then back from B to A. Theoretically the levelling should close without any error but this will very seldom happen. However, the error should be within acceptable limits, certainly not more than 1 mm per 50 m for levellings less than 1 kilometre.

2. *Constant errors*
These errors are due to instrumental defects and will always be of the same sign.

(a) *Non-verticality of the staff*   This is a serious source of error. Instead of being held vertically the staff may be leaning forward or backward. In Fig. 6.22 the staff is 3° out of vertical. If a reading of 4.000 m is observed, it will be in error by 5 mm. The correct reading is

$$4 \times \cos 3° = \underline{3.995 \text{ m}}$$

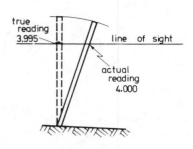

Figure 6.22

The error can be eliminated by fitting the staff with a circular spirit level. The staffholder must ensure that the bubble is centred when the staff is being read. A second method of eliminating the error is for the staffholder to swing the staff slowly backwards and forwards across the vertical position during the observation. The observer then reads the lowest reading.

(b) *Collimation error in the instrument*   In a properly adjusted level the line of sight must be perfectly horizontal when the bubble of the spirit level is central. If this condition does not prevail there will be an error in the staff reading. In Fig. 6.23 the line of sight is inclined and the resultant error is *e*, increasing with length of sight. The error can be entirely eliminated by making backsights and foresights equal in length. The error *e* will be the same for each sight and the true difference in level will be the difference in the readings. It is fair to say that collimation errors on instruments are fairly common on building sites. Since the resulting errors can be completely eliminated by taking backsights

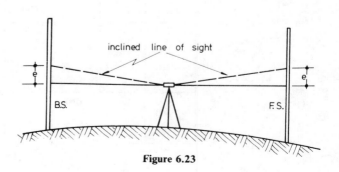

Figure 6.23

and foresights of equal length, this practice should always be adhered to.

(c) *Staff graduation error* This error is more common than is often imagined, particularly at this transition stage when Imperial staffs are being converted to metric by sticking metric graduation strips to the faces of old staffs. Care must be taken to ensure that the zero of the strips coincides with the bottom of the staff and that the various staff sections are properly graduated.

3. *Random errors*

These errors are due to physical and climatic conditions. The resulting errors are small and are likely to be compensatory.

(a) *Effect of wind and temperature* The stability of the instrument may be affected causing the height of collimation to change slightly.

(b) *Soft and hard ground* When the instrument is set on soft ground it is likely to sink slightly as the observer moves around it. When set in frosty earth, the instrument tends to rise out of the ground. Again the height of collimation changes slightly.

(c) *Change points* At any change point the staff must be held on exactly the same spot for both foresight and backsight. A firm spot must be chosen and marked by chalk. If the ground is soft a change plate (Fig. 6.24) must be used. The plate is simply a triangular piece of metal bent at the apices to form spikes. A dome of metal is welded to the plate and the staff is held on the dome at each sighting.

Figure 6.24

(d) *Human deficiencies* Errors arise in estimating the millimetre readings, particularly when visibility is bad or sights are long.

All of the errors in this class tend to be small and compensating and are of minor importance only in building surveying.

Permanent adjustments

In the preceding section on errors, it was pointed out that the line of collimation might not be parallel to the axis of the bubble tube. The levelling is still correct provided the sights taken are of equal length. This is not always possible, however, particularly when many intermediate sights are being taken and the only way in which these errors can be eliminated is to ensure that the instrument is in good adjustment.

Prior to the commencement of any contract the levelling instrument must be tested and if necessary, adjusted.

The adjustments of the various instruments are slightly different, because of their differing constructions.

1. Dumpy level

(a) The vertical axis must be truly vertical when the bubble of the spirit level is central.
Test
(i)   Set up the level and level the spirit level over two footscrews. If the instrument is not in good adjustment the relationship of the vertical axis to the spirit level will be as shown in Fig. 6.25(a), i.e., there is an inclination error, *e*.
(ii) Turn the telescope through 90° and recentralize the bubble.
(iii) Repeat these operations until the bubble remains central for positions (i) and (ii).
(iv) Turn the telescope until it is 180° from position (i). The vertical axis will still be inclined with an error *e* and the bubble of the spirit level will no longer be central. It will, in fact, be inclined to the horizontal at an angle of 2*e*, Fig. 6.25(b). The number of divisions, *n*, by which the bubble is off centre is noted.
Adjustment
(v)   Turn the footscrews until the bubble moves back towards the centre by *n*/2 divisions, i.e., by half the error. The vertical axis is now truly vertical, Fig. 6.25(c).
(vi) Adjust the spirit level by releasing the capstan screws and raising or lowering one end of the spirit level until the bubble is exactly central. The other half *n*/2 of the error is thereby eliminated, Fig. 6.25(d), and the spirit level axis is at right angles to the vertical axis.

(b) The line of collimation must be parallel to the newly adjusted spirit level, in other words, the line of collimation must be horizontal.

The question is often asked 'How can the line of sight be in error in a dumpy level?' The answer is, that the instrument can receive a knock thereby displacing the reticule vertically. Even without any knocks the displacement can take place through natural wear and tear, effects of temperature, etc.
Test
The test called the 'two-peg' test is as follows:
(i)   Select two points A and B 60 m apart and

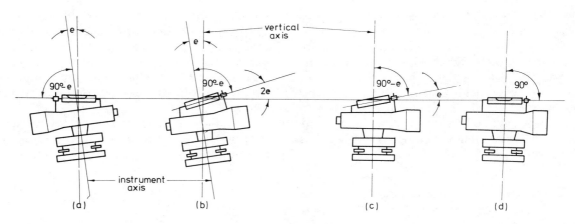

Figure 6.25

hammer two pegs firmly into the ground.

(ii) Set up the instrument exactly mid-way between the pegs and level carefully.

(iii) Sight staff A and note the reading.

(iv) Remove the staff to B. Read the staff and adjust the height of peg B until exactly the same reading as at A is obtained.

(v) The two pegs A and B will be at exactly the same height irrespective of whether the line of collimation is horizontal or not. In Fig. 6.26 the line of collimation is inclined upwards by an angle (α) and causing the staff reading at A to be in error by an amount e. Since peg B is exactly the same distance from the instrument as A there will also be an error $\grave{e}$.

(vi) Remove the instrument to B and, using a plumb-bob, set it up so that the eyepiece is exactly over peg B. Level carefully (Fig. 6.26).

(vii) Hold a staff on B and read it against the eyepiece of the instrument.

(viii) Remove the staff to A and read it. The reading should be identical to the staff reading at B in (vii) above. If not, there is a collimation error caused by the reticule being displaced vertically.

Adjustment

(ix) Release the antagonistic adjusting screws and move the reticule up, in this case (Fig. 6.27) until the required reading is obtained at A.

(x) Since pegs A and B are at exactly the same height, the line of sight is now horizontal, i.e., it is parallel to line AB and the instrument is in adjustment.

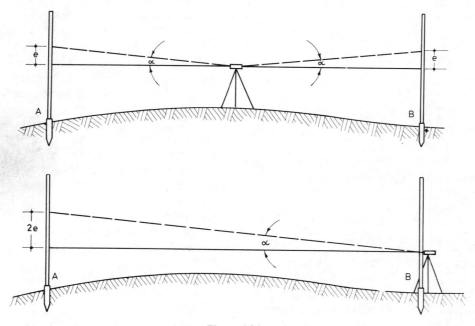

Figure 6.26

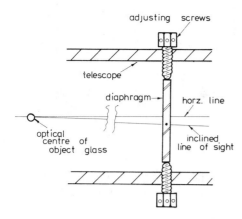

Figure 6.27

2. Tilting level

(a) The tilting level differs from the dumpy in that the vertical axis need only be approximately vertical for the instrument to be in good adjustment. This condition is fulfilled satisfactorily by centring the small circular spirit level at each set-up.

(b) The line of collimation must be horizontal when the bubble of the main spirit level is central.

Test

The test is identical to that for the dumpy level, namely the 'two-peg' test.

Suppose that the test has been carried out and it is discovered that the line of collimation is not horizontal, i.e., staff reading A does not equal staff height B when the instrument is set up over B.

Adjustment

The line of sight can be tilted using the tilting screw until the reading at A agrees with staff height B. The line of sight is now horizontal, since pegs A and B are at the same height, but, since the tilting screw has been moved, the main spirit level will no longer be central.

It is adjusted by means of the spirit level capstan screws until the bubble is exactly central. The main spirit level is therefore parallel to the now horizontal line of sight and the condition is fulfilled.

3. Automatic level

The vertical axis need only be approximately vertical as with the tilting level. Once more this condition is adequately fulfilled using the small circular spirit level.

The line of collimation must be horizontal when the small circular spirit level is central. The 'two-peg' test is carried out as before and if the line of collimation is found to be in error the reticule can usually be adjusted as with the dumpy level. If this is not possible the compensator unit can be adjusted

but this is definitely not a job for the surveyor and the instrument should be returned to the manufacturer.

In the two-peg test the pegs A and B need not be level. It is also common practice to set the instrument beyond peg B instead of over it at the second set-up.

Example 10 In a two-peg test a levelling instrument was set up exactly mid-way between pegs A and B which are 80 m apart. The pegs were adjusted until the same reading was obtained on each.

The instrument was then set over peg B where its height was measured as 1.350 m and a reading of 1.450 m was made to a staff held on peg A.

(a) Calculate the collimation error of the instrument.

(b) Express the error as (i) a percentage (ii) an angle of elevation or depression.

Solution Since the instrument was set up mid-way between the pegs and the same reading was obtained at each peg, the pegs must be level.

(a) Instrument at peg B

$$\text{Apparent difference in level} = 1.450 - 1.350$$
$$= 0.10 \text{ m}$$

i.e., the line of collimation is elevated by 0.10 m over a length of 80 m.

(b) (i) percentage error $= \dfrac{0.10}{80} \times 100$

$$= 0.125 \text{ per cent}$$

(ii) Tan elevation angle $= \dfrac{0.10}{80}$

$$= 0.00125$$

$$\text{angle} = +00° \, 04'$$

In the two-peg test, the pegs A and B need not be set level as is evidenced in the following example.

Example 11 A levelling instrument was set up exactly mid-way between two pegs P and Q lying 70 m apart and the following readings were obtained to a staff held vertically on the pegs in turn (Fig. 6.28):

Reading to P—0.765 m; reading to Q—1.395 m

When the instrument was set over peg P, the following information was obtained:

instrument height P—1.305 m; staff reading Q—1.855 m

Calculate the collimation error of the instrument and express it as an angle of elevation or depression.

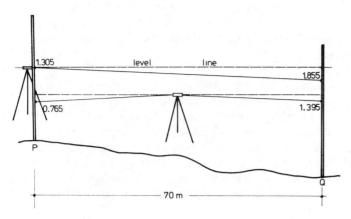

Figure 6.28

Solution Instrument mid-way between pegs P and Q.

True difference in level QP = 1.395 − 0.765

$$= 0.630 \text{ m}$$

Instrument set over peg P

Apparent difference in level QP = 1.855 − 1.305

$$= 0.550 \text{ m}$$

Since the apparent difference is less than the true difference in level, the line of collimation is depressed by

$$0.630 - 0.550 \text{ m} = 0.080 \text{ m over } 70 \text{ m}$$

$$\text{Angle of depression} = \tan^{-1} \frac{0.080}{75.0}$$

$$= \tan^{-1} 0.001\ 07$$

$$= 00° 03' 56''$$

More examples of instrument adjustment are given in Chapter 21.

Curvature and refraction

Throughout this chapter, reference has been made to a horizontal line of sight as distinct from a level line. If the earth is considered to be a perfect sphere (Fig. 6.29) a level line would be, at all points, equidistant from the centre. However, the line of sight through a levelling instrument is a horizontal line tangential to the level line. If a staff were held on B, the staff reading observed from A would therefore be too great, by BB_1. This is the curvature correction, c, and it can be calculated as follows.

In triangle AB_1O, L is the length of sight in

kilometres and R = mean radius of the earth (6370 km).

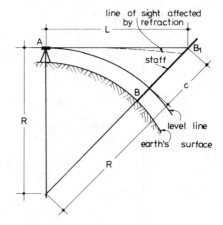

Figure 6.29

By Pythagoras's Theorem:

$$(R + c)^2 = R^2 + L^2$$

That is, $R^2 + c^2 + 2Rc = R^2 + L^2$

Therefore $c(c + 2R) = L^2$

$$c = L^2/(c + 2R)$$

Since c is so small compared with R, it can be ignored.

So $c = (L^2/2R)$ kilometres

That is, $c = \left(\dfrac{L^2}{12\ 740} \right)$ km

But c is required in metres while L remains in kilometres.

$$c = \left(\frac{L^2 \times 1000}{12\ 740} \right) \text{ metres}$$

$c = 0.0785L^2$ metres (where L is in kilometres)

The line of sight is not really horizontal. It is affected by refraction in such a manner that the line of sight is bent downwards towards the earth. Refraction is affected by pressure, temperature, latitude, humidity, etc., and its value is not constant. Its value is taken as $\frac{1}{7}$ curvature and is opposite in effect to that of curvature. Thus

$$\text{combined correction} = 0.0785L^2 - \tfrac{1}{7}(0.0785L^2)$$

$$= \tfrac{6}{7}(0.0785L^2)$$

$$= 0.0673L^2 \text{ metres}$$
$$\text{(where } L \text{ is in}$$
$$\text{kilometres)}$$

Example 12 Calculate the corrected staff reading for a sight of 1500 metres if the observed reading is 3.250.

$$\text{Length of sight} = 1.5 \text{ km}$$

$$\text{Correction for curvature} = (0.0673 \times 1.5^2) \text{ m}$$
$$\text{and refraction} = (0.0673 \times 2.25) \text{ m}$$

$$= 0.151 \text{ m}$$
$$\text{Observed staff reading} = 3.250$$
$$\text{Correction} = -0.151$$
$$\text{Correct staff reading} = 3.099 \text{ m}$$

Example 13 Calculate the correction due to curvature and refraction over a length of sight of 120 metres.

$$c = (0.0673 \times 0.12^2) \text{ m}$$
$$= (0.0673 \times 0.0144)$$
$$= 0.001 \text{ m}$$

Since 0.001 m is negligible, the correction can be neglected for lengths of sight less than 120 metres. It is good practice, when levelling, to restrict the length of sight to about 50 metres. Furthermore, at this distance the staff reading should never be allowed to fall below 0.5 m because of the variation in refraction caused by fluctuations in the density of the air, close to the ground.

Reciprocal levelling

The importance of having sights of equal length has already been stressed. Briefly, collimation errors are eliminated by this technique. If the length of sight does not exceed 120 metres curvature and refraction errors are negligible. However, there are occasions when a long sight must be taken and collimation and curvature errors are present.

For example, in Fig. 6.30, the difference of level between two points A and B on opposite banks of a wide river is required.

The instrument is set at A and the instrument height, h_1, is measured. The staff is held on B and the staff reading s_1 recorded. In Fig. 6.30 r, the refraction error and c, the curvature error are clearly shown. Since AA_1 is the level surface through A the difference in level between A and B is the distance A_1B.

$$\text{Now } A_1B = h_1 + c - r - s_1$$

The instrument is then removed to B, the height h_2 measured and the staff reading s_2 recorded: r and c are the refraction and curvature errors as before. The difference in level between B and A is the distance B_1A.

$$B_1A = s_2 + r - c - h_2$$

Note that since the sights are of equal length, the collimation error is the same and is therefore cancelled.

The difference in level is the mean of A_1B + B_1A

$$= \tfrac{1}{2}(h_1 + c - r - s_1 + s_2 + r - c - h_2)$$
$$= \tfrac{1}{2}(h_1 - h_2 + s_2 - s_1)$$

Strictly, this is not the true difference of level since the value of r will be slightly different in the two equations. However, it will be sufficiently close for building surveys provided the observations are made as soon after each other as possible.

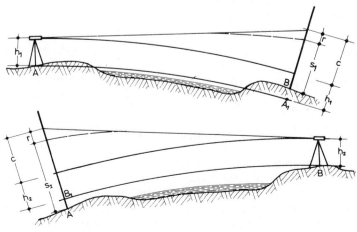

Figure 6.30

Example 14 Observations were made between points X and Y on opposite sides of a wide water-filled quarry as follows:

Level at X — Instrument height 1.350
Staff reading Y   1.725

Level at Y — Instrument height 1.410
Staff reading X   1.055

Calculate the true difference in level between the stations and the reduced level of Y if X is 352.710 AOD.

Solution

True difference in level $= \frac{1}{2}(h_1 - h_2 + s_2 - s_1)$

$$= \frac{1}{2}(1.350 - 1.725$$
$$+ 1.055 - 1.410)\text{ m}$$
$$= \frac{1}{2}(2.405 - 3.135)\text{ m}$$
$$= \frac{1}{2}(-0.730)\text{ m}$$
$$= \underline{-0.365 \text{ m}}$$

Reduced level X $= 352.710$ m AOD

Therefore fall X − Y $= -0.365$ m

and reduced level Y $= \underline{352.345 \text{ m AOD}}$

Test questions

1 The following list of readings was taken in sequence during a levelling survey.

Reading (m)	Remarks
1.250	BM, 1.435 m AOD
1.285	Peg A
1.125	Peg B
0.810	Change point
1.555	
1.400	Inverted staff reading taken to the underside of a bridge
1.235	Peg C
0.665	Change point
1.905	
0.070	BM, 4.600 m AOD

Adopt a standard form of booking and reduce the levels to Ordnance Datum.

(OND, Building)

2 The table below shows the results of a levelling along the centre line of a roadway where settlement has taken place. The road was initially constructed at a uniform gradient rising at 1 in 75 from A to B. Assuming no settlement has taken place at station A:

(a) reduce the levels
(b) calculate the maximum settlement along the line AB.

BS	IS	FS	Remarks	Chainage (m)
3.540			OBM 78.675	
0.410		3.665		
0.525		2.245		
	2.840		Station A	0
	2.440			30
	2.045			60
2.475		1.655		90
	2.090			120
	1.700			150
	1.315			180
	0.900			210
2.465		0.485		240
	2.055			270
	1.645		Station B	300
		2.040	TBM 78.000	

(IOB, Site Surveying)

3 Table 6.20 shows the readings taken to determine the clearance between the river level and the soffit of a road bridge. Reduce the levels and determine the clearances between the river level and the soffit of the bridge.

BS	IS	FS	RL	Remarks
0.872			21.460	OB mark
0.665		3.980		
	2.920			River level at 'A'
	−1.332			Soffit of bridge at 'A'
	−1.312			Soffit of bridge at 'B'
	−1.294			Soffit of bridge at 'C'
	−1.280			Soffit of bridge at 'D'
	2.920			River level at 'D'
4.216		0.597		
		1.155		OB mark

Table 6.20

(City and Guilds of London Institute)

4 Table 6.21 shows a page of a level book in which certain entries are missing. Complete the missing entries and carry out the normal checks.

5 Table 6.22 shows the staff readings obtained from a survey along the line of a proposed roadway from A to B.
(a) On the table, reduce the levels from A to B applying all the appropriate checks.
(b) Given that the finished roadway has to be evenly graded from A to B, calculate the depths of cut or fill at 20 metre intervals between A and B.

BS	IS	FS	Rise	Fall	Reduced levels
3.786					36.642
	—	2.474			—
	1.960			0.648	38.468
—		3.560		—	36.868
	3.698			—	34.042
	0.670		—		—
—		2.180		—	35.560
	1.052		1.186		—
2.874		—		—	34.992
	—		1.158		—
	0.950		0.766		36.916
		1.412		—	—

Table 6.21

(City and Guilds of London Institute)

BS	IS	FS	HPC or		Surface red. level	Grade red. level	Fill	Cut	Remarks
			Rise	Fall					
0.824					39.220				TBM No.1
	1.628								A
	0.790								20 m from A
	0.383								40 m from A
2.154		1.224							60 m from A
	2.336								80 m from A
	2.757								100 m from A
2.555		0.461							Change point
	2.275								120 m from A
	0.436								140 m from A
	0.227								160 m from A
	0.716								180 m from A
	0.652								B, 200 m from A
		0.233							TBM No. 2

Table 6.22

(SCOTVEC, Ordinary National Diploma in Building)

7. Vertical sections

One of the most important, and certainly one of the most common applications of levelling, is sectioning.

Whenever narrow works of long length, e.g., roads, sewers, drains, etc., are to be constructed, it is necessary to draw vertical sections showing clearly the profile of the ground.

Two kinds are necessary:
1. Longitudinal sections, i.e., a vertical section along the centre-line of the complete length of the works.
2. Cross-sections, i.e., vertical sections drawn at right angles to the centre-line of the works.

The information provided by the sections provides data for:
(a) Determining suitable gradients for the construction works.
(b) Calculating the volume of the earthworks.
(c) Supplying details of depth of cutting or height of filling required.

Figure 7.1 is the plan of a small building site compiled from the notes of a linear survey.

The proposed architectural development shows a roadway XY serving several new houses. The roadway is to be constructed on a regular gradient of 1 in 20 rising from the existing formation level of point X (chainage 0 m) towards point Y (chainage 80 m). A sewer is to be constructed along the centre line of the roadway and is to connect into the manhole X (where the existing invert level is 91.20 m AD) at a gradient of 1 in 40.

In dealing with drains and sewers, reference is made to the invert level rather than the formation level. The invert level is the inside of the bottom of the pipe, but for practical purposes it may be considered to be the bottom of the excavation.

Occasionally the pipes are laid on a bed of concrete, in which case the thickness of the concrete must be subtracted from the invert level to obtain the level of the bottom of the excavation.

The volume of material required to build the roadway and the volume of material to be removed in the sewer construction will be required sometime during the construction period. In order to calculate

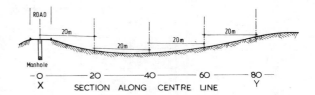

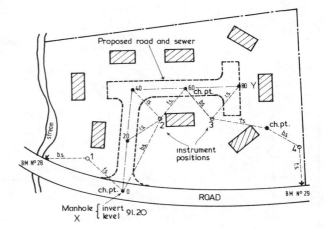

Figure 7.1

these volumes, a longitudinal section along line XY and cross-sections at 20 m intervals will have to be drawn.

Longitudinal sections

1. Field work

Before the longitudinal section can be plotted the following field work is necessary:
(a) A levelling must be made along the centre line with levels taken at all changes of gradient. A level is also taken at every tape length whether or not it signifies a change in gradient.
(b) Horizontal measurements must be made between all the points at which levels were taken. The measurements are accumulated from the first point such that all points have a running chainage. Pegs are left at every tape length to enable cross-sections to be taken later, Fig. 7.1.

It should be noticed that the line of the section must firstly be laid out accurately on the ground by using a theodolite. Setting out by theodolite is described in a subsequent chapter.

Procedure

Generally a surveyor and three assistants are required if the section is long. The surveyor takes the readings and does the booking; one assistant acts as staffman, while the other two act as chainmen taking all measurements, lining-in ranging poles along the previously established centre line and leaving pegs at all tape lengths.

A flying levelling is conducted from some nearby bench mark to the peg denoting zero chainage. Thereafter the levelling is in series form with intermediate sights taken as necessary.

The tape is held at peg zero chainage and stretched out along the line of the section. The chainmen and staffman work together and while the latter holds the staff as a backsight at peg zero, the chainman marks the changes in gradient and calls out the chainages of these points to the observer. The staffman follows up and holds the staff at all of the changes of gradient. When one tape length has been completed, a peg is left and the next length is observed. The procedure is repeated until the complete section has been levelled.

As in all surveying work a check must be provided. In sectioning this can be done by flying levelling from the last point of the section to the commencing bench mark, or to some other, closer, bench mark.

Example 1 Table 7.1 shows the readings obtained during the levelling of the proposed works along line XY. The levelling commenced on BM No. 28 and finished on BM No. 29.

2. Plotting

(a) The field work is reduced and all checks applied. At this stage the reader should complete Table 7.1 thereby obtaining the reduced levels of all the points along the section.

The result should be checked against Table 7.2 which shows the reduced levels.

Chainage	Reduced length	Remarks
0 m	92.420	X Start of section
20 m	92.175	
40 m	93.450	
60 m	94.650	
80 m	96.420	Y End of section

Table 7.2

It will be noticed that the levelling does not close exactly onto bench mark 1, the discrepancy being 0.010 m. This closure error is acceptable and the reduced levels are considered to be satisfactory.

(b) The scales are chosen for the section drawing, such that the horizontal scale is the same as the scale of the plan view of the site. Compared to the length of the section, the differences in elevation of the section points will always be comparatively small: consequently the vertical scale of the section is exaggerated to enable the differences in elevation to be readily seen. The horizontal scale is usually enlarged five to ten times, producing the following scales in Fig. 7.2.

Horizontal Scale   1:500

Vertical Scale   1:100

(c) The reduced levels are examined and the lowest point is found. A horizontal line is drawn to represent an arbitrary datum some way below the lowest point of the section. The line is also a multiple of ten metres above the true datum. In Fig. 7.2 the arbitrary datum of 210.000 metres above Ordnance Datum has been chosen and clearly annotated to this effect.

(d) The horizontal chainages are carefully measured along the arbitrary datum line and perpendiculars are erected at each chainage point.

(e) The reduced level of each station is carefully scaled off along the perpendicular. Each station so established is then joined to the next by a straight line to produce an exaggerated profile of the ground along the section. The lines joining the stations

BS	IS	FS	Rise	Fall	Reduced level	Distance	Remarks
1.370					91.085		TBM No. 28 (RL 91.085)
2.795		0.035				0	X Start of section
	3.040					20	
	1.765					40	
3.865		0.565				60	
	2.095					80	Y End of section
2.240		1.075					
		0.385					TBM No. 29 (RL 99.285)

Table 7.1

must *not* be drawn as curves since the levels have been taken at all changes of slope and the gradient is therefore constant between any two points.

(f) The proposed works are added to the drawing. The formation level of any construction works is the level to which the earth is excavated or deposited to accommodate the works. In Fig., 7.2 the formation level of the new roadway at X is 91.42 m. The roadway is to rise at a gradient of 1 in 20 from X at chainage 0 m to Y at chainage 80 m.

Since the profile is exaggerated true gradients cannot be shown on the section.

Formation level X is plotted and a second point, at chainage 20 metres, is plotted 1 metre higher than X. When joined, the line between the points represents the gradient. The depth of the excavation required to accommodate the roadway is scaled from the section, the depths being the distances between the surface and the line representing the roadway formation level. Since all scaling and calculations are performed on this section, it is called the working drawing.

(g) The presentation drawing is then compiled by tracing the working drawing onto a plastic sheet or piece of tracing linen in black ink only. From this tracing any number of prints can be obtained.

Calculation of cut or fill

In addition to being scaled from the section, the depths of cut or height of fill are also calculated. The method of calculation is similar for all vertical sections. Generally, it consists of:

(a) Calculating the reduced level at each chainage point.

(b) Calculating the proposed level of the new works at each chainage point.

(c) Subtracting one from the other. Where the surface level is higher than the proposed level, there must be cutting and where the proposed level exceeds the surface level, filling will be required.

In Example 1 (Fig. 7.2) the complete calculation is as follows:

(a) Surface levels—obtained from Table 7.2 i.e., the field book reduction.

(b) *Proposed levels (roadway)*

Gradient of roadway XY = 1 in 20 rising from X to Y

Rise over any length = (1/20) of length

Therefore Rise over 20 m length = $(1/20) \times 20$ m = 1.000 m

Formation level at 0 m (X) = 92.420 m
Formation level at 20 m = 92.420 + 1.000 = 93.420 m
Formation level at 40 m = 93.420 + 1.000 = 94.420 m
Formation level at 60 m = 94.420 + 1.000 = 95.420 m
Formation level at 80 m = 95.420 + 1.000 = 96.420 m

Check:

Total rise is (96.420 − 92.420) m over 80 m = 1 in 20

(c) Proposed levels (sewer)

Gradient of sewer XY = 1 in 40 rising from X to Y

Therefore Rise over 20 m = $1/40 \times 20$ = 0.500 m
Invert level 0 m (X) = 91.20 m
Invert level 20 m = 91.20 + 0.5 = 91.70 m

etc.

(d) Cutting or filling (roadway)

At each chainage point the proposed level exceeds or equals the surface level. There will therefore be filling at each chainage point, e.g., at chainage 20 m the depth of filling = (formation level − surface level) = 93.42 − 92.175 = 1.245 m.

(e) Cutting or filling (sewer)

At each chainage point there will be cutting since the surface level is higher than the sewer level, e.g., at chainage 20 m the cutting is (92.175 − 91.700) = 0.475 m.

Since the calculation is always presented in tabular fashion, the remainder of the calculation is shown with Fig. 7.2. The table showing the cutting and filling always accompanies the longitudinal section in presentation work.

Cross-sections

1. Level cross-sections

It may not be necessary actually to observe the levels in the field. The ground across the centre line at any chainage point may be level or nearly so, in which case the centre line level is assumed to apply across the line of the section.

Plotting level cross-sections The plotting is very similar to the plotting of longitudinal sections. One essential difference, however, is that the cross-section is plotted to a natural scale, that is, the horizontal and vertical scales are the same.

In Example 1 (Fig. 7.2) the sewer is to be 0.8 metre wide and the excavation is to have vertical sides. The sections are plotted as follows:

(a) A line representing the arbitrary datum of 90.0 m is drawn lightly above the longitudinal section.

(b) The construction lines of the chainage points of the longitudinal section are extended vertically to cut this datum and act as centre lines for the cross-sections.

(c) The ground level and the sewer invert level are plotted from the datum on the centre line and horizontal lines drawn lightly through both points.

(d) The width of the sewer is marked and the vertical sides of the drain drawn to cut the horizontals in (c) above.

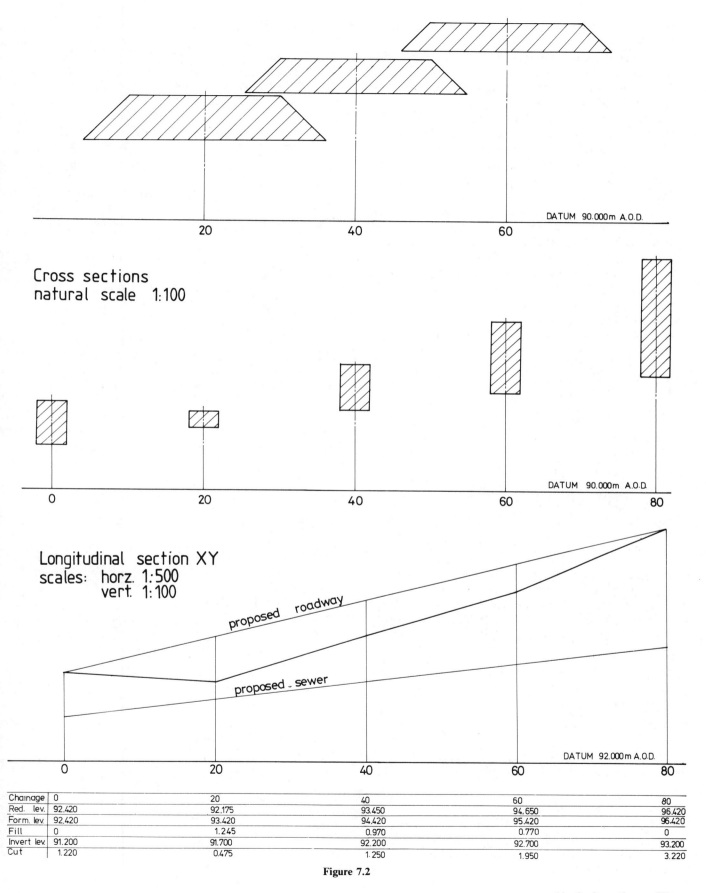

Cross sections
natural scale 1:100

DATUM 90.000m A.O.D.

Longitudinal section XY
scales: horz. 1:500
       vert. 1:100

proposed roadway

proposed sewer

DATUM 92.000m A.O.D.

Chainage	0	20	40	60	80
Red. lev.	92.420	92.175	93.450	94.650	96.420
Form. lev.	92.420	93.420	94.420	95.420	96.420
Fill	0	1.245	0.970	0.770	0
Invert lev.	91.200	91.700	92.200	92.700	93.200
Cut	1.220	0.475	1.250	1.950	3.220

Figure 7.2

(e) The rectangle denoting the excavation is drawn boldly.

The proposed roadway is to be four metres wide and any earthworks are to be formed with sides sloping at 45° (i.e., 1 unit vertically to 1 unit horizontally). The cross-sections are drawn as follows:

(a) A line representing the arbitrary datum of 90.0 m is drawn above the longitudinal section.

(b) The construction lines of each chainage point of the longitudinal section are extended vertically to cut this datum and act as centre lines for the cross-sections.

(c) The ground level and the formation level of the roadway are plotted from the datum on the centre lines and horizontal lines drawn lightly through both points.

(d) The width of the road is marked on the line representing the formation level and the sloping sides of the embankment drawn at 45° to cut the line representing ground level.

(e) The trapezium representing the road embankment is drawn boldly.

2. Non-level cross-sections

In cases where the ground across the centre line at any chainage point is obviously not flat, the following field work is required to obtain levels for plotting the cross-sections.

(a) Field work

Cross-sections are taken at right angles to the longitudinal section at every point observed on the latter. Generally, this rule is not strictly observed and cross-sections are taken usually at every tape length. The following field work is necessary:

(a) Right angles are set out, using a simple hand instrument, e.g., prism square or optical square. Where the ground is relatively flat, the right angle may be judged by eye. A ranging pole is inserted at either side of the centre line on the line of the cross-section.

(b) A levelling must be made from the peg previously established on the centre line of the longitudinal section, to every point where the gradient changes on the line of the cross-section (Fig. 7.3). In addition, levels are always taken on either side of the centre line at points C_1 and C_2 denoting the formation width.

Where the ground is relatively flat, one instrument setting is usually sufficient (Fig. 7.3(a)). If the cross-gradient is steep, a short series levelling is required (Fig. 7.3(b)).

Each cross-section is independent of every other. The peg on the centre line at each tape length is a temporary bench mark for its particular cross-section.

(c) Horizontal measurements must be made between all points at which levels are taken to cover the total width of the proposed works.

Procedure The procedure is much the same as that for longitudinal sections. One surveyor and three assistants is the ideal number of personnel required. However, since the distances involved are short, the surveyor and one assistant frequently take the levels and measurements without further assistance.

Example 2   Table 7.3 shows a part of a field book with the readings at cross-sections 80 m and 100 m of a proposed 6 m wide roadway recorded.

(b) Plotting

The plotting is very similar to the plotting of longitudinal sections. One essential difference, however, is that the cross-section is plotted to a natural scale, that is, the horizontal and vertical scales are the same. Briefly, the plotting is carried out in the following manner:

(a) The levels are reduced. Table 7.4 shows the field work of cross-section 80 m reduced. The reduction of cross-section 100 m is left to the reader.

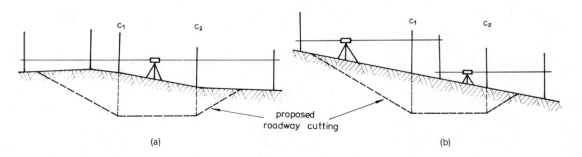

(a)   (b)

proposed roadway cutting

Figure 7.3

BS	IS	FS	Rise	Fall	Reduced level	Distance	Remarks
1.450					214.210	80	Peg on centre line
	0.800					80	6.5 m left of centre line
	0.850					80	11.8 m left of centre line
		1.070				80	16.2 m right of centre line
1.320					214.100	100	Peg on centre line
	0.510					100	8.0 m left of centre line
	0.920					100	14.1 m left of centre line
	2.420					100	6.2 m right of centre line
		1.980				100	17.6 m right of centre line

Table 7.3

(b) A line representing the arbitrary datum of 210.000 m AOD is drawn for each cross-section and the measurements to left and right of the centre line are scaled off accurately (Fig. 7.4).

(c) Perpendiculars are erected at each point so scaled and the reduced levels of each point plotted.

(d) The points are joined to form a natural profile of the ground.

(e) The formation level of the new works is obtained for each cross-section from the longitudinal section. The respective formation levels at 80 m and 100 m are 211.800 m and 212.000 m. These formation levels are plotted on the cross-sections. The finished width of the new works, called the formation width, is drawn and the side slopes taken up by material are plotted. In the example in Fig. 7.4 the formation width of the road is 6 metres and the sides slope at 1 unit vertically to 2 units horizontally.

The area enclosed by the surface profile, the formation width and the side slopes form the area of cutting or filling as the case may be.

Examples on levelling and sectioning

Example 3 The following readings were taken on the surface, along the route of a proposed main drain:

B. Sight 4.128 on Bench mark 1 (Red. lev. 63.107 m)
F. Sight 0.995 on Manhole cover A (Chainage 0 m)
B. Sight 1.540 on Manhole cover A

BS	IS	FS	Rise	Fall	Reduced level	Distance	Remarks
1.450			0.650		214.210	80	Peg on centre line
	0.800				214.860	80	6.5 m left of centre line
	0.850			0.050	214.810	80	11.8 m left of centre line
		1.070		0.220	214.590	80	16.2 m right of centre line
1.450		1.070	0.650	0.270	214.590		
−1.070			−0.270		−214.210		
0.380			0.380		0.380		

Table 7.4

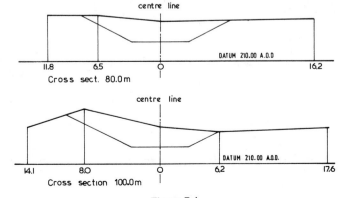

Figure 7.4

I. Sight 1.880 (Ch. 20 m); 2.340 (Ch. 40 m); 1.940 (Ch. 60 m)

F. Sight 1.710 (Ch. 80 m); B. Sight 2.310 (Ch. 80 m)

I. Sight 2.040 (Ch. 100 m); 2.240 (Ch. 120 m); 1.010 (Ch. 140 m); 0.670 (Ch. 160 m)

F. Sight 0.127 on Bench mark 2 (Red. lev. 68.253)

(a) Book and reduce the levels applying the appropriate arithmetical checks.

The drain (100 mm diameter) is to begin at an invert level of 63.440 m at the manhole (chainage 0 m) and rise towards chainage 160 m. In order to keep the excavation to a minimum the drain is to rise at a gradient of 1 in 60 from the manhole for part of its length and at 1 in 100 for the remainder such that the minimum cover to the top of the pipe is 1 metre.

(b) Draw a vertical section along the proposed drain track at a horizontal scale of 1 in 1000 and a vertical scale of 1 in 100, to show clearly the surface and top and bottom of the pipe.

(c) From the section determine:
(i)   the point of minimum cover;
(ii) the cover at 160 m chainage.

Solution

(a) See Table 7.5.

(b) The completed vertical section is shown in Fig. 7.5. Briefly the section is drawn up thus:
(i)   Choose arbitrary datum 60.000 m.
(ii) Plot chainages and their reduced levels.
(iii) Invert level of pipe = 63.440 m. Pipe thickness (100 mm) = 0.100 m therefore level of top of pipe = 63.540 m.

BS	IS	FS	Rise	Fall	Reduced level	Distance	Remarks
4.128					63.107		Bench mark 1
1.540		0.995	3.133		66.240	0	Manhole cover
	1.880			0.340	65.900	20	Surface
	2.340			0.460	65.440	40	Surface
	1.940		0.400		65.840	60	Surface
2.310		1.710	0.230		66.070	80	Surface
	2.040		0.270		66.340	100	Surface
	2.240			0.200	66.140	120	Surface
	1.010		1.230		67.350	140	Surface
	0.670		0.340		67.710	160	Surface
		0.127	0.543		68.253		Bench mark 2
7.978		2.832	6.146	1.000	68.253		
−2.832			−1.000		−63.107		
5.146			5.146		5.146		

Table 7.5

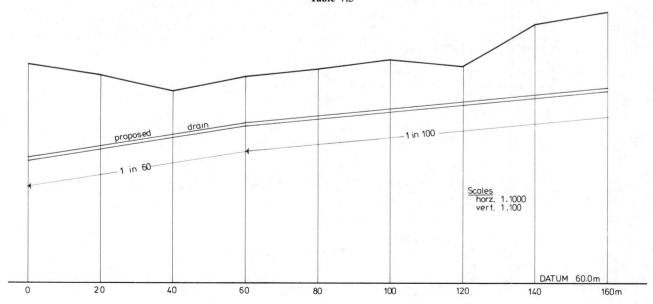

Figure 7.5

(iv) Draw two lines representing top and bottom of pipe rising at 1 in 60 from these levels.

(v) Draw a construction line anywhere on the section rising at 1 in 100. Using set-squares make a parallel line representing the top of the pipe such that it lies a maximum distance of 1 metre below the surface at only *one* point. This point occurs at chainage 120 m.

(vi) Continue the 1 in 100 gradient till it intersects the 1 in 60 gradient.

(vii) Draw a parallel line (100 mm) below this line to represent the bottom of the pipe.

(c) (i) The point of minimum cover is 120 m.

(ii) Cover at 160 m chainage = 2.17 m.

Example 4 The following reduced levels were obtained along the centre line of a proposed road between two points A and B.

Chainage (m)	Reduced level (m)	Chainage (m)	Reduced level (m)
'A' 0	83.50	50	82.45
10	83.84	60	82.20
20	84.06	70	82.41
30	83.66	80	82.70
40	83.30	'B' 90	83.05

The roadway is to be constructed so that there is one regular gradient between points A and B.

(a) Draw a longitudinal section along the centre line at a horizontal scale of 1:1000 and a vertical scale of 1:100.

(b) Determine from the section the depth of cutting or height of fill required at each chainage point, to form the new roadway.

(c) Check the answers by calculation.

Solution

(a) The longitudinal section is shown in Fig. 7.6.

(b) *Gradient of AB*

Reduced level of A = 83.50 m
Reduced level of B = 83.05 m
Difference = 0.45 m
Distance A to B = 90 m
therefore Gradient A to B falling = 0.45 in 90
= 1 in 200

Proposed levels (Formation levels):

Fall over 10 metres = 0.05 m

therefore proposed level
at ch. 10 m = 83.50 − 0.05 = 83.45 m
proposed level at ch. 20 m = 83.45 − 0.05 = 83.40 m
etc.

Chainage (m)	Reduced level (m)	Proposed level (m)	Cut (m)	Fill (m)
A 0	83.50	83.50		
10	83.84	83.45	0.39	
20	84.06	83.40	0.66	
30	83.66	83.35	0.31	
40	83.30	83.30	—	—
50	82.45	83.25		0.80
60	82.20	83.20		1.00
70	82.41	83.15		0.74
80	82.70	83.10		0.40
90 B	83.05	83.05		—

Example 5 Cross-sections are required at chainages 20 m, 40 m, and 60 m in Example 4 above. Levellings were made, and the results tabulated in the field book are shown in Table 7.6.

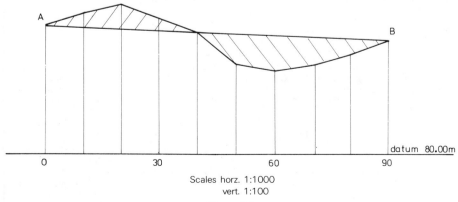

datum 80.00m

0   30   60   90

Scales horz. 1:1000
vert. 1:100

Figure 7.6

BS	IS	FS	HP collimation	Reduced level	Distance	Remarks
1.52				84.06	20	Peg on centre line
	0.88				20	5 m left of centre line
		2.02			20	5 m right of centre line
1.61				83.30	40	Peg on centre line
	0.56				40	5 m left of centre line
		2.34			40	5 m right of centre line
1.47				82.20	60	Peg on centre line
	0.67				60	5 m left of centre line
		1.66			60	5 m right of centre line

Table 7.6

(a) Calculate the reduced levels using the height of collimation system.

(b) Using a natural scale of 1:100 draw cross-sections at 20 m, 40 m, and 60 m chainages, given that the formation width of the roadway is 5 m and that the sides of any cuttings or embankments slope at 30° to the horizontal.

Solution

(a) See Table 7.7.

(b) The cross-sections are shown in Fig. 7.7. The method of plotting is briefly as follows:

(i) Draw arbitrary datum lines of 82.00 m.

(ii) Plot centre line, 5 m left and 5 m right along the datum lines.

(iii) Plot the reduced levels of these points and join them to produce the surface profiles.

(iv) Plot the formation levels at 20 m, 40 m, and 60 m. These are 83.40, 83.30, and 83.20 respectively from the calculations of Example 4.

(v) Plot the formation widths, 2.5 m on both sides of the centre line.

(vi) Draw the side slopes at 30° to the horizontal. Where the reduced levels exceed the formation level, a cutting is produced as in Fig. 7.7(a). In Fig. 7.7(c) the formation level exceeds the reduced level producing an embankment, while in Fig. 7.7(b) the cross-section is part cutting and part embankment.

Test questions

1 Table 7.8 shows levels taken along the centre line of a proposed sewer.

The sewer at chainage 0 m has an invert level of 88.900 and is to fall towards chainage 50 m at 1 in 100.

(a) Reduce the levels and apply the appropriate checks.

(b) Draw a vertical section along the centre line of the sewer on a horizontal scale of 1:500 and vertical scale of 1:50.

(c) From the section determine the depth of cover at each chainage point.

(d) A mechanical excavator being used to form the sewer requires a minimum working height of 4.50 m. What clearance (if any) will it have when working below the bridge?

(e) Given that the sewer track is 0.60 m wide with vertical sides, draw cross-sections to a scale of 1:100 to show the excavation at each chainage point.

2 (a) Table 7.9 shows notes made during a levelling for the construction of a new road. Complete the table and show all checks.

(b) The roadway is to be regraded from A to C, both of whose present levels are to remain unaltered. Calculate the gradient AC and the depth of cut or fill required at B.

(c) Choose any suitable scales and draw a vertical section from A to C.

Solution

(a)

BS	IS	FS	HP collimation	Reduced level	Distance	Remarks
1.52			85.58	84.06	20	Peg on centre line
	0.88			84.70	20	5 m left of centre line
		2.02		83.56	20	5 m right of centre line
1.61			84.91	83.30	40	Peg on centre line
	0.56			84.35	40	5 m left of centre line
		2.34		82.57	40	5 m right of centre line
1.47			83.67	82.20	60	Peg on centre line
	0.67			83.00	60	5 m left of centre line
		1.66		82.01	60	5 m right of centre line

Table 7.7

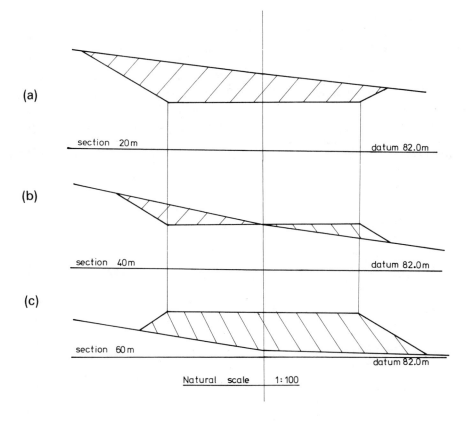

(a)

section 20m

datum 82.0m

(b)

section 40m

datum 82.0m

(c)

section 60m

datum 82.0m

Natural   scale   1:100

Figure 7.7

BS	IS	FS	Rise	Fall	Red. lev.	Dist.	Remarks
1.670					92.550		BM
1.520		3.870				0	Ground level
	0.910					10	Ground level
	1.590					20	Ground level
	1.770					30	Ground level
	1.660					40	Ground level
	−4.200					50	Underside of bridge
		0.720				50	Ground level

Table 7.8

BS	IS	FS	Rise	Fall	Red. lev.	Dist.	Remarks
2.580						0 m	At A
	2.340					15	
	2.640					30	
	2.130					45	
0.830		2.580				60	
	0.930				57.750	62.5	At B
	0.340					75	
	1.790					90	
1.300		2.530				105	
	0.660					120	
	1.920					135	
		1.100				150	At C

Table 7.9

(d) Draw cross-sections at the points of maximum cut and maximum fill to show the road construction given that the road is 5 m wide with 45° side slopes.

3 (a) Table 7.10 shows the results of a levelling survey along the line of a proposed trench, which is to be 0.8 m wide with vertical sides.

Reduce the levels to Ordnance Datum and show all the appropriate checks.

(b) On the worksheet, draw a section of the surface from A to B using:

> Horizontal Scale 1:500
> Vertical Scale 1:50

(c) The trench is to rise at a gradient of 1:40 from A to B and the minimum cover below the surface is to be 1 metre. Add the trench to the longitudinal section and calculate the depth of cut at each chainage point.

(d) Draw cross-sections at chainages 0 m, 50 m, and 90 m to a natural scale 1.50.

BS	IS	FS	HPC	Reduced level	Distance (m)	Notes
1.326				62.580		OSBM
	0.994				0	A
2.356		0.782			10	
	2.289				20	
	1.753				30	
	2.025				40	
	2.230				50	
	1.649				60	
1.928		1.467			70	
	1.635				80	
		0.968			90	B

Table 7.10

8. Contouring

When planning a housing estate, the builder will always try to take advantage of the natural slope of the ground. Houses built on different levels always give the estate a more interesting and pleasant appearance. Site plans must therefore have means of depicting the undulations of the surface, that is, they must show the surface relief.

Several means are available, namely spot levels, hill shading, hachuring, and contouring of which the latter is by far the most common method.

Contour line

A contour line is a line drawn on a plan joining all points of the same height above or below some datum. The concept of a contour line can readily be grasped if a reservoir is imagined. If the water is perfectly calm, the edge of the water will be at the same level all the way round the reservoir forming a contour line. If the water level is lowered by say, five metres the water's edge will form a second contour.

Further lowering of the water will result in the formation of more contour lines (Fig. 8.1). Contour lines are continuous lines and cannot meet or cross any other contour line nor can any one line split or join any other line, except in the case of a cliff or overhang.

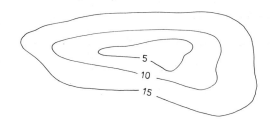

Figure 8.1

Figure 8.2 shows the contour plan and section of an island. The tide mark left by the sea is the contour line of zero metres value. If it were possible to pass a series of equidistant horizontal planes 10

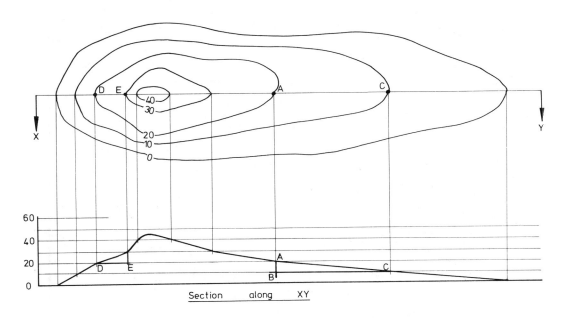

Figure 8.2

metres apart through the island, their points of contact with the island would form contours with values of 10 m, 20 m, 30 m, and 40 m.

Gradients

The height between successive contours is called the vertical interval or contour interval and is always constant over a map or plan. On the section the vertical interval is represented by AB. The horizontal distance between the same two contours is the distance BC. This is called the horizontal equivalent.

The gradient of the ground between the points A and C is found from:

$$\text{Gradient} = \frac{AB}{BC} = \frac{\text{Vertical Interval}}{\text{Horizontal Equivalent}}$$

Since the vertical interval is constant throughout any plan the gradient varies with the horizontal equivalent, for example

$$\text{Gradient along AC} = \frac{10}{100} = \frac{1}{10} = 1 \text{ in } 10$$

$$\text{Gradient along DE} = \frac{10}{30} = \frac{1}{3} = 1 \text{ in } 3$$

Contour characteristics

It should be clear from the above examples that the gradient is steep where the contours are close together and conversely flat where the contours are far apart.

In Fig. 8.3 three different slopes are shown. The contours are equally spaced in Fig. 8.3(a) indicating that the slope has a regular gradient. In Fig. 8.3(b) the contours are closer at the top of the slope than at the bottom. The slope is therefore steeper at the top than at the bottom and such a slope is called a concave slope. Conversely Fig. 8.3(c) portrays a convex slope.

A river valley has a characteristic 'V-shape' formed by the contours (Fig. 8.4(a)). The 'V' always points towards the source of the river, i.e. uphill. In contrast, the contours of Fig. 8.4(b) also form a 'V' shape but point downhill forming a 'nose' or spur.

Example 1 Figure 8.5 shows contours at vertical intervals of 2 metres over a building site drawn to a scale of 1:2500.
(a) What are the site dimensions?

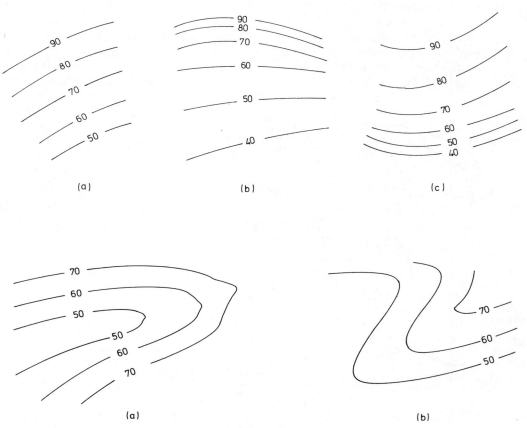

(a)   (b)

Figure 8.4

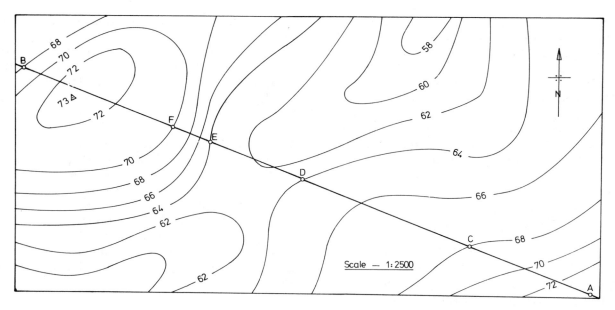

Figure 8.5

(b) Describe briefly the terrain along the line AB.

(c) Calculate the gradient between:
    (i) points D and C
    (ii) points E and F

Solution

(a) 380 m × 190 m.

(b) From A the ground falls as a concave slope to the centre of the area where it flattens out at the saddle formed between the two shallow valleys lying to the north and south. It then rises steeply in the form of a convex slope flattening onto a small hillock approximately 73 metres above datum. From the summit there is a gentle descent to B.

(c) *Line DC*   Rise = 4 m Horiz. Dist. = 120 m
                therefore Gradient = 1 in 30
    Line EF   Rise = 6 m Horiz. Dist. = 27 m
                therefore Gradient = 1 in 4.5

Methods of contouring

Once he has completed an accurate survey the surveyor knows the planimetric position of all points on the site relative to each other. His second task is to make a levelling to enable him to draw the accurate positions of the contour lines over the site.

Choice of vertical interval

The vertical interval of the contour lines on any plan depends on several factors, namely:

(a) The purpose and extent of the survey. Where the plan is required for estimating earthwork quantities or for detailed design of works a small vertical interval will be required. The interval may be as small as 0.5 metre over a small site but 1 to 2 metres is more common, particularly where the site is fairly large.

(b) The scale of the map or plan. Generally, on small-scale maps the vertical interval has to be fairly large. If not, some essential details might be obscured by the large number of contour lines produced by a small vertical interval.

(c) The nature of the terrain. In surveys of small sites, this is probably the deciding factor. A close vertical interval is required to portray small undulations on relatively flat ground. Where the terrain is steep, however, a wider interval would be chosen.

The methods of contouring can be divided into two classes.

1. Direct method

Using this method, the contour lines are physically followed on the ground. The work is really the reverse of ordinary levelling for whereas, by the latter operation the levels of known positions are found, in contouring it is necessary to establish the positions of known levels.

Figure 8.6 shows a small site where contours are required at vertical intervals of 2 metres. A bench mark (RL 94.070 m) has been established previously at the road junction. In order to fix the contour positions two distinct operations are necessary.

(a) *Levelling*

A level is set on site at some position from which the bench mark can be observed comfortably. In the figure, the level is set at position A and a backsight of 1.630 m observed on the BM.

Height of collimation = 94.070 + 1.630 = 95.700 m. From this instrument position, the contours at 94.000 m and 92.000 m can be observed. The staffman holds the staff facing the instrument and backs slowly downhill. When the observer reads 1.700 m, the bottom of the staff is at 94.000 m, since

$$\begin{array}{r} \text{Height of collimation} = 95.700 \text{ m} \\ \text{94 metre contour} = \underline{94.000 \text{ m}} \\ \text{Staff reading} = \underline{\underline{1.700 \text{ m}}} \end{array}$$

The staffman marks the staff position by knocking a peg into the ground. He then proceeds roughly along a level course, stopping at frequent intervals, where he is directed to move the staff slowly up or downhill until the 1.700 m reading is observed from the instrument. A peg is inserted at each correct staff position. In Fig. 8.6, pegs 1 to 9 have been established on the 94.000 m contour.

The 92.000 m contour is similarly established at pegs 10 to 21 using a staff reading of 3.700 m.

In order to set out the 90.00 m contour a staff reading of 5.700 m would be required. Since the staff is only 5 metres high, this is an impossibility, and a foresight must be taken to a changepoint. Table 8.1 shows the booking where it can be seen that the foresight reading is 4.830.

By carefully following the reduction, the reader will observe that staff readings of 2.180 and 4.180 enable the setting out of the 90 m and 88 m contours to take place. These contours are marked on the ground by pegs 22 to 28 and 29 to 31 respectively. Finally the levelling is checked back to the Bench mark.

(b) *Survey of the pegs*

The plan position of the pegs has now to be established to allow plotting to take place. On a small site, such as this, the pegs are surveyed by a system of auxiliary chain lines and offsets, all of which are shown on Fig. 8.6.

On a larger site, instrumental methods would be employed, e.g., compass or tacheometric traverse.

The plan positions of the contours are plotted directly onto the site plan and smooth curves drawn through them.

2. Indirect method

When using this method, no attempt is made to follow the contour lines. Instead, a series of spot levels is taken and the contour positions are interpolated. Three distinct operations are involved.

(a) *Setting out a grid*

On the site shown in Fig. 8.7 the longest side is chosen as baseline and ranging rods or small wooden pegs are set out along the line at 20-metre intervals. The interval depends on the contour vertical interval; the factors affecting the latter have been previously discussed. Briefly, the baseline intervals should be 5 or 10 metres where

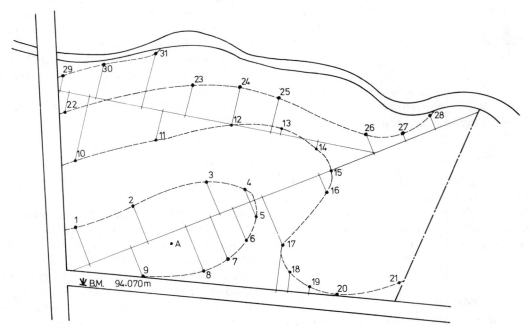

Figure 8.6

BS	IS	FS	Collimation	Reduced level	Distance	Remarks
1.630			95.700	94.070		Bench mark
	1.700			94.000		94 m Contour
	3.700			92.000		92 m Contour
1.310		4.830	92.180	90.870		CP
	2.180			90.000		90 m Contour
	4.180			88.000		88 m Contour
3.820		0.310	95.690	91.870		CP
		1.620		94.070		Bench mark
6.760		6.760		0.000		
	Difference = 0.000					

Table 8.1

the vertical interval is to be close (say 0.5 m or 1 m) and should be a maximum of 20 metres in all other cases.

At both ends of the baseline, right angles are erected by either tape or prism square and pegs are established at 20-metre intervals along the lines. In Fig. 8.7 the last peg on each line is at 60 metres. The distance between these two pegs is measured parallel to the baseline to form a basic rectangle and to check the setting out.

The distance should equal the baseline length. Further pegs are set along this line at 20-metre intervals, enabling a network of 20-metre squares to be established by simple alignment from the basic rectangle. Pegs are placed at every intersection.

The numerous pegs must be easily identified at a later stage so each is given a grid reference. There are many reference systems in common use, each

with its own advantages and disadvantages. Though slightly cumbersome, an excellent system is formed by giving all distances along and parallel to the baseline, an 'x' chainage and all distances at right angles to the baseline, a 'y' chainage. In Fig. 8.7, peg A has a reference ($x100\ y20$) while the reference of peg B is ($x20\ y60$).

The grid values of any point on the boundary can be obtained by measuring from the basic rectangle along the appropriate line to the boundary. The point X would have the identification of ($x116\ y40$) while Y would be ($x60\ y75$). This is the main advantage of the identification system.

(b) *Levelling*

A series levelling is conducted over the site and the reduced level of every point on the grid is obtained.

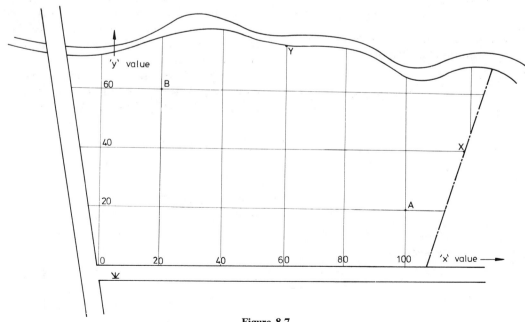

Figure 8.7

Figure 8.8 shows a part of the grid with its appropriate reduced levels.

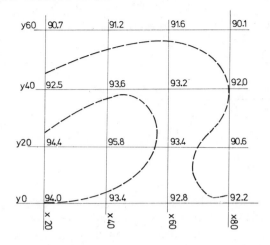

Figure 8.8

(c) *Interpolating the contours*

(i) *Mathematically*   The positions of the contours are interpolated mathematically from the reduced levels by simple proportion, e.g., the 92 m contour passes somewhere between the stations (*x*20 *y*40) and (*x*20 *y*60) (Fig. 8.8).

Referring to Fig. 8.9, the plan position of the contour is calculated thus:

Horizontal distance between stations   = 20 m

Difference in level between stations   =
(92.5 − 90.7) m = 1.8 m

Difference in level between station
(20, 60) and 92 m contour   =
= (92 − 90.7) m = 1.3 m

therefore horizontal distance from station
(20, 60) to 92 m contour (by simple
proportion)

$$= \left(\frac{1.3}{1.8} \times 20 \right) \text{ m } = 14.5 \text{ m}$$

The calculation is performed on a calculator and is the most commonly used method of the several available.

When a great many points are to be interpolated, a hand calculator is sometimes used, particularly when levels are taken to two decimal places. Use is made of a calculation table (Table 8.2). The interpolation of the 92 m contour, shown above, appears in the first line of the table while the remainder of the table is devoted to the calculations necessary for plotting the 92 m contour line completely.

The contour positions are plotted on the plan (Fig. 8.8) and joined by a smooth curve. The 94 m contour is also plotted on the figure and the reader should satisfy himself that the plotted position is correct.

(ii) *Graphically*   Figure 8.10 is an enlarged portion of the grid whereon the position of the 94 m contour is to be plotted between the points (*x*20 *y*20) and (*x*20 *y*40). The respective levels of the points are 94.4 m and 92.5 m.

Using any suitable scale on a scale rule, the 2.5 graduation mark is placed against the 92.5 level and the scale positioned as shown in Fig. 8.10. The 4.0 graduation representing the 94.0 contour line and the 4.4 graduation representing the 94.4 m level are marked on the plan. The latter mark is joined to the 94.4 m level and a parallel line drawn through the 4.0 m graduation to meet the grid line at M. From the similar triangles, formed by the scale, grid line and construction lines, point M is the true position of the 94 m contour on the grid line.

An alternative graphical device is the radial interpolation graph. The graph is drawn on tracing paper and is constructed as in Fig. 8.11.

Two mutually perpendicular lines AB and CD are drawn. The shorter line CD is divided into twenty equal parts and radial lines are drawn from A to each of the divisions.

In Fig. 8.12, the graph is shown in use. The 92 m contour position is required between the stations (*x*60 *y*40) and (*x*60 *y*60) where the reduced levels are

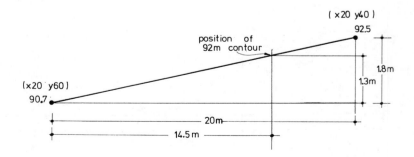

Figure 8.9

93.2 m and 91.6 m respectively. The difference in level between the points is therefore 1.6 m. The overlay is laid on the grid with AB and CD parallel to the grid lines, until 16 divisions are intercepted between the two grid stations.

Each division represents 0.1 m in this case and the 92 m contour will therefore be 4 divisions from

station ($x60\ y60$). A pin-hole is made through the overlay at this point and the mark denoted by a small circle on the site plan. It should be noted that the overlay can be manoeuvred until any convenient number of divisions is intercepted by the grid stations. If, for example, 8 divisions were chosen, each would represent 0.2 metre.

Col. 1	Col. 2	Col. 3	Col. 4	Col. 5	Col. 6	Col. 7
Grid Station	Grid Station	Contour value (m)	Dist. between Col. 1 & Col. 2	Diff. in level Col. 1 & Col. 2	Diff. in level Col. 1 & Col. 3	Dist. between Col. 1 & Col. 3
$x20\ y60$	$x20\ y40$	92	20	1.8	1.3	$(1.3/1.8) \times 20 = 14.5$
$x40\ y60$	$x40\ y40$	92	20	2.4	0.8	$(0.8/2.4) \times 20 = 6.7$
$x60\ y60$	$x60\ y40$	92	20	1.6	0.4	$(0.4/1.6) \times 20 = 5.0$
$x60\ y20$	$x80\ y20$	92	20	2.8	1.4	$(1.4/2.8) \times 20 = 10.0$
$x80\ y00$	$x80\ y20$	92	20	1.6	0.2	$(0.2/1.6) \times 20 = 2.5$

Table 8.2

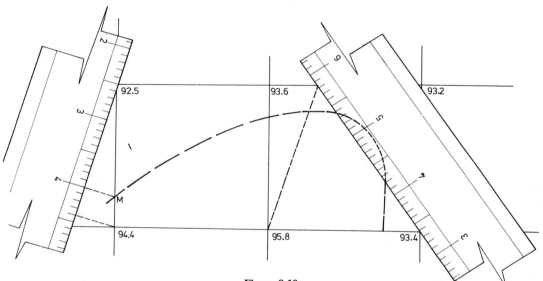

Figure 8.10

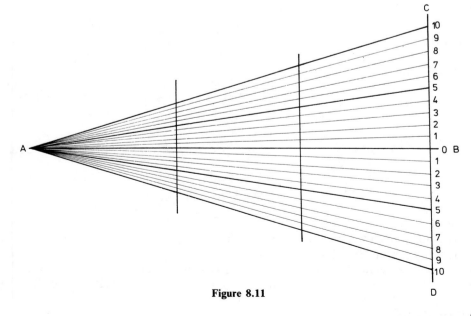

Figure 8.11

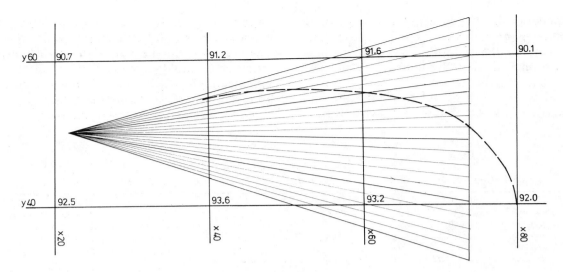

Figure 8.12

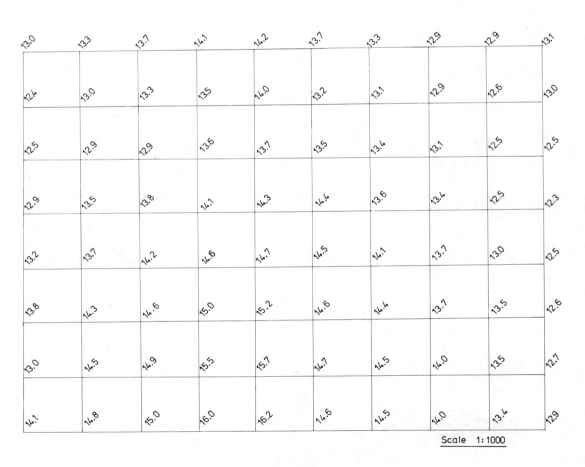

Scale 1:1000

Figure 8.13

Since the overlay is not drawn to any scale, it can be kept and used for any contour interpolation. It is probably the fastest method of interpolating and it is therefore worth while spending a few minutes'

time in preparing a really good overlay.

Example 2 The network of spot levels shown in Fig. 8.13 was observed during the contouring of a building site. Using any interpolation method, draw the contour lines at 0.5 metre vertical intervals.

The solution to this problem is given in Fig. 8.14.

Use of contour maps

Of the many uses to which contour maps are put, the following are the most relevant to building surveying.

1. Vertical sections

Figure 8.14 shows a contoured area where vertical sections are required along the lines XX and YY.

Vertical sections from contours are produced in a manner very similar to that for drawing sections from reduced levels. Briefly the section XX is drawn as follows:

(a) The scale for verticals is chosen. It may be a natural scale, where surface undulations vary greatly, or a five to ten times exaggeration of the horizontal scale on relatively flat ground. A scale of 1:100, i.e., horizontal scale × 10 is chosen in Fig. 8.14.

(b) The datum is drawn parallel to the line of the section XX either in the position shown or clear of the plan. On occasion the vertical section may be drawn on the line XX itself.

(c) The contour positions along the line XX are projected at right angles to the datum line.

(d) The contour values are scaled along the perpendiculars and joined to form the section.

The construction of the vertical section along line YY is left to the reader.

2. Intersection of surfaces

Whenever earth is deposited it will adopt a natural angle of repose. The angle will vary according to whether the material is clayey or sandy, wet or dry, but will almost certainly be between 45° and 26½°,

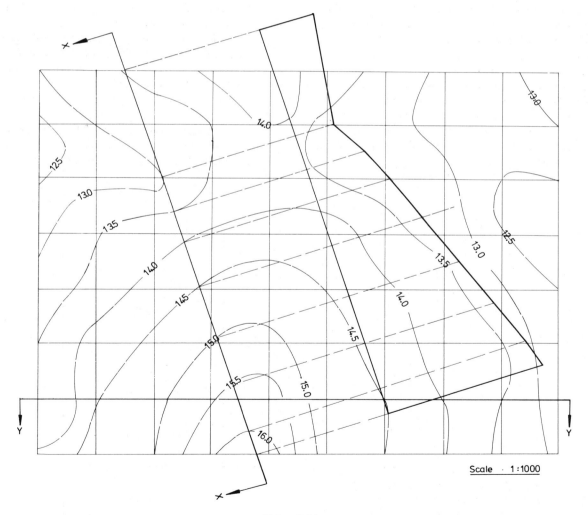

Scale · 1:1000

Figure 8.14

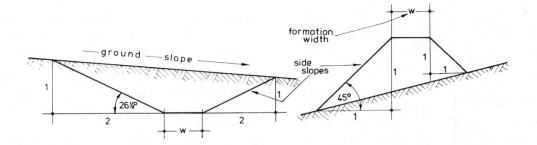

Figure 8.15

that is, the material will form side slopes of 1 in 1 to 1 in 2. In Fig. 8.15 the cutting has side slopes which batter at 1 in 2 while the side slopes of the embankment batter at 1 in 1.

New construction work is often shown on plans by a system of contour lines. The portion of the embankment shown in Fig. 8.16 has a formation width of 10 metres, is 10 metres high and is to accommodate a level roadway 100 metres long. The sides of the embankment slope at 1 unit vertically to 1½ units horizontally and the embankment is being constructed on absolutely flat ground.

Contour lines at 1 metre vertical intervals are drawn parallel to the roadway at equivalent 1½ metre horizontal intervals. Since the embankment is formed on level ground the outer limits, denoting the bottom of the embankment will be perfectly straight lines parallel to the top of the embankment. These lines will be 15 metres on either side of the formation width.

Very seldom will the ground surface be perfectly level. Figure 8.17 shows a series of contour lines portraying an escarpment, the southern scarp slope of which rises steeply from 0 m to 9 m, while the dip slope slopes gently northwards from 9 m to 6 m. The embankment already described is to be laid along the line AB.

The isometric view in Fig. 8.17 illustrates the situation.

Approximately halfway along the embankment, the ground surface is at a height of 9 m, that is, 1 m below the formation level of the embankment. It follows therefore that the width of the embankment will be very much narrower than at the commencement of the embankment, that is, at point A. The plan position of the outer limits will no longer be a straight line.

The actual position is determined by superimposing the embankment contours on the surface contours.

The intersection of similar values forms a point

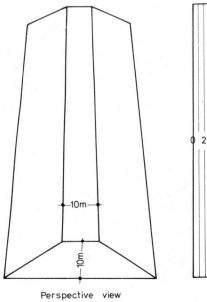

Perspective view

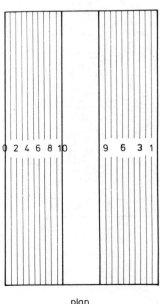

plan

Figure 8.16

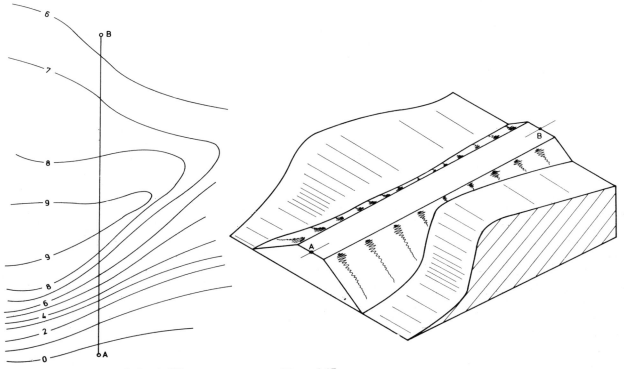

Scale 1 : 1000

Figure 8.17

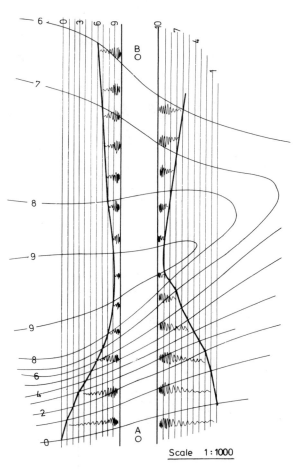

Scale 1 : 1000

Figure 8.18

on the tail of the embankment (Fig. 8.18). When all of the intersection points are joined a plan position of the outer limits is obtained.

When cuttings are to be constructed, the procedure for drawing the outline is identical to that described above. However, when the contours along the cutting slopes are drawn, values increase with distance from the formation whereas the contour values decrease on embankments.

Example 3 Figure 8.19 shows the position of a proposed factory building and its service roads, together with ground contours at 1-metre vertical intervals.

The formation level over the whole site is to be 0.5 m below finished level, that is, 83.000 m AOD and any cuttings or embankments are to have side slopes of 1 unit vertically to 2 units horizontally.

Draw on the plan the outline of the required earth works.

Solution (Fig. 8.20)
(a) There is no excavation or filling at points A and B since the ground level and formation level are the same.
(b) All earthworks northwards are in cutting. Contours are drawn parallel to the roadway or building at 2-metre horizontal intervals and numbered from 83 m to 87 m.

All earthworks southwards from AB are filling. Contours are drawn parallel to the roadway and

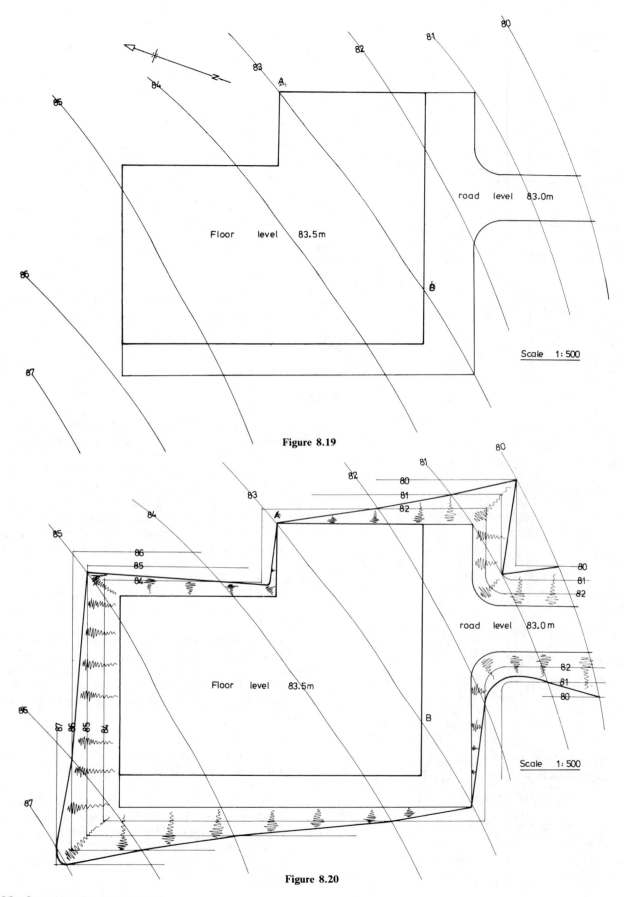

Figure 8.19

Floor   level   83.5m

road   level   83.0m

Scale   1:500

Figure 8.20

Floor   level   83.5m

road   level   83.0m

Scale   1:500

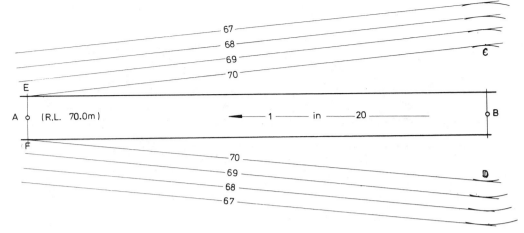

Figure 8.21

building and numbered from 83 m to 80 m. (Note the contours at the roadway curves.)

(c) The intersections of ground contours and earth works contours of similar value form the edge of the embankments or cuttings.

Sloping earthworks

When earthworks are to be formed to some specific gradient, the drawing of the contours is not so straightforward as with level works. Figure 8.21 shows an embankment being formed on flat ground the level of which is 68.000 m. The gradient of the embankment is 1 in 20 rising from a formation level of 70.000 m at A towards B. The length of the embankment is 60 m, and the sides slope at 1 in 2.

The embankment contours are formed by joining two points of the same level on the embankment sides.

The reduced level of point A is 70.000 m, therefore the reduced level of point B is:

$$70.000 + (60/20) \text{ m} = 73.000 \text{ m}$$

A point of 70.000 m level will therefore be 3 metres vertically below point B. The horizontal equivalent of 3 m vertical at a gradient of 1 in 2, is 6 m. Points C and D are drawn 6 m from the formation edge and joined to points E and F respectively to form the 70 m embankment contours.

All other contours at vertical intervals of 1 metre are parallel to these two lines, and are spaced at 2-metre horizontal intervals.

Example 4 Figure 8.22 shows ground contours at vertical intervals of 2 metres and the centre line of a proposed roadway AB. The following data refer to the roadway:

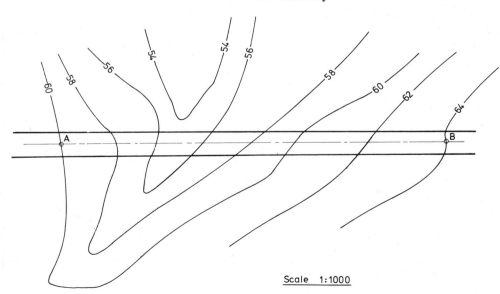

Scale 1:1000

Figure 8.22

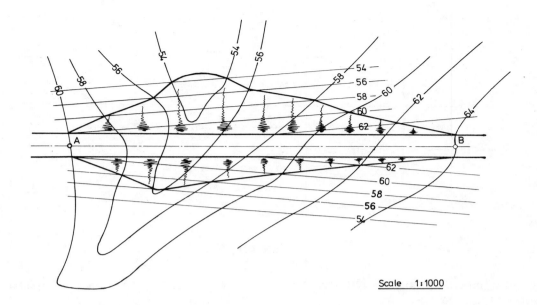

Scale   1:1000

Figure 8.23

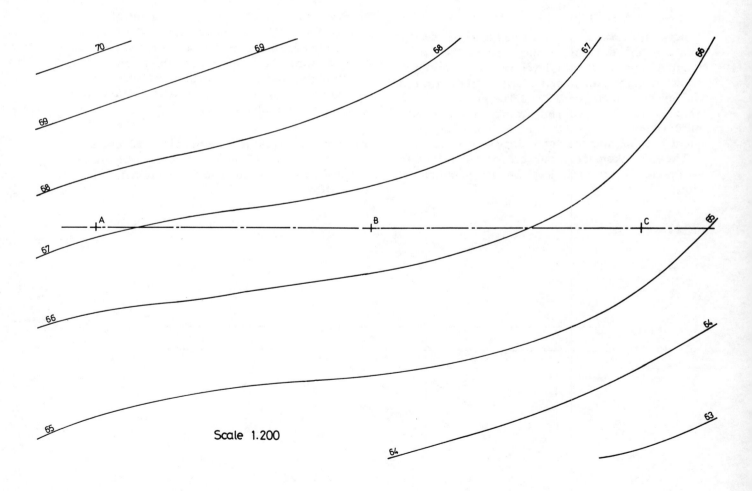

Scale   1.200

Figure 8.24

(a) Formation width   6 metres
(b) Formation level 'A'   60.00 metres
(c) Gradient A–B   1 in 25 rising
(d) Side slopes   1 in 2

Show clearly the outline of any earthworks formed between A and B.

The solution is shown in Fig. 8.23.

Test questions

1 Sketch contour patterns to illustrate the following ground characteristics:
(a) Convex slope
(b) Cliff
(c) Steep-sided river valley.

2 Figure 8.24 shows ground contours at 1 metre vertical intervals.

ABC is the line of a proposed level roadway having the following specification:
(a) Formation level   69.0 m
(b) Formation width   5.0 m
(c) Side slopes   1 in 1 (45°)
(i) Show on the plan the outline of the earthworks required to form the roadway.
(ii) Draw natural cross-sections at A, B, and C to show the roadway embankment and the ground surface.
(iii) From the plan or section, determine the ground levels at the base of the side slopes and on the centre line of the embankment across the sections at A, B, and C.

3 The following readings were taken on a 4-metre staff during a contour survey by the direct method:
(a) 3.120 m on a BM (Red. lev. 96.300)
(b) 0.220 m on a foresight at Change point 1.

The level was then moved to a new position and a reading of 1.350 m taken to the change point. Determine the staff readings required to locate the 100 m and 98 m contour lines.

9. Bearings

In Chapter 4, a detailed description of linear surveying was given and Fig. 4.16(a) shows the framework of a linear survey of a small area. This is the most elementary method of surveying, relying purely on the accurate determination of length. No attempt is made to obtain angular measurements.

It is true to say that in all other forms of surveying, the bearing or direction of the various lines must be obtained by using some form of angular measuring instrument, the simplest of which is a compass.

Magnetic bearings

On all compasses, there is a magnetic needle, which, when suspended freely, or when allowed to swing freely on a pivot, will settle in the magnetic meridian, that is, it will point to magnetic north. The scientific reasons for this do not really concern the building surveyor. Some theories suggest that it is due to the iron core of the earth's centre while others ascribe it to the electrical currents in the atmosphere caused by the earth's rotation. The fact is that the needle does settle in the magnetic meridian.

Figure 9.1 shows the earth with the magnetic meridians meeting at the north and south magnetic poles.

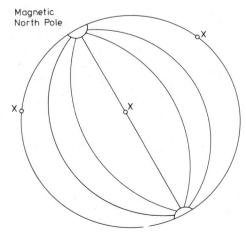

Figure 9.1

If a compass were held at any point X the magnetic needle would settle with its north-seeking end pointing to the magnetic north pole. If a protractor were laid below the needle and rotated so that zero pointed along the magnetic meridian (Fig. 9.2) the direction of any line radiating from X could be determined relative to magnetic north.

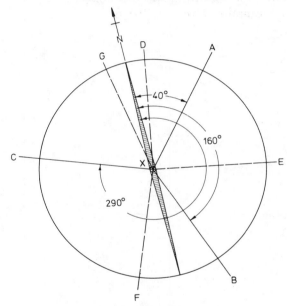

Figure 9.2

The line XA makes an angle of 40° with the magnetic meridian while lines XB and XC make angles of 160° and 290° respectively. The maximum value of any angle is 360° that is, the angle can have any value on the whole circle of graduations provided it is measured in a clockwise direction. The angle of 160° formed by the meridian XN and the line XB is the whole circle magnetic bearing of line XB and is defined as the angle measured in a clockwise direction from the magnetic meridian to the line.

Example 1 Refer to Fig. 9.2 and estimate the whole circle magnetic bearings of lines XD, XE, XF, and XG.

Solution XG, XD, XF, and XE have whole circle bearings of 350°, 10°, 200°, and 100° respectively.

True bearings

Figure 9.3 shows the earth with the geographical meridians coinciding at the true north and true south poles. Any angle measured in a clockwise direction from the true meridian to a line is a true whole circle bearing (WCB).

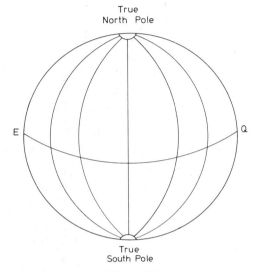

Figure 9.3

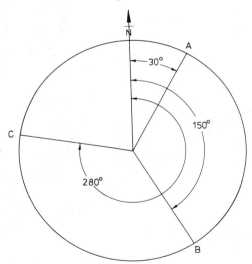

Figure 9.4

The true whole circle bearings of lines XA, XB, and XC (Fig. 9.4) are 30°, 150°, and 280° respectively.

Grid bearings

In Chapter 3 it was pointed out that on Ordnance Survey maps only the Central Meridian of 2° west longitude points to True north. All other north grid lines are parallel to the Central Meridian and point to Grid north.

The grid whole circle bearing of any survey line is the clockwise angle between a grid north line and the survey line.

Figure 9.5 shows a section of an OS map where the grid lines are at 1 km intervals. The grid bearing of the line between Cardross station (Grid ref. 345774) and Craigendoran station (Grid ref. 315812) is 318° 20′ when measured by protractor. Its true bearing however is known to be 316° 05′. The difference is due to the fact that the grid lines deviate westwards from true north by 2° 15′; consequently this value must be subtracted from the grid bearings in this locality to produce true bearings.

Example 2 (Fig. 9.5) Using a protractor, measure the GRID bearings of the lines between:
(a) Trig Station (369805) and Cardross Station (345774).
(b) Trig Station (369805) and Ben Bowie (340829).
(c) Trig Station (313788) and Trig Station (372783).

Solution   (a) 216° 30′   (b) 309°   (c) 96°

This deviation between true and grid north varies with longitude and would be zero at the 2° west central meridian. To the east of the central meridian the grid lines deviate eastwards from true north, hence the angular difference must be added to the measured grid bearings in order to produce true north bearings.

Magnetic declination

When Figs 9.2 and 9.4 are combined, the result in Fig. 9.6 shows that the true and magnetic whole circle bearings of any line are different. The difference is called magnetic declination. Magnetic declination is said to be westerly when the magnetic meridian lies to the west of the true meridian.

Since the three quantities, magnetic whole circle bearing, true whole circle bearing, and magnetic declination are related, it is possible to calculate the third quantity if any two of them are known.

Example 3 Supply the missing quantities in the table below:

	Column 1	Column 2	Column 3
Magnetic WCB	60°	230°	—
True WCB	—	221°	166°
Magnetic Declination	10° west	—	4° west

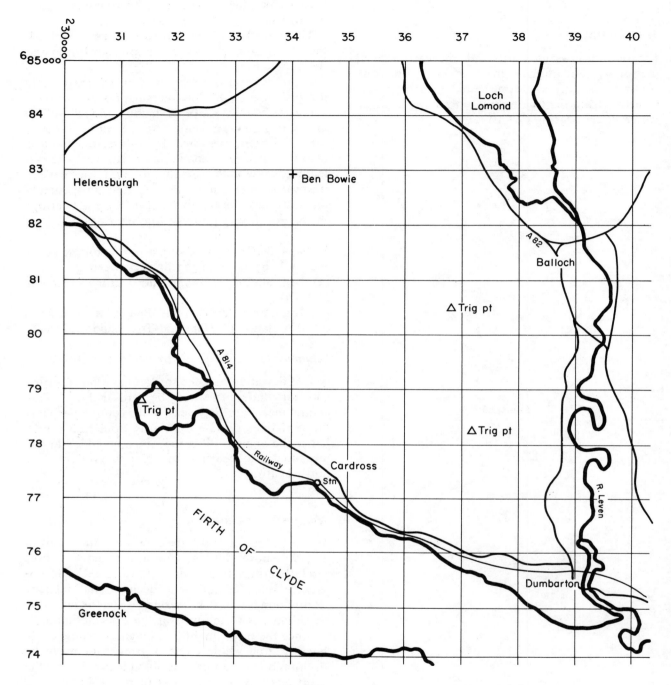

Figure 9.5

Solution

True WCB = 60° − 10°   Mag. Decl. = 230° − 221°
      = 50°   = 9° W.

Mag. WCB = 166° + 4°
      = 170°   (see Fig. 9.7)

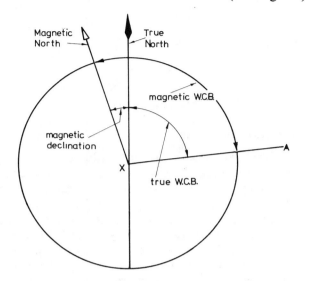

Figure 9.6

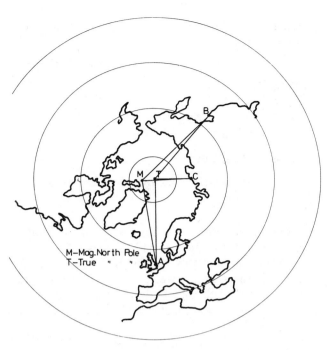

Figure 9.8

This simple example has been used principally to show that the value of magnetic declination varies, the reasons for the variation being as follows.

1. Geographical location

Figure 9.8 shows a plan view of the earth with the angle of declination drawn at three different locations. Clearly the value of the angle changes with geographical position. At point A the value is say 9° west; at point B it is 6° east, while at point C the value is almost zero.

2. Annual variation

The magnetic poles of the earth are constantly changing their position relative to the true north and south poles, with the result that the value of declination at any point on the earth slowly changes its value throughout the year. The mean annual change in the value of declination, called the secular variation, is about one-tenth of a degree in Great Britain. At the present moment, i.e., 1988, the value of declination is about 7½° west in the Midlands. Because of the secular variation, this value will decrease to zero by the year 2055.

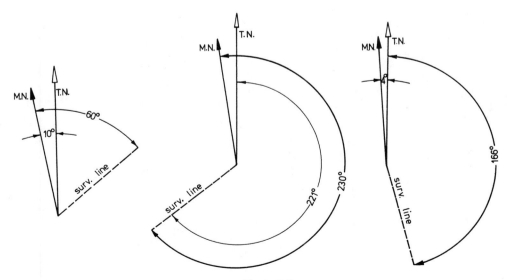

Figure 9.7

Example 4 Supply the missing quantities in the table below:

	Column 1	Column 2	Column 3
Magnetic WCB	40°	—	330°
True WCB	—	55°	315°
Magnetic Declination	10° W	5° E	—

Solution (Fig. 9.9)

Column 1	Column 2	Column 3
30°	50°	15° W

Example 5 Refer to Fig. 9.5. Given that the magnetic declination is 11° 15′ west of true north and that the deviation of the grid north lines from true north is 2° 15′ west, calculate the magnetic whole circle bearings between the points
(i) Cardross Station (345774) and Trig Station (369805)
(ii) Trig Station (313788) and Ben Bowie (340829)
(iii) Trig Station (372783) and Trig Station (369805)

Solution
(a) The relationship between the various north directions is shown in Fig. 9.10.
(b) In order to obtain true north bearings measure grid bearings by protractor and subtract 2° 15′.
(c) To obtain magnetic north bearings add 11° 15′ to the *true* bearings since the declination is west.

	Grid bearing	True bearing	Magnetic bearing
(i)	36° 30′	34° 15′	45° 30′
(ii)	33° 00′	30° 45′	42° 00′
(iii)	351° 00′	348° 45′	0° 00′

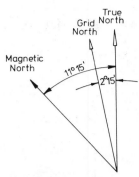

Figure 9.10

Assumed bearings

On local surveys of small building sites and engineering works, it may not be necessary to relate the survey to either magnetic or true north. Some arbitrary point is chosen as reference object and treated as being the equivalent of the north pole. Common reference objects (ROs) are tall chimneys, church steeples, and pegs hammered into the ground at some points where they can be easily found. Whole circle bearings are therefore the clockwise angles measured from the RO to any point. Alternatively two points on the ground may be pinpointed on a large scale OS map where the grid bearing of the line joining the points may be measured by protractor.

Forward and back bearings

It should be clear that any survey line can have four whole circle bearings, namely magnetic, true, grid, or arbitrary. With the addition of certain other information these bearings can be interrelated as required.

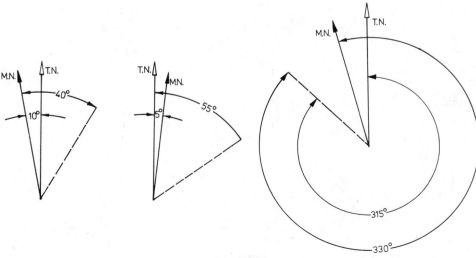

Figure 9.9

Figure 9.11 shows a line joining two stations A and B where the whole circle bearing from A to B is 63°.

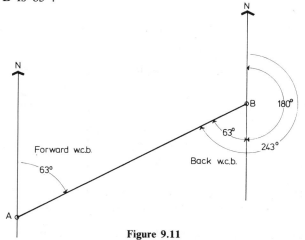

Figure 9.11

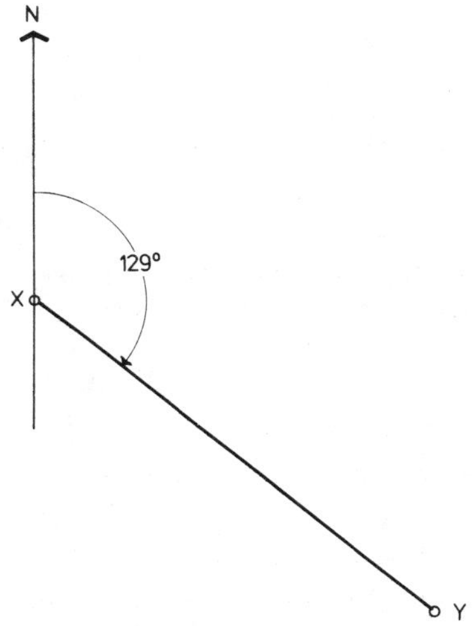

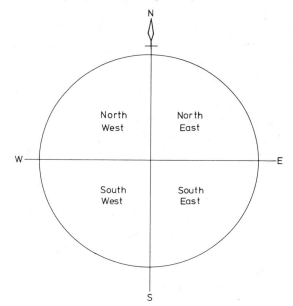

Figure 9.12

If an observer were to stand at station A and look towards station B he would be looking in a direction 63° relative to the chosen north point. In surveying terms, he is looking *FORWARD* towards station B from station A, and the forward whole circle bearing of the line AB is 63°. If he now stood at station B and looked towards station A he would be looking *BACK* along the line, i.e., in exactly the reverse direction.

This direction BA is known as the *BACK* whole circle bearing. The angular value is the clockwise angle from the north point to the line BA, namely (180° + 63°) = 243°.

From this simple example an equally simple rule can be deduced.

Rule 'To obtain back bearings from forward bearings, or vice versa, add or subtract 180°.'

In Fig. 9.12 the forward whole circle bearing of line XY is 129°. Using the rule above the back whole circle bearing YX is (180° + 129°) = 309°. The

reader should now complete the diagram by drawing the north point through station Y and proving that 309° is the correct answer.

Example 6 The forward bearings of four lines are as shown below:

AB—31°   BC—157° 30′   CD—200°   DE—347° 15′

Calculate the back bearings.

Solution

Line	Forward		Back
AB	31° 00′	+180°	211° 00′
BC	157° 30′	+180°	337° 30′
CD	200° 00′	−180°	20° 00′
DE	347° 15′	−180°	167° 15′

Quadrant bearings (reduced bearings)

In subsequent calculations, particularly rectangular coordinates, the whole circle bearings require to be converted to quadrant bearings.

If the cardinal points of the compass are drawn and labelled north, east, south, and west respectively the whole 360° circle will have been divided into four quadrants of 90° (Fig. 9.13).

Figure 9.13

The quadrants are known as the north-east, south-east, south-west, and north-west quadrants. The quadrant bearing of any line is the angle which it makes with the north–south axis. This angle is given the name of the quadrant into which it falls. In Fig. 9.14 the whole circle bearings of four lines AB, AC,

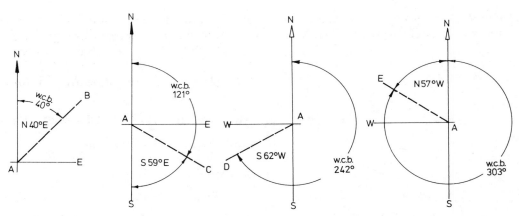

Figure 9.14

AD, and AE are 40°, 121°, 242°, and 303° respectively.

The angles which the lines make with the north-south axis are respectively:

AB	40°	Quadrant Bearing	N 40° E
AC	59°	Quadrant Bearing	S 59° E
AD	62°	Quadrant Bearing	S 62° W
AE	57°	Quadrant Bearing	N 57° W

The conversion of whole circle to quadrant bearings, sometimes called reduced bearings, can be readily carried out by the following rules.

1. When the whole circle bearing lies between 0° and 90°, its quadrant bearing has the same numerical value and lies in the NE quadrant.
2. When the whole circle bearing lies between 90° and 180° the quadrant bearing is (180° − WCB) and lies in the SE quadrant.
3. When the whole circle bearing lies between 180° and 270° the quadrant bearing is (WCB − 180°) and lies in the SW quadrant.
4. When the whole circle bearing lies between 270° and 360° the quadrant bearing is (360° − WCB) and lies in the NW quadrant.

Example 7 Convert the following circle magnetic bearings into quadrant bearings:

60° 30'   240° 10'   352° 10'   131° 00'

Solution

Whole circle bearing   Quadrant bearing

60° 30'		N60° 30' E
240° 10'	(240° 10' − 180°)	S 60° 10' W
352° 10'	(360° − 352° 10')	N07° 50' W
131° 00'	(180° − 131° 00')	S 49° 00' E

Test questions

1 Convert the following whole circle true bearings into quadrant bearings:

311° 10'   247° 30'   060° 10'   093° 00'   270° 00'
167° 50'   111° 10'   264° 50'   359° 10'   179° 00'

2 Convert the following quadrant bearings into whole circle bearings:

N 30° 10' W   S 60° 30' W   S 07° 45' E   N 10° 00' E
East   S 79° 10' W   S 89° 50' E   N 64° 30' E

3 The following true whole circle bearings are known in a certain area:
(a) 162° 10'   (b) 350° 00'   (c) 348° 30'   (d) 210° 10'
At a certain time the magnetic declination in the locality is found to be 10° 30' west.

Calculate the magnetic whole circle bearings of the lines then convert them to magnetic quadrant bearings.

4 Calculate the back bearings corresponding to the following whole circle magnetic bearings.
001° 00'   179° 00'   181° 20'   359° 30'   92° 10'

Convert the back bearings into true quadrant bearings given that the declination is 9° 30' west.

5 From a survey station A, bearings are taken to two points B and C and the following results obtained:

AB   S 84° 30' E
AC   S 79° 20' W

The Ordnance Survey grid bearing of line AB is known to be N 84° 30' E. Calculate the grid bearing of line AC and express it as a quadrant bearing.

6 (a) The bearing of a line AB was measured from the grid lines of an Ordnance Survey plan in 1960 and was found to be 60° 00'. At this particular longitude, the grid lines diverge westwards from the true meridian by 2°. A bearing of AB taken by compass was found to be 70° 00'.

Calculate the value of magnetic declination in 1960.

(b) The declination was shown to be 10° 30' on a 1968 revised edition of the OS map.

Calculate the average secular variation over the period 1960–68.

7 What is the clockwise angle between whole circle bearings 30° 30' and 320° 30'?

8 If the bearing of a line AB is N 17° 45' W, calculate the whole circle bearing of another line at right angles to the line AB.

9 (a) What is the distinction between the terms declination and variation?

(b) Define true meridian, magnetic meridian, assumed meridian.

10 In Fig. 9.5 a yacht is to sail from the pier at ref. 308813 to the harbour at 399752. Draw on the figure a suitable course and measure by protractor the grid whole circle bearings of the various course alterations.

11 Complete the following table.

Magnetic WCB (forward)	26°	88°	233°	315°	8°
Magnetic declination	8°W	2°E	5°E	5°W	10°W
True WCB (forward)					
True WCB (back)					
True quad. brg (back)					

12 Figure 9.15 shows a typical compass survey following the route of a drain. Using a protractor, measure the whole circle magnetic bearings (forward) of the survey lines from A to E. Hence convert these bearings to back bearings.

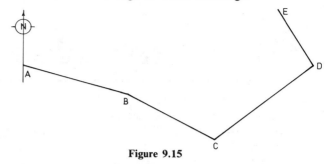

Figure 9.15

13 Figure 9.16 shows a typical theodolite survey following the course of a river. The clockwise angles between the survey lines have been measured and the forward bearing of the first line AB is known to be 60°. Draw the North meridian through each survey point and determine the forward bearing of each line using the geometrical theorems associated with parallel lines.

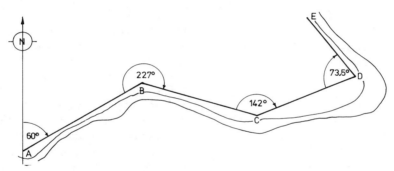

Figure 9.16

10. Principles of traversing

In Chapter 1 it was shown that a survey of any area of ground can be made by obtaining the bearings of a number of well-chosen lines to form the framework of a traverse survey.

A theodolite is used to measure accurately the angles of such a survey and from these angles the bearings of the lines are calculated (Chapter 9, question 13).

The bearings of a survey framework can be obtained directly and speedily though less accurately by using a compass.

This chapter deals with compass traverses and provides a base on which subsequent coordinate calculations and theodolite traverses may be built.

The two most commonly used compasses are:

    1. The prismatic compass.
    2. The surveyor's compass.

Prismatic compass

The Hilger & Watts prismatic compass is shown in Figs 4.5 and 10.1.

It consists essentially of the following parts.
(a) A lightweight telescopic tripod with miniature ball and socket levelling joint. The stem from the ball is externally screwed to receive the compass box.
(b) The compass box which is a non-magnetic glass-topped circular metal box 115 mm in diameter. Two sighting vanes are attached to the outside of the box and are hinged to enable them to fold down against it.
(c) A steel pivot attached to the centre of the box.
(d) A broad magnetic needle and compass ring having a fine jewelled centre, recessed to rest upon the steel pivot.
(e) A damping device in the form of a push button brake pin.
(f) Various accessories such as folding mirror, sun glasses, etc.

Sighting arrangements

Two sighting vanes are mounted on the compass box as shown in Fig. 4.5.

The object vane is a window containing a sighting wire which is accurately aligned onto the target by the observer whose eye is placed at the slit in the eye vane.

In order to read the compass graduations, the observer has only to lower his eye until he is looking directly into the 45° prism. His line of sight is thereby deflected by 90° and enables him to read the appropriate graduation on the compass ring.

Compass graduations

The compass ring and magnetic needle together, form one unit. When the needle comes to rest in the magnetic meridian, the circle graduations are oriented to magnetic north.

The circle is graduated at intervals of half a degree and is figured in a clockwise direction every five degrees as illustrated in Fig. 10.1(b).

It might be expected that the zero graduation will occur at the north end of the needle but closer examination will reveal that the zero in fact is found at the south end. The reason for the transposition is the fact that the observer reads the graduations through the prism. When observing due north, his reading must be 0° and the only way in which this can be achieved is to have the zero reading at the south end.

The 45° reading prism has convex sides and acts as a lens magnifying the compass graduations. Since it is a lens, it inverts the figures and the compass ring is therefore manufactured in such a way that the figures are reversed. When read through the prism they appear right way up.

Surveyor's compass

The Brunton pocket transit is a good example of a surveyor's compass. It is manufactured by Hilger & Watts and is shown in Fig. 4.5. The instrument

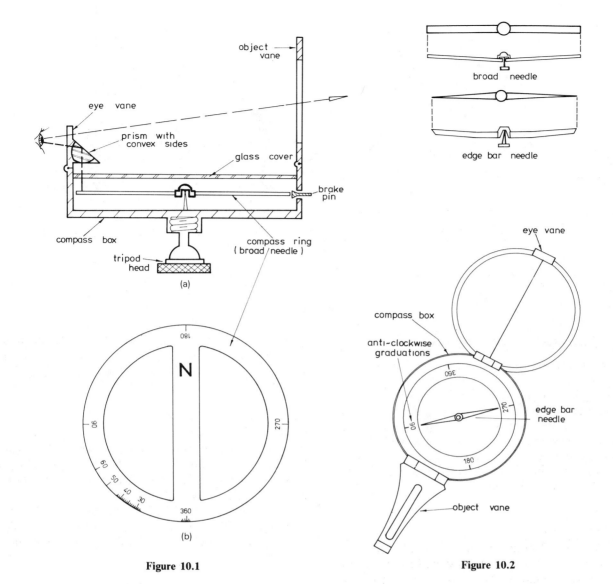

Figure 10.1

Figure 10.2

measures only 70 mm in diameter by 25 mm deep. It is a combined clinometer, Abney level, and compass, the latter function only being the one under consideration.

The instrument, shown diagrammatically in Fig. 10.2, comprises the following parts.

(a) A tripod fitted with a ball and socket joint similar to the prismatic compass.

(b) A non-magnetic metal compass box 70 mm in diameter. The folding lid opens to act as the eye vane of the sighting arrangement. The object vane is a simple window without sighting wire, which folds down against the lid when closed.

(c) An edge-bar needle with a jewelled centre recessed to balance on the fine steel pivot in the centre of the compass box.

The edge-bar needle differs from the broad needle in that it is flattened in a plane at right angles

to the plane of rotation of the needle, while the latter is flattened in the plane of rotation.

(d) Compass graduations attached to the compass box. The graduations therefore rotate around the centre of the instrument with the compass box and sighting vanes. The graduated circle is read against the pointed north end of the needle to give the bearing of the line being observed.

Since the needle remains stationary and the circle moves, the graduations must be anti-clockwise.

The smallest graduation on this instrument is one degree and the figuring is at five-degree intervals.

Example 1 Make a sketch showing a reading of 210° as it would appear on:
(a) The prismatic compass.
(b) The surveyor's compass.

Solution

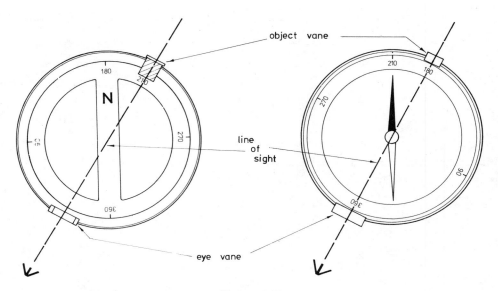

Figure 10.3

Traverse surveying with the compass

The principal use of the compass is in traverse surveying.

A compass traverse is a series of adjoining lines, the bearing and length of each of which must be obtained. Figure 10.4 is an example of an open traverse, following a stream. Stations A, B, C, D, and E are the traverse stations chosen such that the end stations of each line are clearly intervisible and the lines are easily measured.

Linear measurements

The lines do not require to be measured with great accuracy but certainly for construction surveying purposes the measurements should be commensurate with chain survey measurements. Cognisance should therefore be taken of steep slopes, and checks should be made to detect gross errors.

Measurement of bearings

The compass is set up over the first station 'A' of the traverse, the needle is released, allowed to settle and a foresight is taken to station B. The whole circle forward bearing of line AB is noted in the field book as 39° (Table 10.1).

The compass is then removed to station B and a backsight taken to station A, that is, the back bearing of line AB is obtained and noted in the field book.

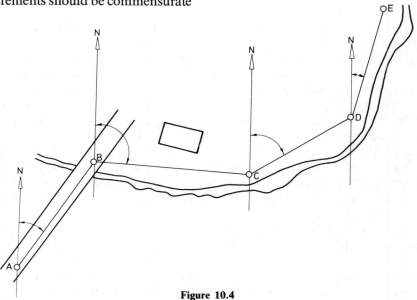

Figure 10.4

| | Observed WCB | | Corrections | | Corrected WCB | | Distance | |
Line	Forward	Back	Forward	Back	Forward	Back	(m)	Remarks
AB	39°	215°					79	
BC	94°	274°					59	
CD	58°	239°					52	
DE	18°	195°					44	

Table 10.1

Forward and back bearings of each line are obtained in the same manner until the traverse is complete. The bearings may be noted in tabular form as in Table 10.1 where is should be noted that the first station in each line is the station from which the forward bearing is taken.

Alternatively a sketch may be made and the bearings simply written along the survey lines.

Local attraction

The back bearing of any line should differ from the forward bearing by ± 180° and back bearing BA should read 219°. However, in the field book (Table 10.1) it is shown as being 215°.

This 4° difference is due to the phenomenon known as local attraction. The presence of overhead power cables, nearby iron railings, piles of metal, etc., will cause the magnetic needle to be deflected from the magnetic meridian and give rise to wrong bearings.

If the backsight and foresight differ by exactly 180° they may be considered to be correct. Should the difference not be 180° however, local attraction must exist at one or both of the stations and neither reading can be depended upon until further readings are taken along the traverse.

The forward and back bearings of the next line BC exhibit a difference of exactly 180° so it can be assumed that there is no local attraction at either station.

Examination of the remaining lines CD and DC reveals that local attraction is affecting the results.

Correcting local attraction

Before the survey can be plotted, attraction must be eliminated. It has already been shown that stations B and C are free from local attraction; therefore any bearing taken from either station will be correct.

The only other bearing taken from B is the back bearing BA. The value of 215° for this bearing must therefore be correct. By similar reasoning forward bearing CD must also be correct, the value being 58°.

Since these four lines are correct, no correction is required and a zero is inserted in the correction columns of Table 10.2. The corrected bearings of the four lines are therefore the same as the observed bearings.

Since the correct back bearing of line BA is 215°, the correct forward bearing of line AB is (215° − 180°) = 35°. This value is inserted in the corrected WCB (forward) column, Table 10.3. The observed forward bearing is 39°, therefore a correction of −4° is required to produce the correct forward bearing. The correction value of −4° is therefore inserted in the correction column.

Similarly, since the correct forward bearing of line CD is 58° the correct back bearing is 58° + 180° = 238°. This value is inserted in the appropriate column (Table 10.3).

The correction to the observed back bearing is therefore 239° − 238° = −1°. This simply means that station D has a local attraction of −1° and all bearings observed from D have to be corrected by this amount. Forward bearing DE is the only line in this category and the value −1° is inserted in the correction (forward) column. The corrected WCB (forward) is therefore (18° − 1°) = 17°.

As a result of these corrections the correct back bearing ED = (17° + 180°) = 197°. The observed back bearing ED is 195° and a correction of +2° is needed to produce the value of 197°. In other words station E has a local attraction of +2°.

Example 2 The following bearings were noted during an open compass traverse. Correct the bearings for local attraction.

| | WC bearings | |
Line	Forward	Back
PQ	69°	249°
QR	82°	260°
RS	75°	258° 30'
ST	172° 30'	354°
TU	153° 30'	331°
UV	354° 30'	172°

Solution

1. Examine the table and find the line or lines where there is no local attraction—line PQ.

Line	Observed WCB		Corrections		Corrected WCB	
	Forward	Back	Forward	Back	Forward	Back
AB	39°	215°		0°		215°
BC	94°	274°	0°	0°	94°	274°
CD	58°	239°	0°		58°	
DE	18°	195°				

Table 10.2

Line	Observed WCB		Corrections		Corrected WCB	
	Forward	Back	Forward	Back	Forward	Back
AB	39°	215°	−4°	0°	35°	215°
BC	94°	274°	0°	0°	94°	274°
CD	58°	239°	0°	−1°	58°	238°
DE	18°	195°	−1°	+2°	17°	197°

Table 10.3

Line	Observed WCB		Corrections		Corrected WCB	
	Forward	Back	Forward	Back	Forward	Back
PQ	69°	249°	0	0	69°	249°
QR	82°	260°	0	+2°	82°	262°
RS	75°	258° 30′	+2°	−1° 30′	77°	257°
ST	172° 30′	354°	−1° 30′	−3°	171°	351°
TU	153° 30′	331°	−3°	−0° 30′	150° 30′	330° 30′
UV	354° 30′	172°	−0° 30′	+2°	354°	174°

Table 10.4

2. Bearings from Q are correct. Therefore forward WC bearing line QR is correct at 82°.

3. The remaining corrections are shown in Table 10.4.

Detailing by compass

The most commonly employed method in detailing is offsetting from the measured chain lines just as in linear surveying. Figure 10.5 shows a set of field notes to accompany the open traverse of Table 10.1. The complete survey is plotted later in this chapter.

Closed traverse

Whenever possible any traverse should be made to close. This procedure allows gross errors to be

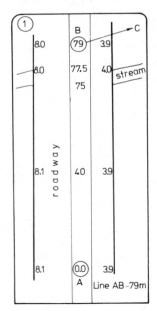

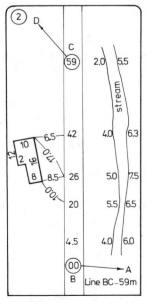

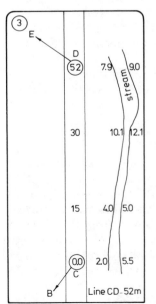

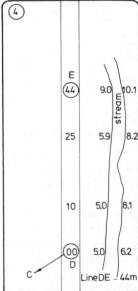

Figure 10.5

detected and also forms a check on the accuracy of the survey.

A closed traverse begins and ends on the same station and is illustrated in Fig. 10.6. The bearings and distances are obtained in the previously described manner and corrections are applied for local attraction.

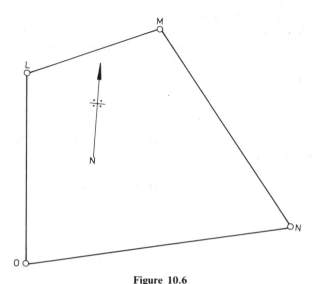

Figure 10.6

It should be noted that some surveyors prefer the tabular method of booking shown alongside Fig. 10.6, although the system has no particular advantage.

Example 3   As a further exercise, the reader should attempt to eliminate local attraction from the traverse shown in Fig. 10.6. The logic is identical to that used in the previous open traverses, though it is set out differently.

Solution

Line	Observed bearing	Correction	Corrected bearing
LO	175° 30′	0°	175° 30′
LM	66° 30′	0°	66° 30′
ML	246° 30′	0°	246° 30′
MN	142° 00′	0°	142° 00′
NM	320° 00′	+2°	322° 00′
NO	255° 30′	+2°	257° 30′
ON	79° 00′	−1° 30′	77° 30′
OL	357° 00′	−1° 30′	355° 30′

Accuracy and limitations of compass surveying

Compass surveys are specially suitable for reconnaissance or whenever speed is important. They can be used to advantage where small irregular areas have to be surveyed, and where the lengths of the survey lines are short.

Under the best conditions, readings of the compass graduations cannot be made to less than 10 minutes of arc which represents a linear displacement at the end of any line of 1 part in 350 parts.

Line	Observed bearing	Length (m)
LO	175° 30′	
LM	66° 30′	93.0
ML	246° 30′	
MN	142° 00′	160.5
NM	320° 00′	
NO	255° 30′	177.0
ON	79° 00′	
OL	357° 00′	129.5

Since the angular accuracy is of this relatively low accuracy, no attempt is made to incorporate refinements into the linear measurements and an accuracy of 1 in 500 is aimed at.

The main advantages of compass surveys are the speed with which they can be done; the low demand on manpower and equipment; the fact that each line is completely defined since its direction, length, and position relative to the other lines is known.

Plotting and adjustment

Two methods of plotting compass surveys are available:
1. Graphical—using protractor and scale.
2. Mathematical—using the method of rectangular coordinates.

Since this is the method of plotting theodolite surveys, it is itself the subject of Chapter 11.

Graphical method

1. *Equipment*

The equipment required for plotting is as follows:
(a) A large circular protractor graduated in half degrees.
(b) An appropriate scale rule.
(c) An instrument for drawing parallel lines. A set

of roller parallels is best, but for classroom and examination work, two set-squares suffice.

(d) A 4H pencil and a needle.

2. *Procedure*

The open traverse given in Table 10.1 and Fig. 10.5 is used in the following description of plotting a survey, the scale used being 1:1000. The corrected framework details are:

Line	AB	BC	CD	DE
WC bearing (forward)	35°	94°	58°	17°
Length (m)	79	59	52	44

(a) A freehand drawing of the survey is made in order to locate the survey centrally on the drawing paper.

(b) A line representing the magnetic meridian is drawn lightly through the first point 'A' of the survey (Fig. 10.7).

(c) The protractor is laid on this line with the 0° graduation facing to magnetic north and the bearing of line AB, namely 35°, is marked off.

(d) A line is drawn through this latter plotting mark and the point A and the length 79 m is marked, thus establishing point B.

(e) A line is drawn through NB parallel to the magnetic meridian through A to represent the magnetic meridian at B.

(f) Using the bearing and distance of the line BC, steps (c) and (d) are repeated to establish station C.

(g) Subsequent stations D and E are likewise established.

(h) Circles of 2 mm diameter are drawn in ink around each survey station and the stations clearly designated.

(i) The offsets are plotted and the features drawn in ink.

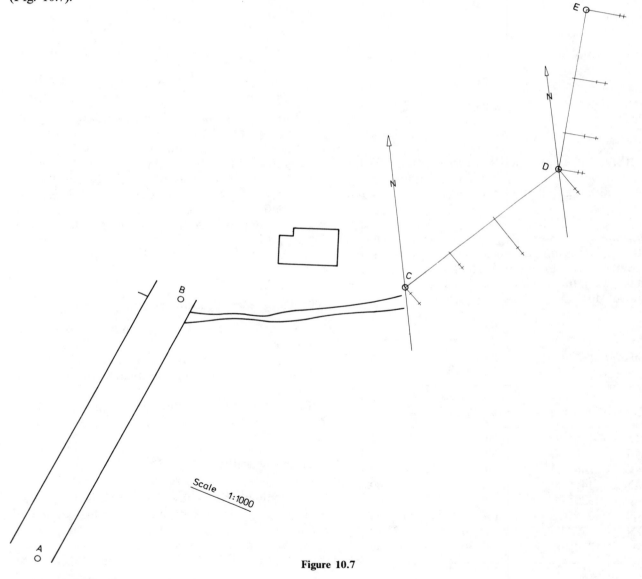

Figure 10.7

(j) The meridians, survey lines, and offsets are erased and a title, scale and north point added in ink to complete the fair drawing.

Test questions

1 The following compass traverse was carried out in the vicinity of overhead power lines.

Line	WC bearing Forward	Back	Distance (m)
PQ	30°	205°	32
QR	97°	272°	75
RS	164°	346°	33
ST	235°	55°	21
TU	261°	81°	88

(a) Calculate the corrected bearings and plot the survey to a scale of 1:500.
(b) From the plotting obtain the shortest distance from point P to line TU.

2 The following notes refer to a closed compass traverse. The bearings have been corrected for local attraction.

Line	AB	BC	CD	DE	EA
WC bearing	29° 30′	100° 45′	146° 30′	242° 00′	278° 45′
Length	83.5	61.8	62.0	51.2	90.4

(a) Plot the survey to a scale of 1:1000.
(b) Scale the closing error.

3 During a compass traverse in an area where the magnetic declination is known to be 10° west of true North, the following data were recorded:

Line	Length (m)	Forward bearing	Back bearing
AB	85	68° 00′	247° 00′
BC	103	146° 30′	326° 30′
CD	58	243° 30′	61° 00′

(a) Adjust the bearings for local attraction.
(b) Plot the survey (scale 1.1000) relative to true North.

(SCOTVEC, Ordinary National Diploma in Building)

4 Table 10.5 shows the results obtained from an open traverse.

Adjust the observed values for local attraction.

(SCOTVEC, Ordinary National Diploma in Building)

Line	AB	BC	CD	DE	EF
Forward bearing	214° 25′	314° 20′	65° 40′	133° 00′	165° 50′
Back bearing	32° 25′	134° 20′	246° 40′	310° 00′	344° 20′

Table 10.5

11. Rectangular coordinates

It is perhaps true to say that if several people plot the same survey using a protractor and scale, very few of the plottings will be identical. This is primarily due to the fact that the protractor cannot be read with the same accuracy by different people.

A much more accurate method of plotting is that of rectangular coordinates which is an entirely mathematical solution to the problem of plotting.

This is the only method of plotting theodolite traverses and while this chapter deals only with compass survey calculations, it must be borne in mind that the calculations of a theodolite traverse are identical in layout and execution and are fully dealt with in Chapter 12.

Basic principle

If it can be imagined for the moment that a gigantic sheet of graph paper could be unrolled over a site, then it would be possible to define the position of any point on the site, by conventional x and y coordinates.

The coordinates of A are (0.0, 0.0), those of B are (+34.6, +20.0), while those of C are (−10.0, −15.0). It should be noted that the x coordinates are given before the y coordinates.

This convention is always followed in surveying.

In Fig. 11.1, A and B are two stations of an open compass traverse, the bearing and distance of the line AB being N 60° E and 40.0 metres respectively. The stations could be plotted by protractor and scale using the field information but if the x and y coordinates could be calculated the protractor could be dispensed with.

If I is the intersection of a parallel to the x axis through B and the y axis through A, then triangle ABI is right angled at I and the quadrant bearing of the line AB is contained within the right-angled triangle, that is:

in triangle ABI

angle IAB = 60°

and side AB = 40.0 m

The x coordinate of point B is the length IB

and IB = AB sin 60°

$\qquad$ = 40.0 × sin 60°

$\qquad$ = +34.6 m

The y coordinate of point B is the length IA

and IA = AB cos 60°

$\qquad$ = 40.0 × cos 60°

$\qquad$ = 20.0 m

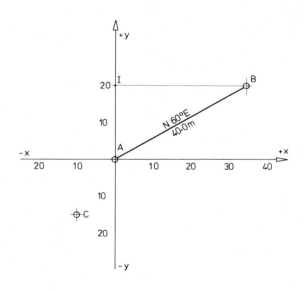

Figure 11.1

Calculation of coordinates

Method 1—Using quadrant bearings

Example 1 The following data refer to an open compass traverse ABCDE. The rectangular coordinates of each station are required:

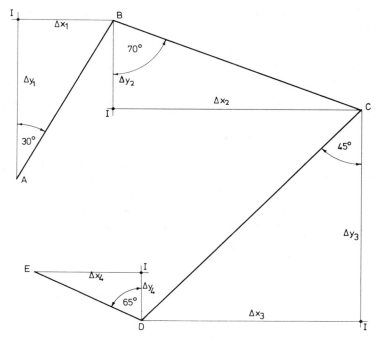

Figure 11.2

Line	AB	BC	CD	DE
WC Bearing	30°	110°	225°	295°
Length	50.0	70.0	82.0	31.2

Solution

1. The first step in any coordinate calculation is to obtain the quadrant bearing of every line. These are as follows:

AB:N 30° E BC:S 70° E CD:S 45° W DE:N 65° W

2. The x and y axes are drawn through two stations at a time such that the quadrant bearing of each line is enclosed in a right-angled triangle (Fig. 11.2), e.g., on line BC the y axis is drawn through B and the x axis through C to enclose quadrant bearing S 70° E.

3. In IAB $\Delta x_1 = $ IB $\Delta y_1 = $ IA

$\qquad\qquad = $ AB sin 30° $= $ AB cos 30°

$\qquad\qquad = 50.0 \times \sin 30.0 = 50.00 \times \cos 30.0$

$\qquad\qquad = \underline{+25.0 \text{ m}} \qquad = \underline{+43.3 \text{ m}}$

Note that both signs are positive since the line travels eastwards (to the right) along the x axis and northwards (upwards) along the y axis.

4. In IBC $\Delta x_2 = $ IC $\Delta y_2 = $ IB

$\qquad\qquad = $ BC sin 70° $= $ BC cos 70°

$\qquad\qquad = 70.0 \sin 70° \qquad = 70° \cos 70°$

$\qquad\qquad = \underline{+65.8 \text{ m}} \qquad = \underline{-23.9 \text{ m}}$

Note that x_2 has a positive sign since the line travels eastwards along the x axis and y_2 has a negative sign, since the line is travelling southwards (down) along the y axis.

5. In ICD $\Delta x_3 = $ ID $\Delta y_3 = $ IC

$\qquad\qquad = $ CD sin 45° $= $ CD cos 45°

$\qquad\qquad = 82.0 \sin 45° \qquad = 82° \cos 45°$

$\qquad\qquad = \underline{-58.0 \text{ m}} \qquad = \underline{-58.0 \text{ m}}$

The signs are both negative since the line travels westwards (to the left) along the x axis and southwards (down) along the y axis.

6. In IDE $\Delta x_4 = $ IE $\Delta y_3 = $ ID

$\qquad\qquad = $ DE sin 65° $= $ DE cos 65°

$\qquad\qquad = 31.2 \sin 65° \qquad = 31.2 \cos 65°$

$\qquad\qquad = \underline{-28.3} \qquad\quad = \underline{+13.2}$

x_4 has a negative sign since the line travels westwards along the x axis, and y_4 has a positive sign since the line travels northwards along the y axis.

Partial coordinates

The differences in x coordinates ($\Delta x_1, \Delta x_2, \Delta x_3, \Delta x_4$) are known as 'departure' differences or more commonly 'partial departures' and are *always* found from the formula:

partial departure = length of line × sin quadrant bearing

'Easterly' departures are always positive and 'westerly' departures are always negative.

The differences in y coordinates (Δy_1, Δy_2, Δy_3, Δy_4) are known as latitude differences or 'partial latitudes' and are *always* found from the formula:

partial latitude = length of line × cos quadrant bearing

'Northerly' latitudes are always positive and 'southerly' latitudes are always negative.

Collectively the partial departures and partial latitudes are known as partial coordinates. These coordinates could be plotted on a rectangular grid, each station being plotted from the previous station.

This system has the distinct disadvantage that if one station is plotted wrongly the succeeding stations will automatically be wrong.

Total coordinates

To overcome this disadvantage, all the stations are plotted from the first station or origin of the survey by converting the partial coordinates to 'total coordinates'.

The total coordinates of any station are obtained by algebraically adding the previous partial coordinates.

Treating station 'A' as the origin of the survey, that is, its coordinates are (0.0, 0.0), the total departure of successive stations, B, C, D, and E are found thus.

Line	Partial departure	Total departure	Station
		0.0	A
AB	Δx_1	$0.0 + \Delta x_1$	B
BC	Δx_2	$0.0 + \Delta x_1 + \Delta x_2$	C
CD	Δx_3	$0.0 + \Delta x_1 + \Delta x_2 + \Delta x_3$	D
DE	Δx_4	$0.0 + \Delta x_1 + \Delta x_2 + \Delta x_3 + \Delta x_4$	E

Numerically the values are:

Line	Partial departure	Total departure		Station
		0.0	= 0.0	A
AB	+25.0	0.0 + 25.0	= +25.0	B
BC	+65.8	0.0 + 25.0 +65.8	= +90.8	C
CD	−58.0	0.0 + 25.0 +65.8 +(−58.0)	= +32.8	D
DE	−28.3	0.0 + 25.0 +65.8 +(−58.0) + (−28.3)	= + 4.5	E

The total latitude of each station is found in an identical manner.

Line	Partial latitude	Total latitude	Station
		0.0	A
AB	Δy_1	$0.0 + \Delta y_1$	B
BC	Δy_2	$0.0 + \Delta y_1 + \Delta y_2$	C
CD	Δy_3	$0.0 + \Delta y_1 + \Delta y_2 + \Delta y_3$	D
DE	Δy_4	$0.0 + \Delta y_1 + \Delta y_2 + \Delta y_3 + \Delta y_4$	E

The numerical values are:

Line	Partial departure	Total departure		Station
		0.0	= 0.0	A
AB	+43.3	0.0 + 43.3	= +43.3	B
BC	−23.9	0.0 + 43.3 +(−23.9)	= +19.4	C
CD	−58.0	0.0 + 43.3 +(−23.9) + (−58.0)	= −38.6	D
DE	+13.2	0.0 + 43.3 + (−23.9) + (−58.0) + 13.2 = +25.4		E

The total coordinates of each station are therefore:

Station	Total departure	Total latitude
A	0.0	0.0
B	+25.0	+43.3
C	+90.8	+19.4
D	+32.8	−38.6
E	+4.5	−25.4

Traverse table

All of the calculations of Example 1 are set out in a traverse table as in Table 11.1(a). It is completed as follows:

1. Enter the station designations in column 10 (i.e., enter letters A to E in successive lines). Otherwise leave line 1 blank.
2. Enter the survey lines AB, BC, CD, and DE in column 1 beginning on line 2.
3. Enter the quadrant bearing of each line in column 2.
4. Enter the plan length of each line in column 3.
5. Calculate the partial coordinates as described.

Enter the partial departures in columns 4 and 5 and the partial latitudes in columns 6 and 7.
6. Calculate the total coordinates from the partial coordinates. Enter the departures in column 8 and latitudes in column 9.
7. A check is provided on the arithmetic thus:

The algebraic sum of the partial departure columns 4 and 5 should equal the difference between the total departures of the first and last stations in column 8.

The algebraic sum of the partial latitude columns 6 and 7 should equal the difference between the total latitudes of the first and last stations in column 9.

Method 2—Calculations of coordinates using whole circle bearings

Although a few surveyors still calculate coordinates using quadrant bearings, the most common practice is to compute the coordinates directly from the whole circle bearings, thus eliminating a potential source of error in calculating the quadrant bearing.

In Figure 11.1 the partial coordinates of line BC are:

1	2	3	4	5	6	7	8	9	10
Line	Quadrant bearing	Dist. (m)	Partial coordinates				Total coordinates		Stn
			E+	W−	N+	S−	Dep.	Lat.	
							00.0	0,00	A
AB	N 30° E	50.0	25.0	—	43.3	—	+25.0	+43.3	B
BC	S 70° E	70.0	65.8	—	—	23.9	+90.8	+19.4	C
CD	S 45° W	82.0	—	58.0	—	58.0	+32.8	−38.6	D
DE	N 65° W	31.2	—	−28.3	13.2	—	+ 4.5	−25.4	E
			+90.8	−86.3	+56.5	−81.9	+ 4.5	−25.4	
			−86.3		−81.9				
			+ 4.5		−25.4				

Table 11.1(a)

Part. dep. = dist. × sin quad brg

= 70.0 × sin 70°

= 70.0 × 0.939 69

= 65.8 m EAST

= +65.8 m

Part. lat. = dist. × cos quad brg

= 70.0 × cos 70°

= 70.0 × 0.342 02

= 23.9 m SOUTH

= −23.9 m

Since whole circle bearing 110° is a second quadrant angle in mathematical terms, then sin 110° = +sin70° and cos 110° = −cos 70°.

The partial coordinates of line BC, using the whole circle bearing are therefore:

Part. dep. = length BC × sin WCB

= 70.0 × sin 110°

= 70.0 × 0.939 69

= +65.8 m

Part. lat. = length × cos WCB

= 70.0 × cos 110°

= 70.0 × −0.342 02

= −23.9 m

Similarly the partial coordinates of line CD are:

Part. dep. = 82.0 × sin 225°

= 82.0 × −0.707 11

= −58.0 m

Part. lat. = 82.0 × cos 225°

= 82.0 × −0.707 11

= −58.0 m

and the partial coordinates of line DE are:

Part. dep. = 31.2 × sin 295°

= 31.2 × −0.906 31

= −28.3 m

Part. lat. = 31.2 × cos 295°

= 31.2 × +0.422 62

= +13.2 m

Using the whole circle bearing method, the traverse Table 11.1(a) is shortened from ten to eight columns in Table 11.1(b). In this table the partial coordinates occupy only two instead of four columns. Using the data of Example 1, the table is completed as follows.

1. Enter the station designations in column 8.

2. Enter the survey lines in column 1 beginning on line 2.

3. Enter the whole circle bearing of each line in column 2.

4. Enter the plan length of each line in column 3.

5. Calculate the partial coordinates as described. Enter the partial departures in column 4 and the partial latitudes in column 5.

6. Calculate the total coordinates by successively adding the partial coordinates. Enter the total departures (commonly called Eastings) in column 6 and enter the total latitudes (commonly called Northings) in column 7.

1	2	3	4	5	6	7	8
	Whole circle		Dep.	Lat.	Total coordinates		
Line	bearing	Dist. (m)	difference	difference	Easting	Northing	Stn
					00.0	00.0	A
AB	30°	50.0	+25.0	+43.3	+25.0	+43.3	B
BC	110°	70.0	+65.8	−23.9	+90.8	+19.4	C
CD	225°	82.0	−58.0	−58.0	+32.8	−38.6	D
DE	295°	31.2	−28.3	+13.2	+ 4.5	−25.4	E
			+90.8	+56.5	+ 4.5	−25.4	
			−86.3	−81.9			
			+ 4.5	−25.4			

Table 11.1(b)

Method 3—Calculation of coordinates using the $\boxed{P \to R}$ function of a calculator

Figure 11.3(a) shows the mathematical concept of polar (r, θ) and rectangular (x, y) coordinates of a line AB, referred to the X and Y axes through the four quadrants numbered 1 to 4.

On all scientific calculators, e.g., CASIO fx, TEXAS TI models, the rectangular coordinates can be found directly from the polar coordinates and vice versa by using the $\boxed{P \to R}$ and $\boxed{R \to P}$ function buttons. These are secondary functions on a calculator and have to be used in conjunction with a button usually marked $\boxed{INV}$ or $\boxed{2nd}$ shown in Fig. 11.3(c).

When the X, Y axes are rotated through 90°, then reflected through 180°, the positive X axis points North and the positive Y axis points East. Thus they form the coordinate axis system for surveying in Fig. 11.3(b).

The polar coordinates (r, θ) become the length and whole circle bearing of the line AB while the x value is the surveying *latitude* and the y value is the surveying *departure*. A slight disadvantage in a $\boxed{P \to R}$ conversion calculation is that the x value (latitude) is produced before the y value (departure) by the calculator.

In Example 1, the polar coordinates of line AB are length 50.0 m and whole circle bearing 30°. The rectangular coordinates, i.e., partial lat. x and partial dep. y are computed on the CASIO fx calculator as follows:

Instruction	*Calculator Operation*
1. Enter length	Input 50.0
2. Operate buttons	$\boxed{INV \ P \to R}$
3. Enter WCB (decimalized)	Input 30.0
4. Operate button	$\boxed{=}$
5. Read answer (part. lat.)	43.3
6. Operate button	$\boxed{X \to Y}$
7. Read answer (part. dep.)	25.0

8. Repeat these steps for line BC (length: 70.0 m; WCB: 110°)

Input 70.0, $\boxed{INV}$ $\boxed{P \to R}$

Input 110°, $\boxed{=}$, read -23.9, $\boxed{X \to Y}$, read $+65.8$

9. Repeat for remaining lines CD and DE

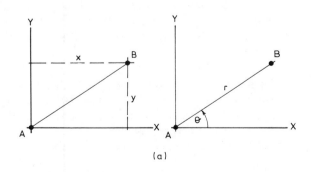

(a)

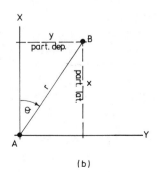

(b)

Figure 11.3

(*Continued*)

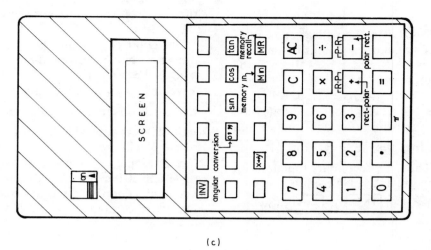

(c)

Figure 11.3 (continued)

Method 4—Calculation of coordinates using a computer program

All of the calculations of the previous sections may be carried out on a microcomputer using a computer program.

Using the $\boxed{P \rightarrow R}$ conversion of the scientific calculator, the length and bearing of a line were input by hand. The calculator then computed the partial coordinates and displayed the answer. The simple computer program works on the same principle.

It is realized that some readers will not wish to or have the facilities to consider further methods of solution, therefore the computer solutions have been placed in Chapter 22. The programs, explanations, and running instructions are provided in programs 1 to 5 where Examples 1 and 2 are fully solved.

Method 5—Calculation of coordinates using spreadsheets

Spreadsheets are commercially available software packages on which tabular solutions for repetitive calculations may be devised. *The spreadsheets, instructions and examples are included with the basic computer programs in Chapter 22.*

Plotting rectangular coordinates

Figure 11.4 is a portion of a rectangular grid drawn to a scale of 1:1000. The lengths of the sides of the squares are 50 m long.

Point A is the origin of the survey and is therefore the intersection of the zero departure and zero latitude lines.

Point C (90.8, 19.4) is plotted thus:
1. The square in which the point will fall is determined by inspection.
2. The total departure, 90.8 m, is scaled along the zero latitude line and the point marked and circled c_1.
3. Similarly the departure is scaled, marked, and circled c_2, along the 50 m latitude line.
4. The points c_1 and c_2 are joined and the line is checked to ensure that it is parallel to the north–south grid lines.
5. The total latitude, 19.4 m, is similarly marked along the 50 m and 100 m departure lines at c_3 and c_4, and the line joining them is checked for parallelism against the east–west grid lines.
6. The intersection of the lines c_1c_2 and c_3c_4 forms the point C.

All other survey points are similarly plotted and clearly marked by a small circle or triangle. The exact position of the survey station is marked by a needle.

Example 2 Calculate and plot the total coordinates of the traverse of Table 10.1 using (a) quadrant bearings, (b) whole circle bearings. Check the plotting against Fig. 10.7.

The relevant traverse data is as follows:

Line	AB	BC	CD	DE
WC Bearing	35°	94°	58°	17°
Length (m)	79.0	59.0	52.0	44.0

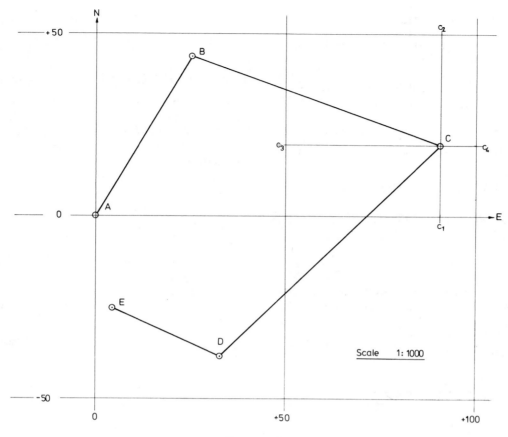

Figure 11.4

Solution

(a) Using quadrant bearings:

Line	Quadrant bearing	Dist. (m)	Partial coordinates				Total coordinates		Stn
			E+	W−	N+	S−	Dep.	Lat.	
							00.0	00.0	A
AB	N 35° E	79.0	45.3	—	64.7	—	+45.3	+64.7	B
BC	S 86° E	59.0	58.9	—	—	4.1	+104.2	+60.6	C
CD	N 58° E	52.0	44.1	—	27.6	—	+148.3	+88.2	D
DE	N 17° E	44.0	12.9	—	42.1	—	+161.2	+130.3	E
			161.2		134.4 −4.1 +130.3	−4.1	+161.2	+130.3	

Table 11.2(a)

(b) Using whole circle bearings:

Line	Whole circle bearing	Distance (m)	Partial departure	Partial latitude	Total dep.	Total lat.	Stn
					0.0	0.0	A
AB	35°	79.0	+45.3	+64.7	+45.3	+64.7	B
BC	94°	59.0	+58.9	−4.1	+104.2	+60.6	C
CD	58°	52.0	+44.1	+27.6	+148.3	+88.2	D
DE	17°	44.0	+12.9	+42.1	+161.2	+130.3	E
			+161.2	+130.3	+161.2	+130.3	

Table 11.2(b)

Calculation of a closed traverse

Basically there is no difference between the calculation of a closed traverse and an open traverse.

The initial and final coordinates of station A should be identical. This will very seldom occur and the resultant closing error must be eliminated.

Example 3 Calculate the total coordinates of all stations in the following traverse given that station A is the origin:

Line   AB   BC   CD   DE   EA
WC
 bearing   29° 30′ 100° 45′ 146° 30′ 242° 00′ 278° 45′
Length (m) 83.50 61.80 62.00   51.20   90.40

Example 3 is an example of a closed traverse which has a closing error. In this example the sexagesimal measure of bearings occurs for the first time in trigonometrical calculations. All of the bearings must be converted to decimal form. Using the scientific calculator a bearing of say 30° 53′ 15″ is decimalized using the converter button, Fig. 11.3(c) as follows.

Press: 30 [° ′ ″] 53 [° ′ ″] 15 [° ′ ″]

The answer appears on the screen as 30.8875°. *The method of converting sexagesimal measure to centesimal measure varies with the type of calculator.*

Closing error of closed traverse (Bowditch's Rule)

The difference between the initial and final coordinates of station A determines the errors in departure and latitude. Figure 11.5 shows the initial and final coordinates plotted on a very much exaggerated scale.

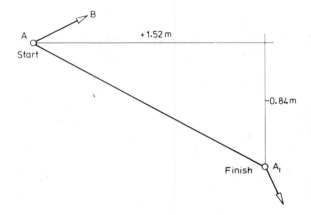

Figure 11.5

Solution   Using quadrant bearings:

Line	Quadrant bearing	Distance (m)	Partial coordinates				Total coordinates		Stn
			E+	W−	N+	S−	Dep	Lat.	
							00.00	00.00	A
AB	N 29° 30′ E	83.50	41.12	—	72.68	—	+41.12	+72.68	B
BC	S 79° 15′ E	61.80	60.72	—	—	11.53	+101.84	+61.15	C
CD	S 33° 30′ E	62.00	34.22	—	—	51.70	+136.06	+9.45	D
DE	S 62° 00′ W	51.20	—	45.20	—	24.04	+90.86	−14.59	E
EA	N 81° 15′ W	90.40	—	89.34	13.75	—	+1.52	+0.84	A
			+136.06 −134.54 +1.52	−134.54	+86.43 −87.27 −0.84	−87.27			

Table 11.3(a)

Solution   Using whole circle bearings:

Line	Whole circle bearing	Distance (m)	Partial dep.	Partial lat.	Total dep.	Total lat.	Stn
					0.00	0.0	A
AB	29° 30′	83.50	+41.12	+72.68	+41.12	+72.68	B
BC	100° 45′	61.80	+60.72	−11.53	+101.84	+61.15	C
CD	146° 30′	62.00	+34.22	−51.70	+136.06	+9.45	D
DE	242° 00′	51.20	−45.20	−24.04	+90.86	−14.59	E
EA	278° 45′	90.40	−89.34	+13.75	+1.52	−0.84	A
			+1.52	−0.84	+1.52	−0.84	

Table 11.3(b)

Table 11.4(a)

Line	Quadrant bearing	Distance (m)	Partial coordinates E+	Partial coordinates W–	Partial coordinates N+	Partial coordinates S–	Corrections Dep.	Corrections Lat.	Corrected partial coordinates E+	Corrected partial coordinates W–	Corrected partial coordinates N+	Corrected partial coordinates S–	Total coordinates Dep.	Total coordinates Lat.	Stn
AB	N 29° 30′ E	83.50	41.12		72.68		−0.364	+0.201	40.76		72.88		0.00	0.00	A
BC	S 79° 15′ E	61.80	60.72			11.53	−0.269	+0.149	60.45			11.38	+40.76	+72.88	B
CD	S 33° 30′ E	62.00	34.22	45.20		51.70	−0.270	+0.149	33.95	45.42		51.55	+101.21	+61.50	C
DE	S 62° 00′ W	51.20		89.34		24.04	−0.223	+0.123		89.74		23.92	+135.16	+9.95	D
EA	N 81° 15′ W	90.40			13.75		−0.394	+0.218			13.97		+89.74	−13.97	E
													0.00	0.00	A
		348.90	+136.06 −134.54 +1.52	−134.54	+86.43 −87.27 −0.84	−87.27	−1.520	+0.840	+135.16	−135.16	+86.85	−86.85			

$$k_1 = \frac{\text{Total correction in departure}}{\text{Total length of survey}}$$

$$= \frac{-1.52}{348.90}$$

$$= -4.36 \times 10^{-3}$$

$$k_2 = \frac{\text{Total correction in latitude}}{\text{Total length of survey}}$$

$$= \frac{+0.84}{348.90}$$

$$= +2.41 \times 10^{-3}$$

Table 11.4(b)

Line	Whole circle bearing	Distance (m)	Partial coords Dep.	Partial coords Lat.	Corrections Dep.	Corrections Lat.	Corrected partials Dep.	Corrected partials Lat.	Total Dep.	Total Lat.	Stn
AB	29° 30′	83.50	+41.12	+72.68	−0.364	+0.201	+40.76	+72.88	0.00	0.00	A
BC	100° 45′	61.80	+60.72	−11.53	−0.269	+0.149	+60.45	−11.38	+40.76	+72.88	B
CD	146° 30′	62.00	+34.22	−51.70	−0.270	+0.149	+33.95	−51.55	+101.21	+61.50	C
DE	242° 00′	51.20	−45.20	−24.04	−0.223	+0.123	−45.42	−23.92	+135.16	+9.95	D
EA	278° 45′	90.40	−89.34	+13.75	−0.394	+0.218	−89.74	+13.97	+89.74	−13.97	E
									0.00	0.00	A
		348.90	+1.52	−0.84	−1.520	+0.840	0.00	0.00			

$$k_1 = \frac{\text{Total correction in departure}}{\text{Total length of survey}}$$

$$= \frac{-1.52}{348.90}$$

$$= -4.36 \times 10^{-3}$$

$$k_2 = \frac{\text{Total correction in latitude}}{\text{Total length of survey}}$$

$$= \frac{+0.84}{348.90}$$

$$= +2.41 \times 10^{-3}$$

The error of closure AA_1 is found by the theorem of Pythagoras to be

$$AA_1 = \sqrt{(1.52^2 + 0.84^2)}$$

$$= \underline{1.74 \text{ m}}$$

The total length of the traverse is 348.90 metres, therefore the fractional accuracy of the survey is:

$$1.74 \text{ m in } 348.90 \text{ m}$$

which equals

$$1 \text{ in } 200 \text{ approximately}$$

Adjustment of closed traverse

The closing error of the survey is due to the accumulation of errors along each draft of the traverse. It should be remembered that the angular and linear errors are assumed to be equally inaccurate. As a result the closing error is distributed throughout the traverse in proportion to the lengths of the various drafts.

In the rectangular coordinate method, it is the separate departure and latitude errors which have to be distributed and not the closing error of distance.

Each error is treated separately and is distributed throughout the traverse in proportion to the lengths of the drafts as follows (Tables 11.4(a) and 11.4(b)):

Using Example 3, the total error in departure = $+1.52$ m, therefore total correction to departures = -1.52 m.

Departure correction at station B

$$= \frac{\text{Length of line AB}}{\text{Total length of survey}} \times \text{total correction}$$

and in fact at any station the departure correction (c_D) is

$$c_D = \left[\frac{\text{length of draft}}{\text{total length of survey}} \right.$$

$$\left. \times \text{ total correction in departure} \right]$$

$$= \frac{l}{L} \times C_D$$

$$= \frac{C_D}{L} \times l$$

Therefore $c_D = k_1 \times l$ (since C_D and L are constants for every line)

The latitude correction (c_L) for every station is derived exactly as above:

$$c_L = \frac{l}{L} \times C_L$$

$$= \frac{C_L}{L} \times l$$

$$= k_2 \times l$$

In example 3 the procedure for correction is as follows:

(a) Total departure correction (C_D) = -1.52 m

Total length of survey (L) $\quad = 348.90$ m

$$\text{Therefore } k_1 = \frac{-1.52}{348.90}$$

$$= \frac{-1.52 \times 10^{-2}}{3.489}$$

$$= \underline{-4.36 \times 10^{-3}}$$

(b) Departure correction at

$$\begin{aligned}
B &= -(4.36) \times 10^{-3} \times 83.50 = -0.364 \\
C &= -(4.36) \times 10^{-3} \times 61.80 = -0.269 \\
D &= -(4.36) \times 10^{-3} \times 62.00 = -0.270 \\
E &= -(4.36) \times 10^{-3} \times 51.20 = -0.223 \\
A_1 &= -(4.36) \times 10^{-3} \times 90.40 = -0.394 \\
&\hspace{5.2cm} \overline{-1.520 \text{ m}}
\end{aligned}$$

(c) Total latitude correction (C_L) = $+0.84$ m

Total length of survey (L) = 348.90 m

$$\text{Therefore } k_2 = +\frac{0.84}{348.90}$$

$$= \underline{2.41 \times 10^{-3}}$$

(d) Latitude correction at

$$\begin{aligned}
B &= +(2.41 \times 10^{-3}) \times 83.50 = +0.201 \\
C &= +(2.41 \times 10^{-3}) \times 61.80 = +0.149 \\
D &= +(2.41 \times 10^{-3}) \times 62.00 = +0.149 \\
E &= +(2.41 \times 10^{-3}) \times 51.20 = +0.123 \\
A &= +(2.41 \times 10^{-3}) \times 90.40 = +0.218 \\
&\hspace{5.2cm} \overline{= +0.840}
\end{aligned}$$

(e) The corrected partial coordinates are obtained from the algebraic addition of the calculated partial coordinates (Table 11.3) and the corrections. For example, corrected partial coordinates of B are

Departure $\qquad\qquad$ *Latitude*

$+41.12 - 0.36 = +40.76 \quad +72.68 - 0.20 = 72.88$

The calculation is more conveniently performed in tabular fashion, as in Table 11.4. It will be immediately apparent that the table is an extension of the original traverse table.

If the arithmetic is performed correctly the sum of the partial eastings should equal the sum of the partial westings. Likewise, the sum of the partial northings and the sum of the partial southings should balance indicating that the errors in departure and latitude have been eliminated.

(f) From the corrected partial coordinates, the total coordinates are obtained by algebraic addition as before. The final coordinates of station A should both be zero if the arithmetic has been done correctly.

Test questions

1 The following results were obtained on an open compass traverse along the south bank of a river:

Line	AB	BC	CD	DE	EF
WC bearing forward	110° 30′	18° 30′	86° 15′	38° 15′	45° 00′
WC bearing back	292° 00′	197° 00′	265° 00′	219° 30′	225° 00′
Distance (m)	90.0	81.5	54.5	100.0	135.0

Line AB		Line BC	
Chainage	Offset	Chainage	Offset
0	4.0(A)	0	(B)
20	1.5	20	14
60	5.5	60	8
74	20.0	81.5	4(C)
90	B		

Line CD		Line DE	
Chainage	Offset	Chainage	Offset
0	6(C)	0	22(D)
30	15	20	11
54.5	19(D)	40	8.2
		60	5.8
		100	6.0(E)

Line EF	
Chainage	Offset
0	(E)
40	14
100	9
135	2(F)

(a) Eliminate local attraction from the observations.
(b) Calculate the coordinates of each station given that 'A' is the origin.
(c) Plot the survey to a scale of 1:2000 showing clearly the south bank of the river.

2 Table 11.5 shows the partial coordinates of a closed traverse.
Calculate:
(a) The closing error.
(b) The accuracy of the traverse.
(c) The corrected final coordinates of all stations, given that station P is the origin.

Line	Distance (m)	East	West	North	South
PQ	122	55.40		108.70	
QR	156	73.20			137.70
RS	65		20.10		61.80
ST	60		56.40		20.50
TP	123		52.90	110.50	

Table 11.5

3 A traverse survey was carried out along the boundaries of a building plot and produced the results as shown in Table 11.6.

Line	Whole circle bearing	Length (m)
PQ	64° 45′	121.25
QR	143° 26′	236.83
RS	234° 15′	179.00
SP	336° 32′	264.97

Table 11.6

Calculate:
(a) The error in closure of the traverse.
(b) The adjusted *TOTAL* coordinates of Q, R, and S with respect to P using Bowditch's Rule.

(SCOTVEC, Ordinary National Diploma in Building)

4 A traverse was run along the boundaries of a building plot and produced the results shown in Table 11.7.

Line	Whole circle bearing	Length (m)
PQ	69° 55′	262.0
QR	166° 57′	155.0
RS	244° 20′	268.0

Table 11.7

Calculate the *TOTAL* coordinates of Q, R, and S taking P as the origin.

(SCOTVEC, Ordinary National Diploma in Building)

12. Theodolites

The theodolite, like the compass, is an angular measuring instrument, but whereas the latter instrument is capable of reading to only about 15 minutes of arc, theodolites can read angles to 1 second of arc. Consequently, the theodolite is of much more intricate design and is more difficult to operate.

At first glance (Fig. 12.1) it appears to be a rather formidable object but a working knowledge can be gained fairly quickly if the instrument is diagrammatically broken into its separate working parts and each examined individually.

Classification

A theodolite is generally classified according to the method used to read the circles. Broadly speaking, the methods are:

1. Vernier
2. Direct reading
3. Optical scale
4. Optical micrometer
5. Opto electronic

Despite this multiplicity of reading systems the basic principles of construction of the theodolite are similar.

Principles of construction

The main components of a theodolite are illustrated in Fig. 12.2. These are:

1. *Tripod*
The purpose of the tripod is to provide support for the instrument. Tripods may be telescopic, that is, they have sliding legs, or may have legs of fixed length.

2. *Theodolite trivet stage*
The trivet stage is the flat base of the instrument which screws on to the tripod and carries the feet of the levelling screws.

3. *Tribrach*
The tribrach is the body of the instrument carrying all the other parts. In the older type of theodolite the tribrach has a hollow, conical-shaped socket into which fits the remainder of the instrument (Fig. 12.2(a)). It has been found that conical axes are prone to wear and lead to inaccuracies with the result that all modern theodolites have hardened steel cylindrical axes (Fig. 12.2(b)). A cylindrical ball race takes the weight of the upper part of the instrument.

4. *Levelling arrangement*
To enable the tribrach to be levelled, levelling screws are fitted between the tribrach and trivet stage. Movement of the footscrews centres the bubble of the plate spirit level, situated on the cover plate of the horizontal circle. The sensitivity of the spirit level is of the order of 2 mm = 40 seconds of arc.

5. *Horizontal circle (lower plate)*
The horizontal circle is in reality an accurately machined protractor graduated in a clockwise direction and numbered from 0° to 360°. On older type instruments the circle is a metal plate of some 110 mm diameter. Modern instruments have glass circles. The horizontal circle is mounted either on a conical axis which fits inside the conical interior of the tribrach (Fig. 12.2(a)) on older instruments or is on a cylindrical axis which fits around the outside of the tribrach (Fig. 12.2(b)).

The circle is therefore free to rotate either within or around the tribrach and can be stopped in any position by applying the lower plate clamp. When tightened, the clamp locks the horizontal circle (lower plate) to the tribrach. A very limited amount of horizontal movement is still possible via the slow motion screw attached to the clamp. The slow motion screw only works when the lower plate clamp is locked.

6. *Alidade*
The alidade is the remainder of the theodolite comprising the uprights ('A' frame of older instruments) which support the telescope and vertical circle and the spirit levels.

(a) Watts ST 20 vernier theodolite

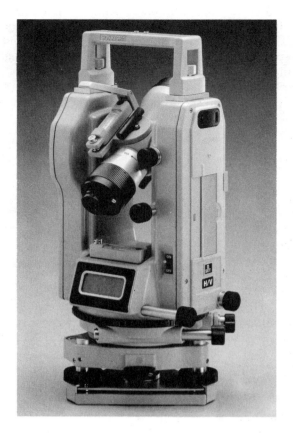

(b) Sokkisha DT 20E electronic theodolite showing upper and lower clamps

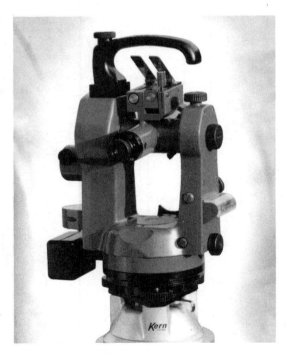

(c) Kern KOS theodolite showing repetition clamp

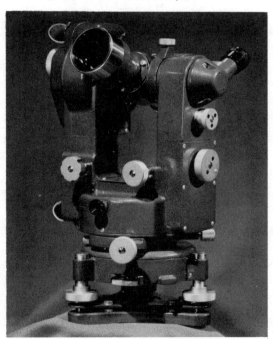

(d) Watts ST 200 theodolite showing circle-setting screw

Figure 12.1

In Fig. 12.2(a) the alidade is carried on a central spindle which fits within the hollow socket of the horizontal circle. It is therefore free to rotate with respect to the circle which is itself free to rotate within the tribrach sleeve as has already been explained.

In Fig. 12.2(b) the alidade fits within the tribrach sleeve and is free to rotate on a precision ball race with respect to it.

7. *Controls for measuring horizontal angles*
(a) *Double centre system* using upper and lower plate clamps. Figure 12.2(a) shows the double centre axis system and figure 12.4(a) shows the two clamps. It is

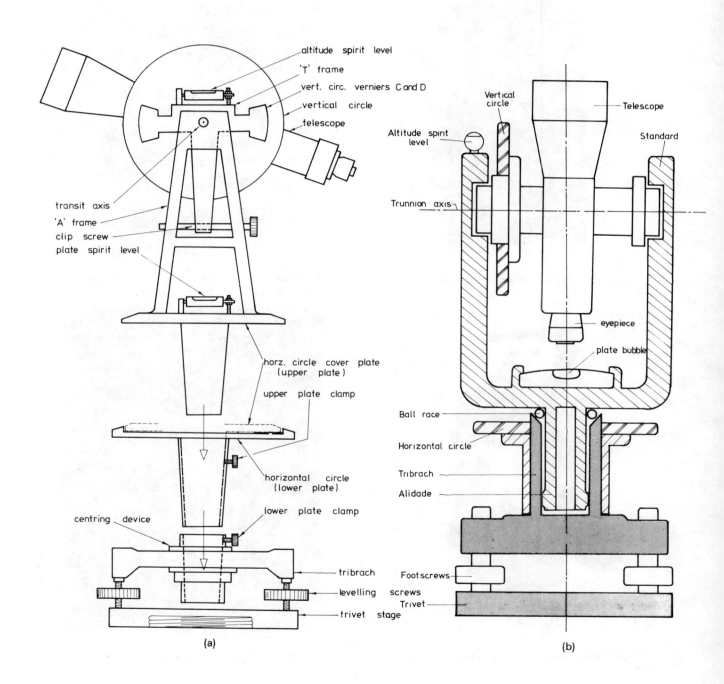

Figure 12.2

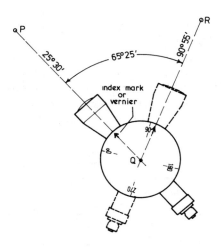

Figure 12.3

essential that the function of the upper and lower plate clamps be understood since they control the entire operation of measuring a horizontal angle.

When both clamps are open the lower plate (horizontal circle) and upper plate (alidade) are free to move in any direction relative to the tribrach and to each other.

When the lower plate clamp is closed, the horizontal circle is locked in position and the alidade is free to rotate over the stationary circle and the reading on the horizontal circle will change continuously.

When both clamps are closed, neither plate can rotate. If the lower plate clamp is now released the upper and lower plates will move together and there will be no change in reading on the horizontal circle. The instrument can be used in the repetition and reiteration measurement of angles (page 149).

(b) *Circle setting screw* Some theodolites do not have a lower plate clamp. The alidade is clamped directly to the tribrach (e.g., Watts No. 1, Kern DKM 1). During measurement the horizontal circle remains stationary while the alidade moves over it in the usual way. However, the circle can be moved by a continuous-drive circle-setting screw and the instrument may therefore be easily set to say 00° 00′ 00″ for any pointing of the telescope, Fig. 12.4(b). The instrument cannot measure angles by the repetition method.

(c) *Repetition clamp system* Some theodolites (Wild T16) are fitted with a repetition clamp instead of a lower plate screw. When the clamp is in the closed position the horizontal circle rests against the tribrach. When the clamp is opened the circle is clamped to the alidade (Fig. 12.4(c)). The upper plate clamp connects the alidade directly to the tribrach and the instrument may be used for either repetition or reiteration measurements (page 149).

In order to measure an angle PQR (Fig. 12.3) the theodolite is set over point Q and the lower plate is locked in any random position. The upper plate clamp is released and the telescope mounted on the alidade is turned to point in turn to stations P and R. A horizontal circle reading is taken for both telescope pointings and subracted to give the horizontal angle.

8. *Index marks*

In order to read the circle for any pointings of the telescope it is convenient to imagine an index mark mounted on the alidade directly below the telescope. As the alidade is rotated, the index mark moves over the horizontal circle. When the alidade is locked, the index mark is read against the circle. Actually,

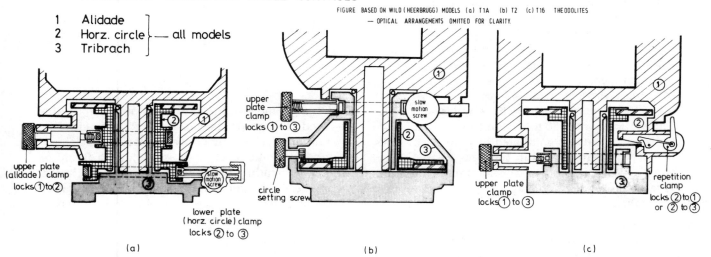

THEODOLITE – HORIZONTAL ANGLE CONTROLS

1 Alidade
2 Horz. circle — all models
3 Tribrach

FIGURE BASED ON WILD (HEERBRUGG) MODELS (a) T1A (b) T2 (c) T16 THEODOLITES — OPTICAL ARRANGEMENTS OMITTED FOR CLARITY.

Figure 12.4

the index mark is really the zero mark of a vernier scale or a mark etched on a glass plate somewhere in the optical train of the theodolite (Figs 12.8 and 12.10).

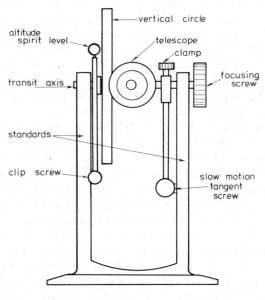

Figure 12.5

9. *Transit axis or trunnion axis*

The transit axis rests on the limbs of the standards and is securely held in position by a locknut. Attached to the transit axis are the telescope and vertical circle. All three are free to rotate in the vertical plane but can be clamped in any position in the plane by a clamp usually known as the telescope clamp (Fig. 12.5). Again, a certain amount of movement is permitted by a slow motion device.

The telescope has been described fully in Chapter 5 and Fig. 5.12 shows the paths of the rays of light through the telescope. A typical specification for a theodolite telescope is

(a) internal focusing—damp and rust resistant;
(b) shortest focusing distance—2 metres;
(c) magnification—× 30;
(d) object glass diameter 42 mm;
(e) field of view—1 degree 12 minutes.

The vertical circle, 80 mm diameter, is attached to the telescope and is graduated in a variety of ways, two of which are shown in Fig. 12.6. Figure 12.6(a) shows the figuring of a vernier theodolite while 12.6(b) shows the figuring of a modern optical theodolite.

10. *Altitude spirit level*

Angles measured in a vertical plane must be measured relative to a truly horizontal line. The line is that which passes through the index arrows of

verniers C and D and is maintained in a horizontal position by the altitude spirit level (Fig. 12.2). It can be seen from the diagram that the spirit level and verniers C and D are attached to a 'T' frame which is made horizontal by activating the clip screw against the standards. The altitude spirit level is more sensitive than the plate spirit level, the sensitivity being 2 mm = 25 seconds.

In a modern theodolite an index mark is provided, which when joined to the centre of the vertical circle forms the horizontal line which is the equivalent of verniers C and D above.

Many modern theodolites are fitted with a coincidence bubble reader which is fully explained in Chapter 5, page 51. This device greatly increases the accuracy of the bubble setting.

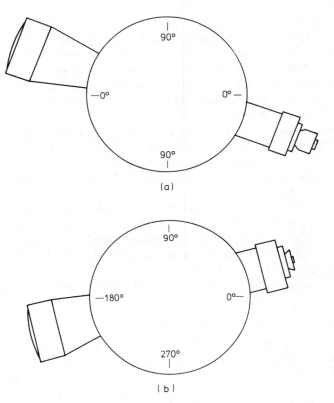

Figure 12.6

On most modern theodolites automatic indexing is used. The spirit level is replaced with either a pendulum device which operates in similar fashion to an automatic level or the surface of a liquid is used. The reading of the vertical circle is reflected from the always true horizontal surface of the liquid and any deflection of the theodolite vertical axis is automatically compensated.

11. *Centering motion*

Since the theodolite must be placed exactly over a survey station, it is fitted with a centring motion fitted usually above the tribrach which allows the whole of the instrument above the tribrach to move relative to the latter. Since the total amount of movement is only 20 mm, the instrument must be placed very accurately over the survey mark before the centring motion is used.

12. *Optical plummet*

On most theodolites an optical plummet is incorporated which greatly aids centring of the instrument particularly in windy weather.

Figure 12.7 is a section through an optical plummet. When the theodolite is properly set up and levelled the observer is able to view the ground station through the eyepiece of the optical plummet, his line of sight being deflected vertically downwards by the 45° prism incorporated in the plummet. Movement of the centring motion allows the theodolite to be placed exactly over the survey station.

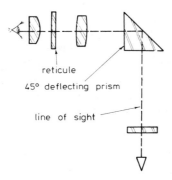

Figure 12.7

Reading the circles

1. *Vernier*

On vernier theodolites (Sokkisha NO10C, Topcon AG20B), the horizontal and vertical circles are divided into 20 minute intervals as in Fig. 12.8. The vernier scale is required to read to 20 seconds so, in accordance with vernier theory, there must be 60 vernier divisions each minutely smaller than any main scale division. In total, the 60 vernier divisions are the same length as 59 main scale divisions.

In Fig. 12.8 the index arrow of the vernier lies between 06° 20′ and 06° 40′ that is, the reading is 06° 20′ + x.

The fractional part x is obtained by finding the point of coincidence of vernier and main scale, that is, at 11′ 40″.

The total reading of the circle is therefore

$$06° \ 20' + 11' \ 40''$$
$$= \underline{06° \ 31' \ 40''}$$

On older instruments (Stanley Eltham), the verniers are read by magnifying microscopes attached to the exterior of the alidade.

The main disadvantage of vernier theodolites is that, in spite of the magnifiers, difficulty is experienced in reading the circles and deciding just which graduations are coincident. If the instrument is old, the circles may be tarnished and the difficulties in reading correspondingly increase.

On modern theodolites the circles are made of glass and the standards supporting the telescope, etc., are hollow. It is therefore possible to pass light through the instrument by a suitable arrangement of prisms.

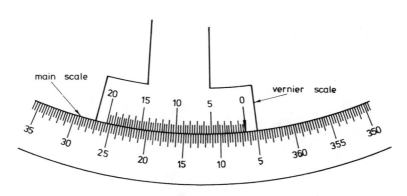

Figure 12.8

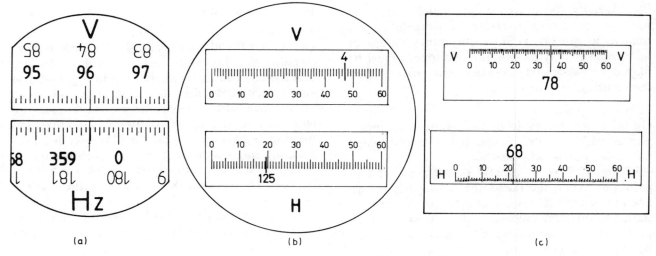

Figure 12.9

2. *Direct reading*

The horizontal circle is read through a reading eyepiece attached to one of the standards or alongside the main telescope eyepiece.

In the lower order theodolites, the horizontal circle is graduated at 5 or 10 minute intervals and read by estimation to the nearest minute. Figure 12.9(a) illustrates the reading of the horizontal and vertical circles of the Zeiss (Jena) Theo 080A minute reading theodolite. The circle has primary graduations at 1 degree intervals and secondary graduations at 5 minute intervals. The respective readings of vertical and horizontal circles are 96° 05′ and 359° 29′.

3. *Direct scale reading*

Intermediate order theodolites employ an optical scale to read the horizontal and vertical circles.

Figure 12.9(b) illustrates the field of view of the Watts minute reading theodolite. The circles are graduated at 1 degree intervals. The image of any degree graduation is seen in the eyepiece super-

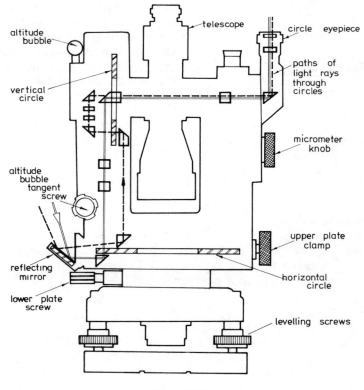

Figure 12.10

imposed on a transparent scale graticule. This scale is graduated at 1 minute intervals and the circle reading is obtained by reading the degree mark against the scale. In the figure the readings of vertical and horizontal scales are 04° 47′ and 125° 19.5′ respectively.

Figure 12.9(c) shows the circle readings of the Kern KIS engineers theodolite. The scales are subdivided into 0.5 minute intervals and reading is by estimation to 0.1 minute. The illustrated readings are vertical 78° 35.6′, horizontal 68° 21.8′.

4. *Micrometer reading*

Figure 12.10 illustrates the optical train through a higher order theodolite, namely Watts No.1.

This is one of the best 20 second theodolites in common use. The optics are simple and the illumination is very clear. In the figure, the double open line follows the optical path through the horizontal circle while the dashed line follows that of the vertical circle.

The readings of both circles are seen through the circle eyepiece mounted on the outside of the standard upright. It can be rotated from one side to the other for comfortable viewing.

The eyepiece contains three apertures, the horizontal and vertical circle graduations appearing in those marked H and V respectively. As with the vernier theodolite, the circles are divided into 20-minute divisions. The horizontal circle, as read against the index arrow in Fig. 12.11(a) is therefore

$$35° \ 20' + x$$

The fractional part 'x' is read in the third aperture by means of a parallel plate optical micrometer inserted into the light path of the instrument.

A parallel plate micrometer is simply a glass block with parallel sides. If the law of refraction is recalled (Chapter 5) it will be remembered that a ray of light striking the glass block at right angles passes through the block unrefracted (Fig. 5.2). If the block is tilted, however, the ray of light will be refracted but the emergent ray is parallel to the incident ray.

In Fig. 12.11(b) the ray of light from the horizontal circle is passing through the parallel plate when in the vertical position. The plate is directly geared to a drum mounted on the standard upright. Rotation of the drum causes the plate to tilt and the main scale reading of 35° 20′ is thereby made to coincide with the index mark. The resultant displacement, x, is read in minutes and seconds on the micrometer scale. The horizontal circle reading shown in Fig. 12.11(c) is therefore

$$
\begin{array}{r}
35° \ 20' \\
+06' \ 40'' \\
\hline
= 35° \ 26' \ 40''
\end{array}
$$

Double reading micrometer

It is frequently necessary in engineering surveying to measure angles with a higher degree of accuracy than can be obtained from a 20-second theodolite. In such cases a theodolite reading directly to 1 second is used.

It is possible to show that if the spindles of the upper and lower plates are eccentric, the measurement of the horizontal angles will be in error. The effects of eccentricity are entirely eliminated if two readings, 180° apart, are obtained and the mean of them taken.

On a 1-second theodolite, the reading eyepiece is again mounted on the upright. In most cases the horizontal and vertical circle readings are not shown simultaneously. On the Hilger & Watts No.2 microptic theodolite, a knob is situated below the reading eyepiece which, when turned to either H or V brings into view the horizontal or vertical circle scale.

The images of divisions diametrically opposed to each other are automatically averaged when setting the micrometer (Fig. 12.12) and the reading direct to 1 second is free from circle eccentricity.

The observer is actually viewing both sides of the circle simultaneously. In order to obtain a reading, the observer turns the micrometer drum, until he sees in the smallest aperture of the viewing eyepiece the graduations from one side of the circle correctly superimposed on those from the other side as in Fig. 12.12.

Both circles are read to 1 second, each circle having its own light path and separate micrometer. The reading of the horizontal circle in Fig. 12.12 is

$$
\begin{array}{l}
183° \ 20' + x \ \text{(main scale)} \\
\underline{ 7' \ 26'' \ \text{(micrometer scale)}} \\
183° \ 27' \ 26''
\end{array}
$$

5. *Opto electronic*

The electronic theodolite is controlled by a microprocessor prompted by the surveyor using a control panel built into one of the standards. The horizontal and vertical circles are etched with alternate light and dark patterns. The patterns are detected by a light source using an incremental encoder system which converts the light into an electrical signal. This signal converted to a pulse signal is received by the microprocessor which converts it into angular

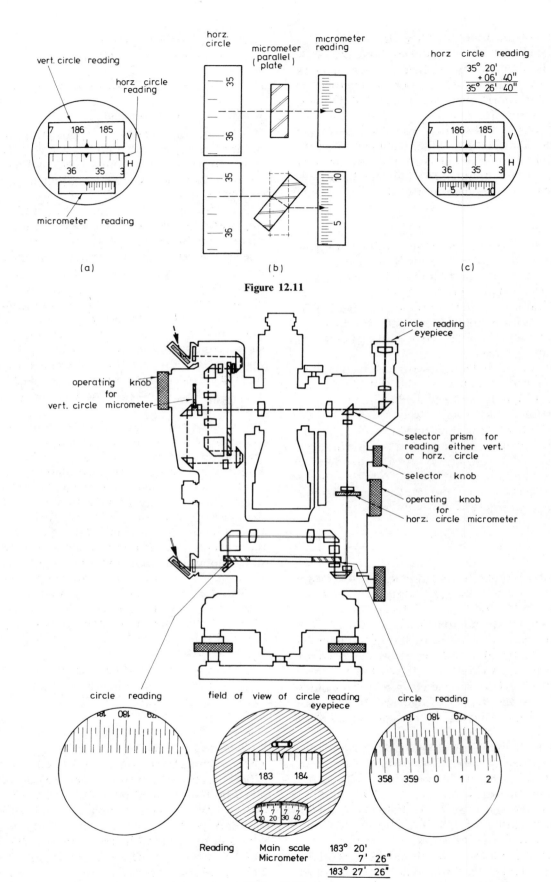

Figure 12.11

Reading   Main scale   183° 20'
           Micrometer   7' 26"
                         ‾‾‾‾‾‾‾‾‾‾‾‾‾
                            183° 27' 26"

Figure 12.12

units and displays them in digital fashion on a liquid crystal display (LCD) shown in Fig. 12.13 (Kern E2 Electronic Theodolite).

The automatic display of the circle readings eliminates the need for verniers, scales and micrometers, and greatly reduces reading and booking errors. The time required for measuring an angle is also reduced because of the zero-set facility on most of these instruments (Topcon DT20, Sokkisha DT20E) whereby the circle can be set to read zero at the touch of a button.

Figure 12.13 Kern E2 electronic theodolite with attached EDM and pocket calculator.

Example 1 Figure 12.14 shows the view through the reading eyepiece of several theodolites in common use:
(i) By inspection determine the system employed to read the circles.
(ii) Determine the readings of the various horizontal and vertical circles shown.

Solution
(a) Optical Scale V—95° 54′ 20″ H—130° 04′ 40″
(b) Optical Scale V—79° 15′ H—224° 54′
(c) Optical Scale V—10° 29′ 00″ H—123° 39′ 20″
(d) Optical Micrometer H—78° 56′ 25″
(e) Vernier H—10° 29′ 40″
(f) Optical Micrometer H—23° 31′ 40″

Setting up the theodolite (temporary adjustments)

The sequence of operations required to prepare the instrument for measuring an angle is as follows

1. *Setting the tripod*
This is probably the most important operation. If the tripod is not set properly, a great deal of time will have to be spent on subsequent operations 2 to 4.

The tripod legs are spread out and rested lightly on the ground around the mark. With the plumb-bob hanging from a pencil laid across the tripod head, the tripod is moved bodily until the plumb-bob is over the mark. The small circular spirit level mounted on the tripod is checked to ensure that the bubble is approximately central. If it is not, one leg of the tripod at a time is moved sideways until it is. Sideways movement of any leg will not greatly affect the position of the plumb-bob.

Only when the bubble is centred and the plumb-bob very close to the survey mark are the legs of the tripod pushed firmly home. This latter action will disturb the position of the plumb-bob and/or bubble and the telescopic sliding arrangements on the legs are used to reestablish the tripod's central position.

2. *Mounting the instrument*
The theodolite is then carefully removed from the box and firmly screwed to the tripod. It should be noted at this point that the theodolite should not be carried on the tripod. Not only is it wasteful of setting-up time but serious damage can be caused should the observer trip. Even discounting mishaps, damage is caused to the instrument's central axis because the weight of the instrument tends to bend it slightly.

3. *Levelling*
The levelling sequence is identical to that of the dumpy level, namely:
(a) Set the plate spirit level over two screws and centralize the bubble.
(b) Turn the instrument through 90° and recentralize the bubble.
(c) Repeat operations (a) and (b) until the bubble remains central for both positions.

4. *Centring*
The centring motion is released and the instrument moved until the plumb-bob is exactly over the survey mark. The centring motion is then tightened. This operation will have resulted in movement of the spirit level bubble from its central position, consequently, the operations of levelling and centring are repeated until both conditions are satisfied.

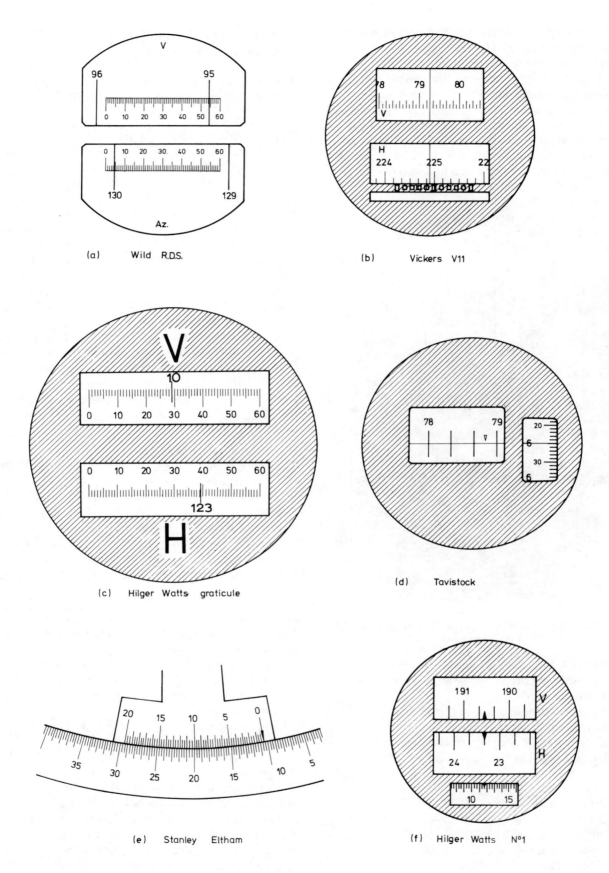

(a)   Wild R.D.S.

(b)   Vickers V11

(c)   Hilger Watts graticule

(d)   Tavistock

(e)   Stanley Eltham

(f)   Hilger Watts No1

Figure 12.14

On a windy day it is very difficult to set the plumb-bob exactly over the survey point.

Most theodolites have an optical plummet which overcomes this disadvantage. Before use of the optical plummet, the theodolite is centred approximately using the plumb-bob and levelled using the plate spirit level. The centring motion is then released and the instrument head moved until the survey point is centred on the cross wires of the optical plummet. Care should be taken to move the shifting head in the two directions used to level the spirit level.

On some theodolites the centring is carried out using a telescope centring rod. The tip of the rod is placed over the survey point and the telescope tripod legs are used to centralize the bubble of a spherical spirit level at right angles to the rod. When the bubble is central, the rod is vertical and the theodolite itself approximately level.

5. *Parallax elimination*

The cross wires are brought clearly into focus by carefully turning the eyepiece until no parallax is present.

Measuring a horizontal angle

When exactly set over a survey mark and properly levelled, the theodolite can be used in two positions, namely:
(a) face left or circle left;
(b) face right or circle right.

The instrument is said to be facing left when the vertical circle is on the observer's left as he sights an object. In order to sight the same object on face right, the observer must turn the instrument horizontally through 180° until the eyepiece is approximately pointing to the target. He then rotates the telescope about the transit axis, thus making the objective end of the telescope face the target. The vertical circle will now be found to be on the observer's right. This operation is known as transitting the telescope.

In figure 12.15 horizontal angle PQR is to be measured.

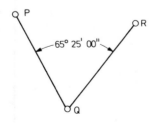

Figure 12.15

(a) *Reiteration method—without setting to zero.*
This method may be used with any type of theodolite and should be mastered before attempting method (b).

The possible horizontal control arrangements of a theodolite are:
(i) upper and lower plate clamps, (ii) repetition clamp, and (iii) circle setting screw (Fig. 12.4).
1. Using the field book (Table 12.1), enter theodolite station Q in column 1 line 1.
2. Enter the left-hand target station P in column 2 line 1.
3. Enter the right-hand target station R in column 2 line 2.
4. Close the lower plate clamp or repetition clamp, if fitted, *and do not touch either of them again.*
5. Set the instrument on face left.
6. Open the upper plate clamp on the alidade and the telescope clamp.
7. Turn the instrument carefully towards the left-hand target P and sight the target using the auxiliary finder sights fitted to the telescope. Lock the upper plate clamp and telescope clamp.
8. Focus the telescope on the target. The cross wires will not be on the target but should be close. Use the slow motion screws on the upper plate clamp and telescope clamp to bisect the target accurately.
9. Read the horizontal circle and note the reading (25° 30′) in column 3 line 1.
10. *REPEAT OPERATIONS 6, 7, 8, and 9* for the right-hand target R, booking the horizontal circle readings (90° 55′) in column 3 line 2.
11. Subtract reading P from reading R (90° 55′ − 25° 30′ = 65°25′) and note in column 3 line 3.

Line	1 Observation station	2 Target station	3 Face left reading	4 Face right reading	5 Accepted mean angle
1 2	Q	P R	25° 30′ 90° 55′	205° 30′ 270° 55′	
3			65° 25′	65° 25′	65° 25′

Table 12.1

In order to measure the angle above, sixteen manipulations of the theodolite controls, two circle readings, and two bookings were required. Clearly an error could easily occur with an inexperienced operator. Besides, even if these operations were conducted perfectly, the theodolite might be in poor adjustment and the angle of 65° 25′ would be incorrect.

All of these possible sources of error are eliminated by remeasuring the angle on face right.

	Col. 1	Col. 2	Col. 3	Col. 4	Col. 5	Col. 6	Col. 7
	Inst. stn	Stn sighted	Face left		Face right		Mean
			Vernier A	Vernier B	Vernier A	Vernier B	
Line 1 Line 2	Q	P R	00° 00′ 00″ 65° 25′ 00″	180° 00′ 00″ 245° 25′ 00″	180° 00′ 00″ 245° 25′ 20″	00° 00′ 00″ 65° 25′ 20″	
Line 3			65° 25′ 00″	65° 25′ 00″	65° 25′ 20″	65° 25′ 20″	65° 25′ 10″

Table 12.2

12. Transit the telescope to set the instrument to face right and make preparations to remeasure the angle.

13. *REPEAT OPERATIONS 6, 7, 8, and 9*, noting the left-hand target reading P (205° 30′) in column 4 line 1. This reading should differ by 180° from that in column 3 line 1 if no errors are present.

14. *REPEAT OPERATIONS 6, 7, 8, and 9* for right-hand target, noting the reading (270° 55′) in column 4 line 2.

15. Subtract reading P from reading R (270° 55′ − 205° 30′ = 65° 25′) and note in column 4 line 3.

16. Calculate the mean value of the angle and note in column 5 line 3.

Note: If a vernier instrument is used, two readings on verniers A and B are taken for each target sighting and booked as in Table 12.2. The initial reading on station P is 00° 00′ 00″ which obviously assists in the angle calculation. The method of setting to 00° 00′ 00″ is described below.

Example 2 Table 12.3 shows the field measurements of four angles of a traverse. Using the table, calculate the values of the angles.

Observation station	Target station	Face left reading	Face right reading	Accepted mean angle
B	A C	89° 16′ 20″ 185° 18′ 40″	269° 16′ 20″ 05° 19′ 00″	
C	B D	185° 39′ 40″ 271° 38′ 20″	05° 39′ 20″ 92° 38′ 40″	
D	C A	275° 18′ 00″ 01° 02′ 20″	95° 18′ 20″ 181° 02′ 40″	
A	D B	00° 00′ 00″ 92° 15′ 30″	180° 00′ 00″ 272° 15′ 30″	

Table 12.3

Solution See Table 12.4.

Observation station	Target station	Face left reading	Face right reading	Accepted mean angle
B	A C	89° 16′ 20″ 185° 18′ 40″	269° 16′ 20″ 05° 19′ 00″	
		96° 02′ 20″	96° 02′ 40″	96° 02′ 30″
C	B D	185° 39′ 40″ 271° 38′ 20″	05° 39′ 20″ 91° 38′ 40″	
		85° 58′ 40″	85° 59′ 20″	85° 59′ 00″
D	C A	275° 18′ 00″ 01° 02′ 20″	95° 18′ 20″ 181° 02′ 40″	
		85° 44′ 20″	85° 44′ 20″	85° 44′ 20″
A	D B	00° 00′ 00″ 92° 15′ 30″	180° 00′ 00″ 272° 15′ 30″	
		92° 15′ 30″	92° 15′ 30″	92° 15′ 30″

Table 12.4

In Example 2 the angle DAB was calculated more easily than the other three angles simply because the initial reading was 00° 00′ 00″. Many surveyors prefer this method of measuring angles and in setting out work it is standard practice.

(b) *Reiteration method—setting to zero*
The actual measurement procedure is the same as for method (a) except that the initial setting of the horizontal circle has to be 00° 00′ 00″. The mechanics of setting the circle varies with the type of theodolite.

(i) Using a double centre theodolite, i.e., one fitted with upper and lower plate clamps, the procedure is:

1. Set the theodolite to face left position.

2. Set the micrometer (if fitted) to 00′ 00″.

3. Release the upper plate clamp only and set the index mark or vernier to zero degrees as closely as possible by eye. Close the clamp and using the slow motion screw, set the circle exactly to zero. The reading is now 00° 00′ 00″.

4. Release the *lower* plate clamp and telescope clamp. Sight the left-hand station P, lock the clamps and using the *lower plate slow motion screw*, accurately bisect the target.

(ii) Using a theodolite fitted with a repetition clamp:

1. Set the theodolite to face left position.
2. Set the micrometer (if fitted) to 00′ 00″.
3. Release the upper plate clamp and set the index mark to zero degrees as closely as possible by eye. Close the clamp and using the slow motion screw, set the circle exactly to zero. The reading is now 00° 00′ 00″.
4. Depress the repetition clamp. This action locks the horizontal circle to the alidade.
5. Release the upper plate clamp and telescope clamp. The alidade and circle will now move together and maintain a zero reading. Sight the left-hand station P, lock the clamps and using the slow motion screws, accurately bisect the target.
6. Release the repetition clamp. The circle will now be free of the alidade.
(iii) Using a theodolite fitted with a circle setting screw:
1. Set the instrument to face left position.
2. Set the micrometer (if fitted) to 00′ 00″.
3. Release the upper plate clamp and telescope clamp. Sight the left-hand station P and using the slow motion screws, accurately bisect the target.
4. Raise the hinged cover of the circle setting screw and rotate the screw carefully until the horizontal circle reads exactly zero.

In all three cases the situation has been reached where the theodolite circle reads zero and the left-hand target P is accurately bisected. The measurement of the angle is completed as follows:
1. In Table 12.5, enter the reading of 00° 00′ 00″ in column 3 line 1.

	1	2	3	4	5
Line	Observation station	Target station	Horizontal FL	Angle FR	Accepted value
1	Q	P	00° 00′ 00″ 65° 25′ 00″	180° 00′ 00″ 245° 25′ 00″	
			65° 25′ 00″	65° 25′ 00″	65° 25′ 00″

Table 12.5

2. Open the upper plate clamp and telescope clamp.
3. Turn the instrument carefully towards the right and sight the right-hand target R.
4. Lock both clamps and using the upper plate slow motion screw and telescope slow motion screw, accurately bisect the target.
5. Read the horizontal circle and enter the reading (65° 25′ 00″) in column 3 line 2.
6. Subtract reading P from reading R (65° 25′ 00″ − 00° 00′ 00″ = 65° 25′ 00″).
7. Transit the telescope to set the instrument on face right.

8. Resight the left-hand target and note the reading which should be 180° 00′ 00″ if no errors have been made and if the instrument is in adjustment. Note the reading in column 4 line 1.
9. Resight the right-hand station and note the reading 245° 25′ 00″ in column 4 line 2.
10. Subtract reading P from reading R (245° 25′ 00″ − 180° 00′ 00″ = 65° 25′ 00″).
11. Calculate the mean angle and enter the value in column 5 line 3.

Repetition method

In order to reduce the number of times that the circle has to be read and thereby reduce a source of error, a method of measuring, known as repeated addition or repetition, is employed. The method is of particular value when small angles such as angle XYZ in Fig. 12.16 are to be measured.

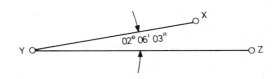

Figure 12.16

Theodolites fitted with a circle setting screw cannot be used to measure angles by repetition.
The method of measuring is as follows:
1. Set instrument to face left and close lower plate clamp. Sight station X and note reading as before.
2. Release UPPER plate clamp, turn the instrument to sight station Z and close clamp. The circle need not be read but usually is to check the final value of the angle.
3. Release the lower plate clamp, resight station X and close the clamp.
4. Release upper plate clamp, sight station Z for the second time and close the clamp.

The reading on the instrument is now double the value of the angle but again it is not noted.
5. Repeat operations 3 and 4 any number of times. The value of the angle is thereby added on the circle. If, after say six repetitions of the measurement, the circle is read, and the value of 12° 36′ 18″ obtained, the mean value of angle XYZ is found by dividing by six:

$$\text{Angle XYZ} = 02° 06′ 03″$$

6. Repeat the angular measurements a further six times on face right and obtain a second mean value

by subtracting the first reading on station X from the final reading on station Z and dividing by six:

$$\text{Say Angle XYZ} = 02°\ 06'\ 07''$$

7. Accepted XYZ $= \frac{1}{2}(02°\ 06'\ 03'' + 02°\ 06'\ 07'')$
$$= 02°\ 06'\ 05''$$

Measuring a vertical angle

It should be remembered that the construction of the theodolite is such that the vertical circle moves with the telescope and the vernier or index marker remain fixed. The vertical angle is measured from the line through the index mark or arrows of the vernier. This line is made horizontal by centring the bubble of the altitude spirit level. It follows therefore that the bubble axis should be parallel to the line through the vernier index marks.

Very often this is not the case and all vertical angles *MUST* be measured on both faces. The procedure is as follows.

Measuring a vertical angle:
1. Set the instrument to face left.
2. Release the telescope clamp and one of the horizontal plate clamps.
3. Sight the target using the finder sights. Lock the telescope clamp and plate clamp.
4. Focus the telescope on the target. The cross wires will not be on the target but should be close. Use the slow motion screws to bisect the target accurately.
5. Set the altitude spirit level (if fitted) to the centre of its run and read the vertical circle.
6. Change the instrument to face right and repeat operations 2, 3, 4, and 5.

Figure 12.6(b) shows the method of graduating the vertical circle of Watts No. 2 theodolite. Because of this system of graduation and since only one reading per face can be obtained, the face left reading of the vertical angle above might be 07° 49′ 56″.

When the face right reading is obtained, it should read 172° 10′ 04″ because the sum of face left and face right should total 180° 00′ 00″. Because of small maladjustments of the instrument, the face right reading will in all probability differ from the above value and the sum will differ from 180° 00′ 00″ by an amount known as the index error. If the actual reading on face right is 172° 09′ 56″, the index error and hence the correct vertical angle are obtained thus:

Face left reading	07° 49′ 56″
Face right reading	172° 09′ 56″
Sum =	179° 59′ 52″
Index error =	−08″

The index error is halved and a correction of +04″ is applied to both angular measurements to bring their sum to 180° 00′ 00″.

Corrected face left reading =	07° 50′ 00″
Corrected face right reading =	172° 50′ 00″
Sum =	180° 00′ 00″

The face left reading is taken as the vertical angle, i.e., 07° 50′ 00″ (elevation).

If an angle of depression is observed using this particular theodolite, the sum of face left and face right readings should total 540° 00′ 00″.

For example:

Vertical angle FL reading =	330° 25′ 10″
FR reading =	209° 35′ 02″
Sum =	540° 00′ 12″
Index error =	+12″

Correct face left reading =	330° 25′ 10″ − 06″
=	330° 25′ 04″
Vertical angle =	360° − 330° 25′ 04″
=	29° 34′ 56″ (depression)

Errors affecting angular measurements

The errors which affect angular measurements can be considered under two headings:
1. Instrumental maladjustments.
2. Human errors.

1. Instrumental maladjustments affecting horizontal angles

If a theodolite is to be in perfect adjustment, the relationship between the various axes should be as shown in Fig. 12.17, namely:
(a) The vertical axis should be truly vertical and at right angles to the plate bubble.
(b) The line of collimation of the telescope should be at right angles to the transit axis.
(c) The transit axis should be at right angles to the vertical axis of the instrument.

The relationships become disturbed through continuous use or misuse, and tests must be carried out before the start of any major contract and at frequent intervals thereafter to ensure that the instrument is in adjustment.

The various tests and adjustments are as follows:
(a) The vertical axis must be truly vertical and at right angles to the plate bubble axis.
Test
(i) Erect the tripod firmly and screw on the theodolite. Set the plate spirit level over two screws

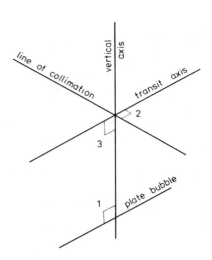

Figure 12.17

and centralize the bubble. If the instrument is not in good adjustment the vertical axis will be inclined by the amount e as shown in Fig. 12.18(a).

(ii) Turn the instrument through 90° and recentralize the bubble.

(iii) Repeat these operations until the bubble remains central for positions (i) and (ii).

(iv) Turn the instrument until it is 180° from position (i). The vertical axis will still be inclined with an error e and the bubble of the spirit level will no longer be central. It will, in fact, be inclined to the horizontal at an angle of $2e$, Fig. 12.18(b). The number of divisions, n, by which the bubble is off-centre is noted.

Adjustment

(v) Turn the footscrews until the bubble moves back towards the centre by $n/2$ divisions, i.e., by half the error. The vertical axis is now truly vertical, Fig. 12.18(c).

(vi) Adjust the spirit level by releasing the capstan screws and raising or lowering one end of the spirit level until the bubble is exactly central. The other

half $n/2$ of the error is thereby eliminated, Fig. 12.18(d), and the spirit level is at right angles to the vertical axis.

Effect of maladjustment The effect of an inclined vertical axis is not serious and in fact the small maladjustment normally encountered has no effect on measurements made with a conventional theodolite.

There is no observational procedure which can be employed to eliminate this error. However, as already implied, the error can be ignored.

(b) The line of collimation of the telescope must be at right angles to the transit axis. The line of collimation is defined as the line joining the optical centre of the object glass to the vertical crosshair of the diaphragm. If the position of the diaphragm has been disturbed, the line of collimation will lie at an angle to the transit axis (Fig. 12.19) with e being the error in the line of collimation.

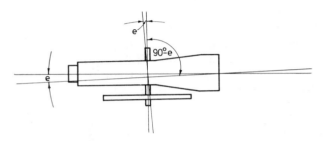

Figure 12.19

Test

(i) After properly setting up the instrument at a point which will be designated I, sight a well-defined mark A about 100 metres away and close both clamps. The line of sight will now be pointing to the target as in Fig. 12.20(a).

(ii) Transit the telescope and sight a staff 'B' laid horizontally about 100 metres away on the other side of the instrument from 'A'.

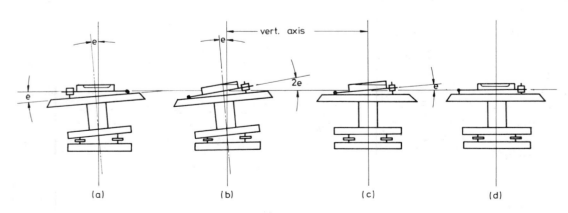

Figure 12.18

Note the reading, 2.100 m in Fig. 12.20(b).

Since the transit axis and line of sight make an angle of $(90 - e)$ and since the transit axis maintains its position when the telescope is transitted, the line of sight must diverge from the straight line by an amount $2e$.

(iii) Change the instrument to face right and again sight mark 'A' Fig. 12.20(c).

(iv) Transit the telescope and sight staff 'B'. The line of sight will again diverge from the straight line AB by an amount $2e$ but on the other side of the line. The staff reading is again noted and in Fig. 12.20(d) is 2.000 m.

(v) If the two staff readings are the same, the instrument is in adjustment and points A, I, and B form a straight line.

Adjustment

(vi) Since the staff readings differ, the difference represents the error $4e$, that is, $4e = (2.100 - 2.000) = 0.100$ m in this case.

At the outset, it was shown that the error was e, therefore $e = (0.100/4) = 0.025$ m.

The error is eliminated by bringing the vertical crosshair to the staff reading $(2.000 + e) = 2.025$ m. This is done by means of the antagonistic adjusting screws situated one on either side of the diaphragm.

In this case the diaphragm has to be shifted to the right (Fig. 12.21). The left screw is loosened slightly and the right screw tightened until the correct staff reading is obtained.

If the crosshair is no longer vertical, it might be necessary to loosen the diaphragm and rotate it

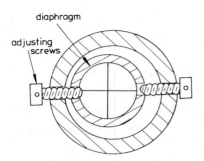

Figure 12.21

slightly until it is perfectly vertical. The verticality is usually checked against a plumb-line.

Effect of maladjustment If the instrument is used in its unadjusted state, every angle will be in error. However, as will be evident from Fig. 12.20, the mean of face left and face right readings is correct. For example the mean staff reading $\frac{1}{2}(2.100 + 2.000) = 2.050$ m is the correct position of B since the line of sight diverges by $2e$ on either side of the mean position for face left and right respectively.

(c) The transit axis must be truly horizontal when the vertical axis is vertical.

If the instrument is not in adjustment, the transit axis will not be horizontal when the instrument is correctly set up. In Fig. 12.22 the error is e. If the telescope is inclined, the crosshair will travel along the plane shown by the dotted line, that is a plane at right angles to the transit axis but *not* a vertical plane.

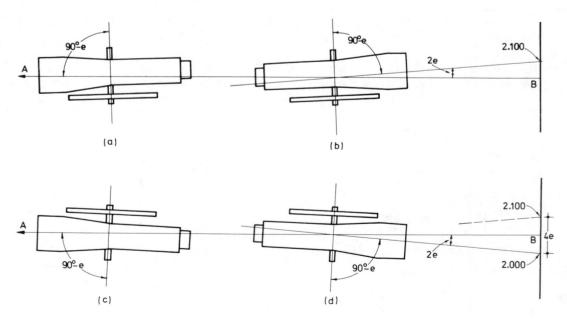

Figure 12.20

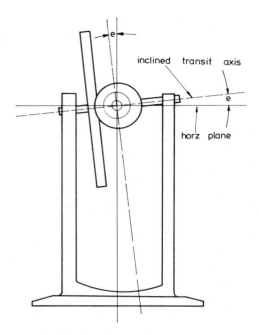

Figure 12.22

Test

(i) Set up the instrument at I, and level it properly. Sight a mark A at an elevation of about 60°. Close both upper and lower plate clamps, Fig. 12.23(a).

(ii) Lower the telescope and sight a horizontal staff or scale laid on the ground at the base of the object A. Note the reading, '*b*'.

(iii) Repeat the operations on face right to obtain a second reading, '*c*', on the scale. If the instrument is in adjustment both readings will be the same.

Adjustment

(iv) If the readings differ, the mean is correct since the telescope will have traversed over planes each inclined at an angle *e* on either side of the vertical.

By means of either upper or lower plate slow motion screw, set the instrument to the mean reading.

(v) Elevate the telescope until the horizontal crosshair cuts mark A. The vertical crosshair will lie to the side of the mark.

(vi) Bring the vertical crosshair onto mark A by means of the transit axis adjustment screw located on the standard immediately below the transit axis.

(vii) Depress the telescope. If the adjustment has been carried out correctly the vertical crosshair should read the mean staff reading.

Effect of maladjustment Angles measured between stations at considerably different elevations will be in error. However, as Fig. 12.23 clearly shows, the mean face left and face right is correct.

It must be noted at this juncture that most modern theodolites make no provision for this adjustment. The complex optical systems make the adjustment very difficult. However, since such instruments are assembled with a high degree of precision this adjustment is usually unnecessary. Besides, the mean of face left and face right observations cancels the error.

Instrumental maladjustment affecting vertical angles

Only one maladjustment materially affects the measurement of vertical angles. It has already been pointed out that vertical angles are measured from a line through the index marks of the verniers. Therefore, when the telescope is horizontal the index arrows should read zero and since they are attached to the altitude spirit level, the bubble of the latter should be central.

Test

(i) Set up the instrument at I on face left, centralize the altitude spirit level and read a vertical levelling

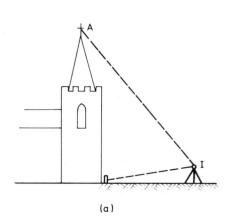

(a)

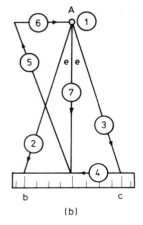

(b)

Figure 12.23

staff at A with the vertical circle verniers reading zero. Note the reading, say 1.500, Fig. 12.24(a).

(ii) Transit the telescope and repeat operation (i) on face right. Note the staff reading, say 1.400 in Fig. 12.24(b).

Adjustment

(iii) The mean staff reading is correct since the face left observation is elevated and the face right reading depressed by equal errors, *e*.

Using the vertical slow motion screw, the telescope is brought to the mean reading of 1.450 m, Fig. 12.24(c).

(iv) The verniers will no longer read zero and must be made to do so by the clip screw which controls the altitude spirit level and verniers.

(v) The movement of the clip screw, however, will move the bubble from its central position. The bubble is recentralized by adjusting the capstan screws situated at one end of the spirit level, Fig. 12.24(d).

Effect of maladjustment   If the instrument is in maladjustment all vertical angles will be in error. However, Fig. 12.24 clearly shows that the mean of face left and face right observations is correct.

2. Human and other errors

Human errors can be considered under two headings.

(a) *Gross errors*

These are mistakes on the part of the observer caused by ignorance, carelessness, or fatigue. They include sighting the wrong target, measuring the anticlockwise angle, turning the wrong screw, opening the wrong clamp, reading the circles wrongly, and booking incorrectly.

These errors can only be avoided by careful observation and by observing each angle at least twice. Only then will any error show up. It is absolutely useless to measure any angle on one face only as it is open to the wildest of errors, as already pointed out.

(b) *Random errors*

Small errors cannot be avoided. They may be due to imperfections of human sight and touch which make it impossible to bisect the targets accurately or read the verniers exactly.

The errors, however, are small, and of little significance. They are minimized by taking several

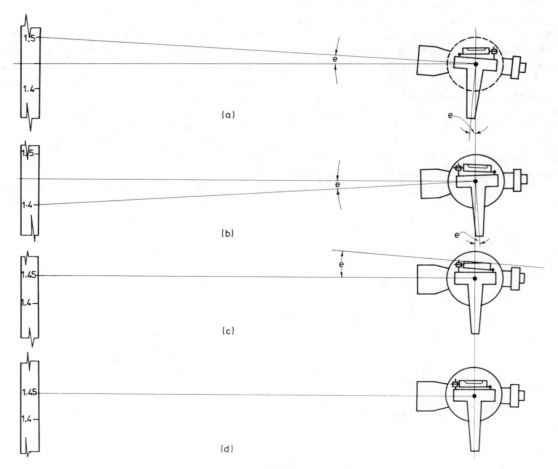

Figure 12.24

observations and accepting the mean.

Other errors arise from such sources as unequal expansion of the various parts of the instrument by the sun, instability of the instrument in windy weather, heat haze or mist affecting the sighting, and lastly inaccurate centring of the instrument.

They can only be avoided by shielding the instrument against the wind or sun and choosing times to observe which are favourable.

Lastly, if the instrument is not correctly centred, nothing can be done to eliminate or minimize the errors which must arise. Great care must be taken to position the instrument over the survey station with accuracy.

Summary

In general, gross errors cannot be eliminated or minimized by any observational system. A compensated measure will only show that there is an error. However, that should be sufficient as the observer should then take steps to trace the error. A mistake in subtraction will not necessitate any repetition of measuring but all other errors under this heading will, and in general another complete compensated measure should be made.

Systematic errors arise from instrumental maladjustments and defects. It has been shown that in almost every case a compensated measure of the angle cancels the error, the exception being that if the vertical standing axis is not truly vertical errors cannot be eliminated. However, they are second order errors and do not affect measurements made by a conventional theodolite.

Small random errors are not eliminated by a compensated measurement but such action minimizes the error and the mean of a larger number of measurements would be very accurate.

It cannot be over-emphasized that a compensated measure must be made of every angle regardless of its importance, otherwise the result is so uncertain as to be meaningless.

Test questions

1 Make a sketch of a theodolite to show clearly the principal parts.

2 Describe briefly the tests and adjustments which should be made to ensure that a theodolite is in good working order.

3 Using a diagram, describe briefly any form of optical plummet built into a theodolite.

4 The following angles were measured during a theodolite traverse:

Instrument station	Target station	Face left reading	Face right reading	Accepted mean angle
B	A	00 24′ 40″	180° 24′ 40″	
	C	65° 36′ 20″	245° 36′ 20″	
C	B	00 00′ 00″	179° 59′ 30″	
	D	66° 34′ 20″	246° 33′ 50″	
D	C	332° 10′ 20″	152° 10′ 40″	
	E	87° 08′ 00″	267° 08′ 20″	

Calculate the mean observed angles ABC, BCD and CDE.

5 Describe briefly any method of measuring a horizontal angle such that errors are minimized.

6 Outline the advantages and disadvantages of the following methods of reading the circles of a theodolite:
(a) Vernier
(b) Optical scale
(c) Optical micrometer.

7 Make a list of errors which may arise in measuring angles, using the following headings:
(a) Gross errors
(b) Systematic errors
(c) Random errors.

8 The following readings refer to a vertical angle measured with a double reading optical micrometer theodolite:

FL 359° 10′ 15″
FR 180° 49′ 51″

Determine the value of the vertical angle.

13. Theodolite traversing

The principle of compass traversing was fully explained in Chapter 11. Briefly the purpose of such a traverse is to establish the bearings and lengths of a series of adjoining lines which together form the framework for the survey of a particular area. The bearings and distances are then plotted by protractor or by rectangular coordinates.

The principle of theodolite traversing is exactly the same. The bearings and distances of every line of the traverse have to be obtained but whereas the compass is capable of directly producing the bearing of every line, the theodolite is not. The theodolite measures clockwise angles only and bearings have to be deduced from these measured angles.

The horizontal length of every line of any traverse must be obtained. Compared with the measurement of angles, the linear measurement of traverse lines generally proves to be the more difficult operation. In order to obtain an accurate traverse, great care must be taken in measuring the lines and applying the appropriate corrections to obtain the horizontal lengths.

The procedure for measuring the lengths and angles is described later in this chapter.

Types of traverse

Theodolite traverses are classified under these headings:

1. Open traverse

Figure 13.1 is an example of an open traverse. The lengths of all lines and the clockwise angles at each station are measured.

The type of traverse is not self-checking and errors in either angular or linear or both measurements may pass unchecked. The only check which can be provided is to repeat the complete traverse or resurvey the traverse in the opposite direction from F to A.

2. Polygonally closed traverse

In this type of traverse, the lines close to form a polygon (Fig.13.2).

The lengths of the lines and the clockwise angles are again measured. The clockwise angles shown by the solid arcs in Fig. 13.2 are the exterior angles of the polygon and are measured when the survey is made in a clockwise direction, that is, the instrument is moved from A to B to C, etc.

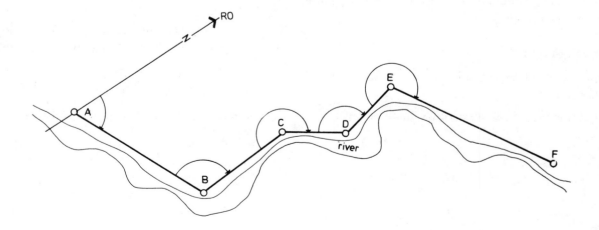

Figure 13.1

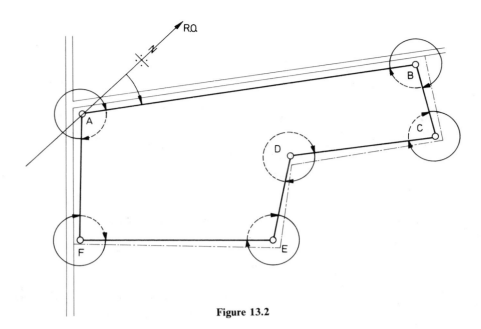

Figure 13.2

The traverse is self-checking to some extent, because the sum of the exterior angles of any polygon should equal $(2n + 4) \times 90°$ where n is the number of instrument stations. In this case the sum of the angles should be

$$(2 \times 6 + 4) \times 90°$$
$$= (12 + 4) \times 90°$$
$$= \underline{1440°}$$

It is very unlikely that the angles will sum to this amount because of the small errors inherent in angular measurements. However, the sum should be very close to this amount and should certainly be within the following limits.

$\pm 40\sqrt{n}$ seconds, in the case of a single-second reading theodolite

and $\pm \sqrt{n}$ minutes, in the case of a twenty-second reading theodolite

In each case, n is the number of instrument stations.

The interior angles of the polygon shown by dotted arcs in Fig. 13.2 are measured when the survey is moving in an anticlockwise direction.

The sum of these angles should be $(2n - 4) \times 90°$ where n is the number of instrument stations. Again the observed sum will differ but should be within the limits shown above.

In all cases, where the sum of the observed angles is not $(2n \pm 4) \times 90°$ the angles must be adjusted so that they do sum to this figure. Examples of such adjustments are given later.

3. Traverse closed between previously fixed points

In this type of survey, the traverse begins on two points A and B of known bearing and ends on two different known points C and D (Fig. 13.3). The survey is once again self-checking in that the bearing of line CD deduced from the traverse angles should agree with the already known bearing of CD. Seldom will the two bearings be identical but they should be within the limits of $\pm \sqrt{n}$ minutes where n is the number of instrument stations occupied during the traverse.

As in the case of a polygonally closed traverse, complete agreement will seldom be obtained and the bearings of the line will require to be adjusted. Future examples illustrate the principle.

Orientation of traverse surveys

It has already been stated that the theodolite in its conventional form does not measure bearings directly. Attachments such as a tubular compass or a gyro can be fitted to enable the theodolite to seek magnetic or true north respectively. However, these attachments are very seldom used in construction surveying and indeed are not included in any of the syllabuses of the various construction institutes.

The probable reason is that a sufficiently close approximation to true north can be obtained readily from Ordnance Survey plans. It is general practice on sites to choose a well-defined point such as a church steeple or tall chimney as a reference for the survey. The chosen point is called the reference object (RO) and the bearing between it and the first survey station can be obtained to within a few minutes of arc from the OS plan.

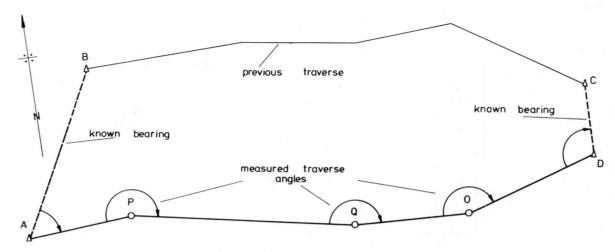

Figure 13.3

If the survey must be oriented to Grid north, it is necessary to begin on two Ordnance Survey points. The bearing, length, coordinates, and reduced levels of the points can be obtained from the local OS office for a small fee.

Theodolite traversing—fieldwork

It is generally accepted that a minimum of four surveyors is required to conduct a theodolite traverse.
Their duties are:
1. To select suitable stations.
2. To measure the distances between the stations.
3. To erect, attend, and move the sighting targets from station to station.
4. To measure and record the angles.
5. To reference the stations for further use.

Factors influencing the choice of stations

On arrival at the site, the survey team's first task is to make a reconnaissance survey of the area with a view to selecting the most suitable stations. Generally the stations have to be fairly permanent and concrete blocks are generally formed *in situ*. A bolt or wooden peg is left in the concrete to act as a centre point (Fig. 4.14). Where the survey mark is to be sited on a roadway a masonry nail is hammered into the surface and a circle painted around it.

The positions of the stations are governed by the following factors:
1. Easy measuring conditions. Since the angles can be measured very accurately with the theodolite, the linear measurements must be of comparable accuracy. Most measurements will be made along the surface using a steel band, therefore the surface conditions must be conducive to good measuring. Roadways, paths, and railways present good measuring conditions since they are smooth and have regular gradients. Conversely, measurements should not be made through long grass or undergrowth, over very undulating ground or heaps of rubble, etc.
2. Avoidance of short lines. Wherever possible the traverse should exclude short lines. The reason is simply that angular errors are introduced if targets at short range are not accurately bisected. For example if the theodolite sights 2 millimetres off target at a range of 10 metres, it is equivalent to sighting 40 millimetres off target at a range of 200 metres.
3. Stations should be chosen so that the actual station mark can be sighted. If a pole has to be erected at a station it must be plumbed exactly above the mark using either a builder's spirit level or a plumb-bob. Failure to observe this simple precaution may result in a fairly substantial angular error. For example in Fig.13.4 a 2-metre pole has been erected on a station and is off vertical by only 20 mm at the top.

If a theodolite at a range of 50 metres sights the top of the pole instead of the ground station, the error in the measured angle will be 80 seconds.
4. Possibility of through bearings. If at all possible the traverse stations should be so located that through bearings may be taken to provide a check on the work. Figure 13.5 illustrates the principle.
5. If the other conditions can be readily satisfied the stations should be chosen near to some permanent objects, such as lamp standards, trees, etc., in order that they may be found readily at a later date by measuring from these objects (Fig. 4.14).

Linear measurement

In the calculation of rectangular coordinates, the true plan, i.e., horizontal length of every line, is

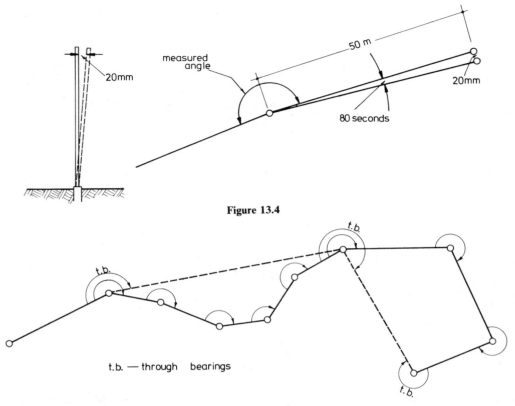

Figure 13.4

t.b.

t.b. — through bearings

t.b.

t.b.

Figure 13.5

required. Cognisance should therefore be taken of the various constant errors which affect linear measurements, and appropriate corrections should be applied to the measured lengths to produce true horizontal lengths. This suggests therefore that:

1. Measurements should be made on the slope and vertical angles should be observed.
2. Each tape length should be aligned accurately using the theodolite.
3. Temperature should be recorded and the appropriate correction applied.
4. The steel band should be standardized before use.
5. The correct tension should be applied to the band and the band should not be allowed to sag.

In engineering and construction surveying it is normal practice to align the tape by eye and dispense with temperature measurements. Frequently the tension is judged but this is not good practice and a spring balance or tension handle should be used.

Assuming that this practice is to be employed, the following equipment is necessary for measuring a traverse line:

1. A steel band graduated throughout in metres and centimetres. The zero marks of the band should be about 200 to 300 millimetres from the ends.

2. A spring balance calibrated in kilogrammes or a BS tension handle.
3. Measuring arrows and two ranging poles.
4. Marking plates to be used on soft gound. A marker plate can be made quite easily from a piece of metal, 100 mm square, if the corners are bent to form spikes. On hard surfaces, the surface is simply chalked and a mark made with a pencil.

Procedure

The line to be measured is divided into bays at each change of gradient as shown in Fig. 13.6. The gradient of each bay is obtained by measuring the vertical angles θ_1, θ_2, θ_3, θ_4 with the theodolite set up at alternate bays only. The vertical angles are entered in the field book as in Table 13.1.

Four persons are required to measure a line with high accuracy. Two are stationed at the rear and two at the forward end of the steel band.

The procedure is as follows:

1. A 30-metre tape is used to lay off marker plates or chalked marks at intervals of approximately 29.8 m.
2. The plates are aligned with the theodolite or by eye and pencil crosses drawn on them.
3. The steel band is laid over the first two plates.

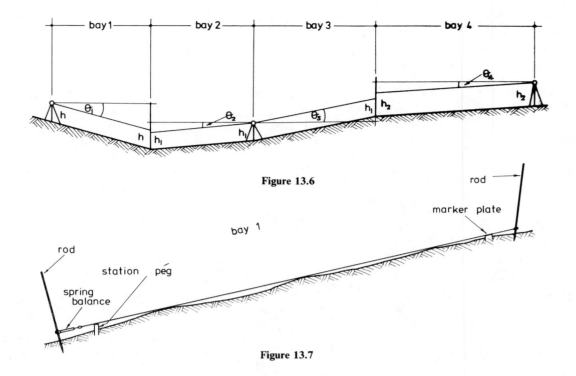

Figure 13.6

Figure 13.7

4. One man anchors the forward end by putting a ranging rod through the handle into the ground (Fig. 13.7).

5. A second man attaches the spring balance to the rear end, tightens the band and anchors it by putting a second rod through the spring balance handle. He levers back on this rod until the correct pull of 5 kg is registered on the spring balance. He then calls 'read' to the third and fourth men stationed at either end of the band.

6. On receiving this command, they read the tape against the pencil crosses. Both readings are entered in the field book on line 1 (Table 13.1).

7. The difference between the readings is worked out.

8. The forward anchor man moves his ranging rod slightly and the whole procedure is repeated to produce a second set of readings (line 2).

9. A third set of readings is obtained (line 3) and the mean of the three sets is calculated.

10. Bay 2 is then measured in identical manner.

11. The total length AB is checked for gross errors against the distance measured by the linen tape.

An alternative method of measuring is commonly used. Complete 30 m band lengths are marked on the plates and summed at the end of the line.

This method is open to more errors than the former and the line should be remeasured in the opposite direction.

12. The field notes are reduced to produce the horizontal length. Examples 1 and 2 show the relevant calculations. If the air temperature is measured during the fieldwork a further correction is required to obtain the plan length of the line.

	Line AB	Tape length 30 m				Standard tension 5 kg	
	Bay No.	Rear end reading	Forward reading	Difference	Mean m	Vertical angle	Remarks
Line 1	1	0.110	29.911	29.801			
Line 2		0.122	29.923	29.801			
Line 3		0.131	29.933	29.802	29.8013	−02° 10′ 50″	
	2	0.082	29.892	29.810			
		0.098	29.908	29.810			
		0.106	29.916	29.810	29.810	+04° 20′ 15″	
	3	0.110	21.926	21.816			
		0.092	21.908	21.816			
		0.099	21.914	21.815	21.8157	+06° 23′ 42″	

Table 13.1

Example 1 Line AB

Bay 1 Horiz.
    length = 29.8013 × cos 02° 10′ 50″ = 29.780 m
Bay 2   = 29.8100 × cos 04° 20′ 15″ = 29.725 m
Bay 3   = 21.8157 × cos 06° 23′ 42″ = 21.680 m
                Horizontal length AB = 81.185 m

13. The tape should be checked against a standard length and, if found to be in error, an appropriate correction must be applied. The correction was fully dealt with in Chapter 4.

Example 2 The following observations were obtained on a theodolite traverse:

Line	Bay	Slope length (m)	Vertical angle
PQ	1	53.220	+03° 25′ 30″
	2	29.610	−00° 30′ 40″
	3	17.325	+04° 19′ 00″

When checked against a standard 30 metre length, the steel tape was found to measure 30.004 m.
Calculate the true length of line PQ.

Solution
Correction for standardization = $(L - l)$ per tape
                                              length
                                    = (30.004 − 30.000)
                                              per tape.

Bay 1 No. of tapes = 53.220/30 = 1.774
       Correction = 0.004 × 1.774
                  = +0.007 m
Correct slope length = 53.220 + 0.007
                     = 53.227 m

Bay 2   Correction = 0.004 × 0.987
                   = +0.004
       Slope length = 29.610 + 0.004
                    = 29.614 m

Bay 3   Correction = 0.004 × 0.5775
                   = +0.002
       Slope length = 17.325 + 0.002
                    = 17.327 m

Plan length = slope length × cos inclination

Bay 1   Plan length = 53.227 × cos 03° 25′ 30″
                        = 53.227 × 0.998 213 7
                        = 53.132 m

Bay 2   Plan length = 29.614 × 0.999 960 2
                        = 29.613 m

Bay 3   Plan length = 17.327 × 0.997 148 7
                        = 17.278 m

So Plan length PQ = 53.132 + 29.613 + 17.278
                  = 100.023 m

Angular observations

For the reasons already explained in Chapter 12, it is necessary to take compensated measures of each angle at the various stations.

The angular observations should be made before the linear measurements then, while the theodolite is still set up, the various vertical angles can be taken and the steel band can be aligned properly.

In Fig. 13.1 the theodolite is set at station A on face left and the angle between the RO and station B is measured. The face right value is then obtained and the mean value calculated. The RO is the back station and B is the fore station. Some form of target must be set at B, and at subsequent stations. Most commonly a ranging rod is used and the point of the rod is sighted. The rod must be plumbed by using a builder's spirit level, in case the point cannot be seen. Many other targets have been devised on site, their form depending on the ingenuity of the surveyor. A common target is shown in Fig. 13.9. It is made by lashing three ranging rods or pieces of timber together and suspending a plumb-bob over the mark.

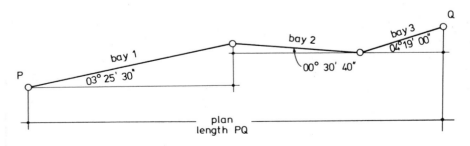

Figure 13.8

Figure 13.9

Inst. stn	Target stn	Face left	Face right	Mean
A	RO	00° 00′ 00″	180° 00′ 00″	
	B	65° 34′ 20″	245° 34′ 20″	
		65° 34′ 20″	65° 34′ 20″	65° 34′ 20″
B	A	00° 05′ 20″	180° 05′ 20″	
	C	110° 10′ 40″	290° 10′ 40″	
		110° 05′ 20″	110° 05′ 20″	110° 05′ 20″
C	B	00° 27′ 00″	180° 27′ 00″	
	D	220° 17′ 40″	40° 17′ 40″	
		219° 50′ 40″	219° 50′ 40″	219° 50′ 40″
D	C	00° 09′ 40″	180° 09′ 20″	
	E	135° 02′ 20″	135° 02′ 20″	
		134° 52′ 40″	134° 53′ 00″	134° 52′ 50″
E	D	00° 02′ 40″	180° 02′ 00″	
	F	251° 01′ 40″	71° 00′ 40″	
		250° 58′ 20″	250° 58′ 40″	250° 58′ 30″

Table 13.2

For the measurement of any angle, three plumbing operations are necessary regardless of the form of target being used, resulting in the expenditure of much time and effort and of course, with such a system small plumbing errors must inevitably occur.

If three tripods are available the errors are eliminated. In the measurement of angle ABC, in Fig. 13.10, the tripods are plumbed and levelled over the stations using a detachable optical plummet. Targets are placed at stations A and C and the theodolite at station B. The plummet, theodolite, and targets are completely interchangeable and are detachable from the levelling head.

The angle ABC is measured in the normal way. Angle BCD is measured by leap-frogging tripod A to station D; target A to B; theodolite B to C; and target C to D. There is no need to centre the theodolite at C since the tripod is already over the mark. Similarly the target at back station B is correctly centred.

The values of the horizontal angles measured on the open traverse A–F in Fig. 13.1 are shown in Table 13.2. The instrument used was a single reading optical micrometer theodolite.

Conversion of angles to bearings

1. Open traverse

The whole circle bearing of every line of the traverse is required to obtain the coordinates of the stations. Since only the first bearing is measured in the field, the others have to be calculated from the measured angles.

In the traverse, the line A–RO is the assumed North meridian. Since a whole circle bearing is, by definition, the angle between the meridian and survey line, the bearing of line AB is 65° 34′ 20″. This is, of course, the forward bearing.

In Fig. 13.11, the meridian is drawn through station B. The back bearing of this line, that is, the direction from B to A is

$$65° 34′ 20″ + 180° 00′ 00″ = 245° 34′ 20″$$

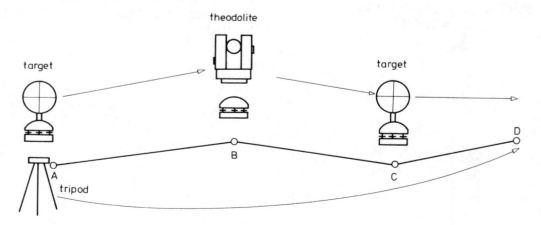

Figure 13.10

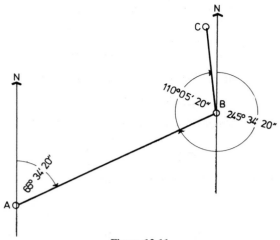

Figure 13.11

The clockwise angle ABC is 110° 05′ 20″ from Table 13.2.

From the diagram, the forward whole circle bearing of line BC is the clockwise angle NBC.

Now angle NBC = angle NBA + angle ABC
= back WCB BA + clockwise measured angle ABC
= 245° 34′ 20″ + 110° 05′ 20″
= 355° 39′ 40″

Frequently it will be found that forward bearings obtained as above will exceed 360°. In these cases 360° is subtracted from the answer to produce the correct bearing.

From these calculations a simple rule for obtaining bearings from angles emerges.

Rule: 'To the back bearing of the previous line, add the next clockwise angle. The sum is either equal to or is 360° greater than the forward bearing of the next line.'

The remaining bearings are calculated thus:

		Whole circle bearing
AB		65° 34′ 20″
Back Brg BA	245° 34′ 20″	
+ABC	110° 05′ 20″	
BC	355° 39′ 40″	355° 39′ 40″
Back Brg CB	175° 39′ 40″	
+BCD	219° 50′ 40″	
	395° 30′ 20″	
	−360°	
CD	35° 30′ 20″	35° 30′ 20″
Back Brg DC	215° 30′ 20″	
+CDE	134° 52′ 50″	
DE	350° 23′ 10″	350° 23′ 10″

Back Brg ED	170° 23′ 10″	
+DEF	250° 58′ 30″	
	421° 21′ 40″	
	−360°	
EF	61° 21′ 40″	61° 21′ 40″

Many surveyors prefer to use an amended version of the rule in which the calculation of the back bearing is omitted. Compensation is made at the end of the calculation for any line by adding or subtracting 180°.

The rule then becomes: 'To the *FORWARD* bearing of the previous line add the next clockwise angle. If the sum is greater than 180°, subtract 180°. If the sum is less than 180°, add 180°. If the sum is greater than 540°, subtract 540°. The result is the forward bearing of the next line.'

In Fig. 13.11 the forward bearing of line BC is calculated as follows:

Forward bearing AB =	65° 34′ 20″
+ ABC	110° 05′ 20″
	175° 39′ 40″
Since the sum is less than 180°	
add 180°	+180° 00′ 00″
Forward bearing BC =	355° 39′ 40″

which agrees with the previous calculation

Using this method the computerized solution of bearings is simplified.

Computer solution

Bearings can be readily calculated using a computer program. Since some readers may not wish, or have the facilities, to consider this method of solution, the program (No. 6) is included in Chapter 22.

Example 3 The following data refer to an open traverse along the route of a proposed roadway (Fig. 13.12).

Angle	Mean observed value	Line	Plan length (m)
DEF	110° 30′	EF	19.67
EFG	120° 15′	FG	41.11
FGH	209° 17′	GH	74.06
GHJ	93° 08′	HJ	59.91

The whole circle bearing of line DE is known from a previous traverse to be 20° 30′. Calculate the whole circle bearings of all lines.

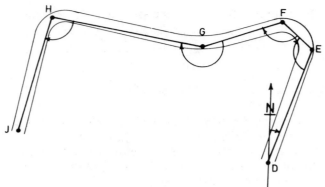

Figure 13.12

Solution

		Whole circle bearing
DE		20° 30′
Back Brg ED	200° 30′	
+ DEF	+110° 30′	
EF	311° 00′	311° 00′
Back Brg FE	131° 00′	
+EFG	+120° 15′	
FG	251° 15′	251° 15′
Back Brg GF	71° 15′	
+ FGH	+209° 17′	
GH	280° 32′	280° 32′
Back Brg HG	100° 32′	
+ GHJ	+ 93° 08′	
HJ	193° 40′	193° 40′

2. Closed traverse

(Angular adjustment and calculation of bearings)

Before the bearings of a polygonally closed traverse can be calculated, the observed angles must be adjusted in order that their sum equals

$(2n + 4) \times 90°$ if the angles are exterior
or $(2n - 4) \times 90°$ if the angles are interior,

where n is the number of angles.

In Fig. 13.2 the values of the exterior angles are:

Angle	Mean observed value
ABC	272° 03′ 10″
BCD	272° 05′ 51″
CDE	104° 50′ 31″
DEF	261° 11′ 06″
EFA	266° 10′ 15″
FAB	263° 38′ 25″

Sum of exterior angles $= (2n + 4) \times 90°$

$$= 16 \times 90°$$

$$= 1440° \ 00′ \ 00″$$

Sum of observed angles $= 1439° \ 59′ \ 18″$

Therefore Angular error $= -42″$

Correction per angle $= + \frac{1}{6}$ of 42″

$$= +07″$$

Corrected Angular values are:

Angle	Mean observed value	Correction	Corrected angle
ABC	272° 03′ 10″	+07″	272° 03′ 17″
BCD	272° 05′ 51″	+07″	272° 05′ 58″
CDE	104° 50′ 31″	+07″	104° 50′ 38″
DEF	261° 11′ 06″	+07″	261° 11′ 13″
EFA	266° 10′ 15″	+07″	266° 10′ 22″
FAB	263° 38′ 25″	+07″	263° 38′ 32″
	1439° 59′ 18″	+42″	1440° 00′ 00″

The bearing of line AB deduced from angle RO–A–B is 43° 40′ 45″ and the bearings of all other lines are calculated in the manner previously described.

		Whole circle bearing
AB		43° 40′ 45″
Back Brg BA	223° 40′ 45″	
+ABC	+272° 03′ 17″	
	495° 44′ 02″	
	−360°	
BC	135° 44′ 02″	135° 44′ 02″
Back Brg CB	315° 44′ 02″	
+BCD	+272° 05′ 58″	
	587° 50′ 00″	
	−360°	
CD	227° 50′ 00″	227° 50′ 00″
Back Brg DC	47° 50′ 00″	
+CDE	+104° 50′ 38″	
DE	152° 40′ 38″	152° 40′ 38″
Back Brg ED	332° 40′ 38″	
+DEF	+261° 11′ 13″	
	593° 51′ 51″	
	−360°	
EF	233° 51′ 51″	233° 51′ 51″
Back Brg FE	53° 51′ 51″	
+EFA	266° 10′ 22″	
FA	320° 02′ 13″	320° 02′ 13″

Back Brg AF	140° 02' 13"
+FAB	+263° 38' 32"
	403° 40' 45"
	−360°
AB	43° 40' 45"

—Agrees with initial bearing

Example 4 Figure 13.13 shows the mean observed angles of closed traverse MNOP.

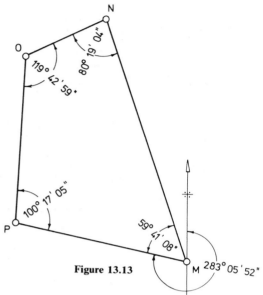

Figure 13.13

Calculate the quadrant bearing of each line.

Solution

$$\text{Sum of interior angles} = (2n - 4) \times 90°$$
$$= 360° \, 00' \, 00"$$

Sum of observed angles = 360° 00' 16"
Angular error = +16"
Correction per angle = −16/4
= −04"

Angle	Observed value	Correction	Correction value
PMN	59° 41' 08"	−04"	59° 41' 04"
MNO	80° 19' 04"	−04"	80° 19' 00"
NOP	119° 42' 59"	−04"	119° 42' 55"
OPM	100° 17' 05"	−04"	100° 17' 01"
	360° 00' 16"	−16"	360° 00' 00"

The bearings are calculated in exactly the same fashion as in the previous examples and the reader should satisfy himself that the answers below are correct.

Line	Whole circle bearing	Quadrant bearing
MN	342° 46' 56"	N 17° 13' 04" W
NO	243° 05' 56"	S 63° 05' 56" W
OP	182° 48' 51"	S 02° 48' 51" W
PM	103° 05' 52"	S 76° 54' 08" E

3. Traverse closed between previously fixed points

In Fig.13.3 the bearings of lines B–A and D–C are known from a previous traverse to be 204° 11' 05" and 02° 10' 47" respectively.

The observed angles of traverse BAPQODC are:

Angle	Observed value
BAP	72° 39' 42"
APQ	187° 40' 12"
PQO	169° 23' 47"
QOD	161° 58' 20"
ODC	106° 17' 21"

Line BA is the opening line of the traverse from which the bearings of all other lines are calculated.

		Whole circle bearing
BA		204° 11' 05"
Back Brg AB	24° 11' 05"	
+BAP	72° 39' 42"	
AP	96° 50' 47"	96° 50' 47"
Back Brg PA	276° 50' 47"	
+APQ	187° 40' 12"	
	464° 30' 59"	
	−360°	
PQ	104° 30' 59"	104° 30' 59"
Back Brg QP	284° 30' 59"	
+PQO	169° 23' 47"	
	453° 54' 46"	
	−360°	
QO	93° 54' 46"	93° 54' 46"
Back Brg OQ	273° 54' 46"	
+QOD	+161° 58' 20"	
	435° 53' 06"	
	−360°	
OD	75° 53' 06"	75° 53' 06"
Back Brg DO	255° 53' 06"	
+ODC	+106° 17' 21"	
	362° 10' 27"	
	−360°	
DC	02° 10' 27"	02° 10' 27"

Known whole circle bearing line DC =	02° 10' 47"
Traverse error =	−20"
Correction per angle =	+20/5"
=	+04"

The observed angles could be recalculated by adding 04" to each and the bearings could be found from the corrected angles as before. The corrected bearing of line AP will therefore be 04″ greater than the original while the bearing of the next line PQ will be 08″ greater. In fact the corrected bearings of each line will increase successively by 04″, until finally bearing DC will be (5 × 04)″ greater than the original uncorrected bearing.

The correction is therefore much more quickly done by simply finding the correction per angle and adding it successively to the uncorrected bearings.

Line	Uncorrected bearing	Correction	Corrected bearing
BA			204° 11′ 05″
AP	96° 50′ 47″	+04″	96° 50′ 51″
PQ	104° 30′ 59″	+08″	104° 31′ 07″
QO	93° 54′ 46″	+12″	93° 54′ 58″
OD	75° 53′ 06″	+16″	75° 53′ 22″
DC	02° 10′ 27″	+20″	02° 10′ 47″

Example 5 AB and FG are two survey lines with bearings fixed from a primary traverse as 39° 40′ 20″ and 36° 18′ 30″ respectively. The bearings are to be unaltered.

A secondary traverse run between the above stations produced the following results.

Angle	Mean observed value
ABC	179° 59′ 40″
BCD	210° 05′ 40″
CDE	149° 44′ 40″
DEF	177° 46′ 40″
EFG	179° 02′ 20″

Calculate the corrected bearings of each line.

Solution

		Whole circle bearing	Correction	Corrected bearing
AB		39° 40′ 20″	—	39° 40′ 20″
Back Brg BA	219° 40′ 20″			
+ABC	+179° 59′ 40″			
	399° 40′ 00″			
	−360°			
BC	39° 40′ 00″	39° 40′ 00″	−10″	39° 39′ 50″
Back Brg CB	219° 40′ 00″			
+BCD	+210° 05′ 40″			
	429° 45′ 40″			
	−360°			
CD	69° 45′ 40″	69° 45′ 40″	−20″	69° 45′ 20″
Back Brg DC	249° 45′ 40″			
+CDE	+149° 44′ 40″			
	399° 30′ 20″			
	−360°			
DE	39° 30′ 20″	39° 30′ 20″	−30″	39° 29′ 50″
Back Brg ED	219° 30′ 20″			
+DEF	+177° 46′ 40″			
	397° 17′ 00″			
	−360°			
EF	37° 17′ 00″	37° 17′ 00″	−40″	37° 16′ 20″
Back Brg FE	217° 17′ 00″			
+EFG	+179° 02′ 20″			
	396° 19′ 20″			
	−360°			
FG	36° 19′ 20″	36° 19′ 20″	−50″	36° 18′ 30″

Correct bearing FG = 36° 18′ 30″
Traverse error   = + 50″
Angular correction = −(50/5)″
                          = −10″

Alternative methods of obtaining bearings

It is possible to find the bearings of traverse lines directly in the field by two methods.

In Fig.13.14 the bearings of lines AB, BC, and CD are required.

The theodolite is set at A on face left and, with the horizontal circle set to read zero, the reference object RO is sighted. The theodolite is in fact oriented along the north line of the survey.

When the upper plate clamp is released and the telescope is directed to station B the reading of the horizontal circle, 70° 00′ 00″, is the bearing of line AB.

The theodolite is then taken to station B and the bearing of line BC is obtained by one of the following methods. In each method, the basic idea is to set up the theodolite so that it is correctly oriented with North.

1. Back bearing method

(a) Calculate the back bearing of line AB.

Back bearing = 70° 00′ 00″ ± 180° = 250° 00′ 00″

Set this bearing on the horizontal circle by releasing the upper plate screw and rotating the upper plate until the circle reads 250° 00′ 00″.

(b) Release the lower plate screw and direct the telescope to station A, on face left; bisect the station and tighten the clamp. The 250° graduation is thereby pointing to station A, that is, on its correct bearing. The zero graduation is therefore pointing along the North meridian and the instrument is correctly oriented.

(c) Release the upper plate, sight station C, tighten the upper clamp, and read the circle. The reading shown in the figure is 130° 00′ 00″ and this is the bearing of line BC.

(d) Transport the instrument to station C, set up and bisect station B, with the circle reading the back bearing of line BC, namely 130° 00′ 00″ ± 180° = 310° 00′ 00″.

(e) Operation (c) above is repeated by sighting D and the circle reading of 40° 00′ 00″ is the bearing of line CD.

2. Direct method with transitting

(a) On finishing the face left observation at station A, retain the reading of 70° 00′ 00″ on the circle by keeping the upper clamp closed.

(b) Transport the instrument to B, release the lower plate clamp and with the circle still reading 70° 00′ 00″, direct the telescope to station A, on face right. The relative positions of telescope and circle are shown in Fig. 13.15(a).

(c) Transit the telescope, whereupon the relative position of telescope and circle will be as shown in

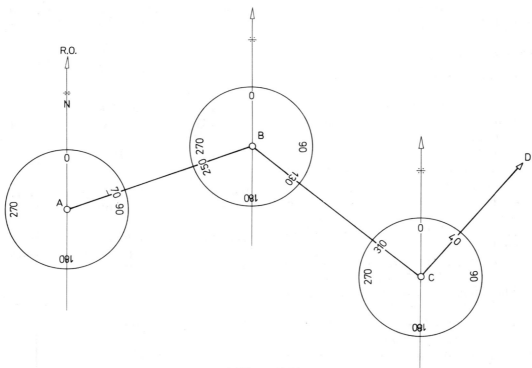

Figure 13.14

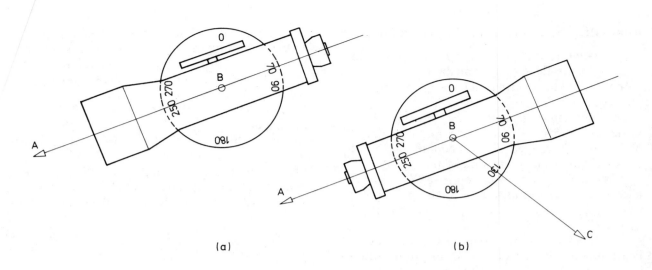

(a)   (b)

Figure 13.15

Fig. 13.15(b). The instrument is thereby correctly oriented with North.

(d) Release upper plate, sight C and read the circle. The reading of 130° 00' 00" is the bearing of line BC.

It should have been noticed that in both of these methods the angles were measured only once and therefore are liable to both gross and constant errors. They are therefore inferior to the method of measuring angles and deducing the bearings. Nevertheless, the methods are used fairly often and when setting out construction works, they are particularly valuable.

Theodolite traversing—plotting

All theodolite traverses are plotted by rectangular coordinates.

Chapter 11, dealing with the plotting of compass surveys, explained the theory of rectangular coordinates. The relevant section of that chapter should be revised.

Briefly, the partial latitude and departure of any survey line are found from the formulae:

Partial latitude = Length of line × cosine of whole circle bearing.

Partial departure = Length of line × sine of whole circle bearing.

The total coordinates of the stations are the algebraic sums of the relevant partial coordinates.

Once the plan lengths and whole circle bearings of the lines of a theodolite traverse have been deduced from the field data the rectangular coordinates of the stations are calculated in the same manner as were the coordinates of any compass traverse.

A scientific calculator should be used, utilizing the angular conversion and polar–rectangular coordinate facilities (Fig. 11.3). All calculations in this chapter were carried out using these facilities and their associated traverse tables.

Example 6 Figure 13.12 (Example 3) shows an open traverse along the route of a proposed roadway. The bearings of this traverse were calculated in Example 4 and are reproduced below, together with the lengths of the traverse lines. Calculate the coordinates of stations F, G, H, and J given that the coordinates of E are 513.21 m East and 494.65 m North.

Line	Whole circle bearing	Length (m)
EF	311° 00'	19.67
FG	251° 15'	41.11
GH	280° 32'	74.06
HJ	193° 40'	59.91

Solution Figure 13.12 shows the layout of the survey and Table 13.3 provides the solution.

Example 7 The corrected bearings of a closed traverse are given below with the horizontal lengths of the lines.

Line	Horizontal length (m)	Corrected W.C.B.
QR	172.200	62° 02' 30"
RS	87.520	332° 24' 50"
ST	93.810	101° 18' 45"
TQ	141.080	172° 02' 00"

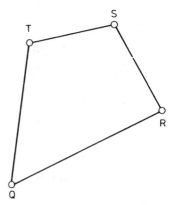

Figure 13.16

(a) Calculate the total coordinates of each station relative to the origin Q; hence calculate:
(b) the closing error,
(c) the accuracy,
of the traverse.

Solution

(a) The coordinates are calculated in Table 13.4.
(b) Closing error = distance between the coordinated positions of station Q.

Difference in departure $= +0.033$ m $= \Delta D$

Difference in latitude $= +0.182$ m $= \Delta L$

Closing error $= \sqrt{(\Delta D^2 + \Delta L^2)}$

(by Pythagoras) $= \sqrt{(0.033^2 + 0.182^2)}$

$= \sqrt{0.034213}$

$= \underline{0.184}$ m

(c) Accuracy of the traverse $= \dfrac{\text{Closing error}}{\text{Total length of traverse}}$

$= \dfrac{0.184}{494.610}$

$= \dfrac{1}{2688}$

Note: The closing error and fractional accuracy of a closed traverse were fully explained in Chapter 11 (page 134) and revision should be made at this point if the reader is at all uncertain of the above calculations.

Adjustment of a closed traverse

Bowditch's method of adjusting a closed traverse was used in Chapter 11 to adjust a compass traverse.

Bowditch assumes that the closing error of the traverse is due equally to errors in the linear and angular measurements. As already explained, this is probably true for compass surveys but is unlikely to be true for theodolite surveys. In spite of the precautions taken in measuring the lengths of the lines, the angular observations are generally more accurate than the linear measurements.

Theodolite traverses are adjusted by the traverse rule. The formulae used have no theoretical foundation; they are purely empirical, that is, they are founded on experience or observation.

Once the errors in latitude and departure have been calculated, they are proportionally eliminated

Line	Whole circle bearing	Distance (m)	Partial dep.	Partial lat.	Total dep.	Total lat.	Stn
					513.21	494.65	E
EF	311° 00′	19.67	−14.85	12.90	498.36	507.55	F
FG	251° 15′	41.11	−38.93	−13.21	459.43	494.34	G
GH	280° 32′	74.06	−72.81	13.54	386.62	507.88	H
HJ	193° 40′	59.91	−14.61	−58.21	372.01	449.67	J
			−141.20	−44.98	372.01	449.67	
					−513.21	−494.65	
					−141.20	−44.98	

Table 13.3

Line	Whole circle bearing	Distance (m)	Partial dep.	Partial lat.	Total dep.	Total lat.	Stn
					00.000	00.00	Q
QR	62° 02′ 30″	172.200	152.102	80.732	152.102	80.732	R
RS	332° 24′ 50″	87.520	−40.529	77.570	111.573	158.302	S
ST	258° 41′ 15″	93.810	−91.987	−18.402	19.586	139.900	T
TQ	187° 58′ 00″	141.080	−19.553	−139.718	0.033	0.182	Q
			+0.033	+0.182	+0.033	+0.182	

Table 13.4

as follows. Correction in departure (c_D) at any station

$$= \frac{\text{Total correction in departure} \times \text{partial departure}}{\text{Sum of departures (neglecting signs)}}$$

$$= \frac{C_D \times \text{dep.}}{\Sigma D}$$

Since C_D and ΣD are constants for every line the correction becomes

$$\text{Correction} = \left(\frac{C_D}{\Sigma D}\right) \times \text{dep.}$$

i.e., $c_D = k_1 \times \text{dep.}$

Similarly the correction in latitude (c_L) at any station

$$c_L = \frac{\text{total correction in latitude} \times \text{partial latitude}}{\text{sum of latitudes (neglecting signs)}}$$

$$= \frac{C_L \times \text{lat.}}{\Sigma L}$$

$$= \left(\frac{C_L}{\Sigma L}\right)$$

i.e., $c_L = \underline{k_2 \times \text{lat.}}$

It is pointless to calculate the total coordinates as in Table 13.4 since they are known to be wrong. The total errors in latitude and departure are evident after calculating the partial coordinates.

Error in departure = algebraic sum of Eastings and Westings columns.

Error in latitude = algebraic sum of Northings and Southings columns.

In the example

$$\Delta D = +0.033 \text{ m} \qquad \Delta L = +0.182 \text{ m}$$

Therefore Correction in dep. = -0.033 m and Correction in lat. = -0.182 m.

Sum of deps = 152.102 (neglecting signs) and Sum of Lats = 158.302 (neglecting signs)

$$+152.069 \qquad\qquad +158.120$$

$$\Sigma D = 304.171 \qquad \Sigma L = 316.422$$

Therefore $k_1 = \dfrac{-0.033}{304.171}$ and $k_2 = \dfrac{-0.182}{316.422}$

The corrections are as follows:

Departure correction to line

$$\text{QR} = -k_1 \times 152.102 = -0.016$$
$$\text{RS} = -k_1 \times 40.529 = -0.004$$
$$\text{ST} = -k_1 \times 91.987 = -0.010$$
$$\text{TQ} = -k_1 \times 19.553 = -0.003$$
$$\overline{-0.033}$$

Latitude correction to line

$$\text{QR} = -k_2 \times 80.732 = -0.046$$
$$\text{RS} = -k_2 \times 77.570 = -0.045$$
$$\text{ST} = -k_2 \times 18.402 = -0.011$$
$$\text{TQ} = -k_2 \times 139.718 = -0.080$$
$$\overline{-0.182}$$

The corrections and partial coordinates are added algebraically to produce corrected partial coordinates, for example

Line RS: Correct part. lat. = $-40.529 - 0.004 = -40.533$.

The algebraic addition of the corrected partial coordinates produces the correct total coordinates.

The complete calculation is much more neatly accomplished in a table and Table 13.5 shows the tabulation of this particular example.

Example 8 The following are the partial coordinates of a short closed traverse

Line	Whole circle bearing	Distance (m)	Partial coordinates Dep.	Partial coordinates Lat.	Corrections Dep.	Corrections Lat.	Corrected partials Dep.	Corrected partials Lat.	Total coordinates Dep.	Total coordinates Lat.	Stn
									00.000	00.000	Q
QR	62° 02′ 30″	172.200	152.102	80.732	−0.016	−0.046	152.086	80.686	152.086	80.686	R
RS	332° 24′ 50″	87.520	−40.529	77.570	−0.004	−0.045	−40.533	77.525	111.553	158.211	S
ST	101° 18′ 45″	93.810	−91.987	−18.402	−0.010	−0.011	−91.997	−18.413	19.556	139.798	T
TQ	172° 02′ 00″	141.080	19.553	−139.718	−0.003	−0.080	−19.556	−139.798	00.000	00.000	Q
	Error =		+0.033	+0.182	−0.033	−0.182	0.000	0.000			
	therefore Correction =		−0.033	−0.182							

Table 13.5

Line	Partial coordinates Dep.	Lat.
AB	+95.34	−55.52
BC	+197.26	+71.79
CD	−22.55	+128.31
DE	−265.95	−46.88
EA	−3.80	−97.90

Adjust the partial coordinates by the traverse rule and calculate the total coordinates of each station given that A is origin.

Solution See Table 13.6.

Miscellaneous coordinate problems

1. Closing bearing and distance
In work involving coordinates it is frequently necessary to calculate the bearing and distance between two points whose coordinates are known. If, for example, the coordinates of the traverse C to H (Fig. 13.17) are calculated, the bearing and distance between the first and last stations would be found as follows.

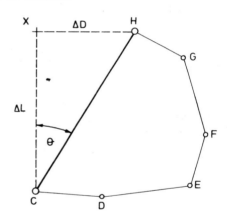

Figure 13.17

Coordinates of Station C:

$$300.20 \text{ E}$$
$$320.00 \text{ N}$$

Coordinates of station H:

$$463.30 \text{ E}$$
$$595.30 \text{ N}$$

In triangle CHX, the difference in departure between the points C and H is ΔD:

$$\Delta D = 463.30 - 300.20$$
$$= +163.10$$

The difference in latitude between the points C and H is ΔL:

$$\Delta L = 595.30 - 320.00$$
$$= 275.30$$

The quadrant bearing of line CH is angle θ:

$$\text{Tan } \theta = \frac{\Delta D}{\Delta L}$$
$$= \frac{163.10}{275.30}$$

Therefore $\theta = 30° 38' 40''$

Therefore Quadrant bearing CH is N 30° 38′ 40″ E. The distance between points C and H is, by Pythagoras

$$= \sqrt{(163.10^2 + 275.30^2)}$$
$$= \sqrt{102\,391.70}$$
$$= 319.99 \text{ m}$$

Alternatively, once the bearing has been calculated, the distance between the points C and H is found in either of the following ways:

Line	Partial coordinates Dep.	Lat.	Corrections Dep.	Lat.	Corrected partials Dep.	Lat.	Total coordinates Dep.	Lat.	Stn
							0.00	0.00	A
AB	+95.34	−55.52	−0.05	+0.03	95.29	−55.49	95.29	−55.49	B
BC	+197.26	+71.79	−0.10	+0.04	197.16	71.83	292.45	16.34	C
CD	−22.55	+128.31	−0.01	+0.06	−22.56	128.37	269.89	144.71	D
DE	−26.95	46.88	−0.14	+0.02	−266.09	−46.86	3.80	97.85	E
EA	−3.80	97.90	0.00	+0.05	−3.80	−97.85	0.00	0.00	A
Error therefore Corr. =	+0.30 −0.30	−0.20 +0.20	−0.30	+0.20	0.00	0.00			

Table 13.6

$$\frac{\Delta L}{CH} = \cos \theta \qquad\qquad \frac{\Delta D}{CH} = \sin \theta$$

$$\text{therefore } CH = \frac{\Delta L}{\cos \theta} \qquad \text{therefore } CH = \frac{\Delta D}{\sin \theta}$$

$$= \frac{275.30}{0.860\ 347} \qquad\qquad = \frac{163.10}{0.509\ 709}$$

$$= \underline{319.99 \text{ m}} \qquad\qquad = \underline{319.99 \text{ m}}$$

Closing bearing and distance using $\boxed{R \rightarrow P}$ *function on calculator*

The calculation may be carried out on a scientific calculator using the rectangular – polar conversion facility (Fig.11.3(b)) as follows. (*Note*: The calculation varies with the type of calculator but fundamentally the method is correct.)

To find the bearing and distance from C to H, subtract the coordinates of C from those of H:

$$N = 595.30 - 320.00 = +275.30 \text{ (as before)}$$
$$E = 463.30 - 300.20 = +163.10 \text{ (as before)}$$

On the calculator:

Enter 275.30, press $\boxed{INV}$ $\boxed{R \rightarrow P}$, enter 163.10, press $\boxed{=}$

Display shows: 319.99 m = length CH

Press $\boxed{X \rightarrow Y}$

Display shows: 30.644 392° = bearing C to H (dec.)

Press $\boxed{INV}$ $\boxed{° \,' \,''}$

Display shows: 30° 38' 40″ = bearing C to H

Using rectangular–polar, angular conversion and memory facilities the complete calculation is as follows.
On the calculator

Enter 463.30 − 300.20

Display shows: +163.10   Press $\boxed{Min}$, i.e., enter into memory

Enter 595.30 − 320.00

Display shows: +275.30   Press $\boxed{INV}$ $\boxed{R \rightarrow P}$

Press $\boxed{MR}$ $\boxed{=}$

Display shows: 319.99   = length CH

Press $\boxed{X \rightarrow Y}$

Display shows: 30.644 392°   = bearing from C to H (dec.)

Press $\boxed{INV}$ $\boxed{° \,' \,''}$

Display shows: 30° 38' 40″   = bearing from C to H

2. Solution of triangles

In Fig. 13.18 ABCD is an open traverse along the route of a sewer. The total coordinates of the stations have been found to be:

A	B	C	D
00.00 E	9.04 E	91.78 E	141.52 E
00.00 N	59.31 S	146.51 S	154.09 S

It is proposed to connect a point X (25.32E, 110.70S) to the sewer as economically as possible.

Calculate the bearing and length of the shortest route.

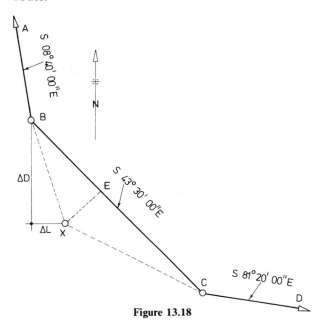

Figure 13.18

The shortest route is the line XE at right angles to the line BC. The length XE can only be found by solving either the triangle BXE or triangle CXE.

Solution of triangle BXE:

(a) Bearing BC = S 43° 30′ 00″ E (given)

$\qquad$ = 136° 30′ 00″

(b) Therefore Bearing EX = 136° 30′ 00″ + 90°

$\qquad$ = 226° 30′ 00″

(c)

Dep. B	9.04 E	Lat. B	59.31 S
Dep. X	25.32 E	Lat. X	110.70 S
ΔD =	16.28	ΔL =	51.39

$$\text{Tan Bearing BX} = \frac{\Delta D}{\Delta L}$$

$$= \frac{16.28}{51.39}$$

$$= \text{S } 17° 34′ 40″ \text{ E}$$

$$= 162° 25′ 20″$$

(d) Distance BX $= \dfrac{\Delta L}{\cos 17° 34′ 40″}$

$$= \frac{51.39}{0.9533\ 079}$$

$$= \underline{53.90 \text{ m}}$$

(e) Angle EBX $=$ Bearing BX − Bearing BE

$\qquad = \quad 162° 25′ 20″$

$\qquad = -136° 30′ 00″$

$\qquad = \quad \underline{25° 55′ 20″}$

(f) Distance EX

$$\frac{EX}{BX} = \sin 25° 55′ 20″$$

therefore EX = 53.90 × 0.437 150 7

$$= 23.56 \text{ m}$$

It should be noted that the majority of problems connected with coordinates finish with an unsolved triangle, in which the coordinates of two points and the bearing of one or more sides is known. Generally the closing bearing and distance between the coordinated points must be found before the triangle can be solved.

Example 9 In Fig. 13.19, two tunnels AB and EF are being driven forward until they meet in order to accommodate telephone cables.

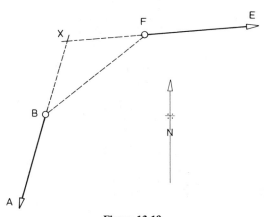

Figure 13.19

Calculate the distance still to be driven in each tunnel given the following information:

Bearing AB = 15° 00′ $\qquad$ Coordinates B $\quad$ 624.30 E
$\qquad\qquad\qquad\qquad\qquad\qquad\qquad\qquad\qquad$ 1300.50 N

Bearing EF = 265° 00′ $\qquad$ Coordinates F $\quad$ 845.90 E
$\qquad\qquad\qquad\qquad\qquad\qquad\qquad\qquad\qquad$ 1482.30 N

Solution

(a) Produce EF and AB until they meet at X. Join BF to form triangle FBX.

(b) Bearing XB = Bearing BA

$$= 195° 00'$$

Bearing XF = Bearing FE

$$= 85° 00'$$

therefore Angle FXB = 195° 00' − 85° 00'

$$= 110° 00'$$

(c) Departure F = 845.90 Latitude F = 1482.30
    B = 624.30   B = 1300.50
    ΔD = 221.60   ΔL = 181.80

$$\text{Tan Bearing BF} = \frac{221.60}{181.80}$$

$$= 50° 38' 05''$$

$$\text{Distance BF} = \frac{221.60}{\sin 50° 38' 05''}$$

$$= 286.63 \text{ m}$$

(d) Angle XBF = Bearing BF − Bearing BX
    50° 38' 05''
    −15° 00' 00''
    = 35° 38' 05''

(e) Angle BFX = Bearing FX − Bearing FB
    = 265° 00' 00''
    −230° 38' 05''
    = 34° 21' 55''

Check angles of triangles FXB

Sum = 110° 00' 00''
      35° 38' 05''
      34° 21' 55''
      180° 00' 00''

(f) In triangle FXB. By Sine Rule:

$$\frac{XB}{\sin F} = \frac{BF}{\sin X}$$

therefore $XB = \dfrac{BF \sin F}{\sin X}$

$$= \frac{286.63 \times \sin 34° 21' 55''}{\sin 110° 00' 00''}$$

$$= \frac{286.63 \times 0.564\ 466\ 9}{0.939\ 692\ 6}$$

$$= 172.17 \text{ m}$$

$$\frac{FX}{\sin B} = \frac{BF}{\sin X}$$

therefore $XF = \dfrac{BF \sin B}{\sin X}$

$$= \frac{286.63 \times \sin 35° 38' 05''}{\sin 110° 00' 00''}$$

$$= \frac{286.63 \times 0.582\ 615\ 6}{0.939\ 692\ 6}$$

$$= 177.71 \text{ m}$$

Test questions

1 The following angles were measured on a closed theodolite traverse

Inst. stn	Target stn	Face left
A	D	00° 10' 20''
	B	89° 26' 40''
B	A	89° 26' 40''
	C	189° 05' 20''
C	B	189° 05' 20''
	D	269° 29' 40''
D	C	269° 29' 40''
	A	00° 09' 00''

Calculate: (a) the measured angles
(b) the angular correction
(c) the corrected angles
(d) the quadrant bearing of each line given that bearing AD is 45° 36' 00''

2 Calculate the closing error and accuracy of the following traverse:

Line	Length	Bearing
AB	110.20	156° 40' 00''
BC	145.31	75° 18' 00''
CD	98.75	351° 08' 00''
DE	163.20	276° 29' 00''
EA	52.34	187° 27' 00''

3 The partial coordinates of a closed traverse are:

Line	Length	Partial dep.	Partial lat.
PQ	252.41	0.00	252.41
QR	158.75	−110.76	−113.82
RS	153.50	−25.24	−151.41
ST	136.74	+136.15	12.67

Distribute the closing error by:
(a) Bowditch's Rule
(b) Traverse Rule
and calculate the corrected partial coordinates in each case.

4 The following data refer to an open theodolite traverse round an old farmhouse which is to be demolished at a future date:

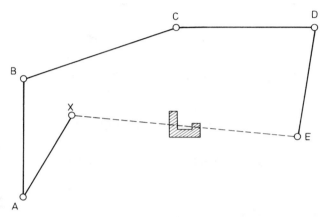

Figure 13.20

Line	Plan length (m)	Angle	Observed value
AX	160.00		
AB	186.40	XAB	330° 00′
BC	234.00	ABC	251° 30′
CD	170.60	BCD	198° 30′
DE	138.00	CDE	280° 45′

Given that station A is the origin and that line AB is north, calculate:
(a) The total coordinates of each station.
(b) The length of and bearing of a proposed sewer which is to be laid between the points X and E.
(c) The values of the angles to be set out at X and E by the theodolite to establish the line of the sewer.

5 The following lengths and angles were obtained on a closed theodolite traverse using a 20 m steel band and theodolite:

Line	Plan length	Quadrant bearing
AB	167.25	North
BC	228.34	N 30° 24′ 00″ E
CD	367.50	S 18° 16′ 40″ E
DA	220.70	N 89° 28′ 40″ W

It is suspected that there is a gross error in one of the linear measurements. Calculate the total coordinates of each station; thence determine the erroneous line and the probable reason for the wrong measurement.

6 Points on the boundary of a proposed building site have the following coordinates:

Station	Departure (m)	Latitude (m)
A	314	762
B	263	801
C	293	849
D	354	871
E	394	793

The points are to be set out by theodolite and tape from the line A to D. Calculate the length and theodolite reading required to set out B, C and E.

(SCOTVEC, Ordinary National Diploma in Building)

7 A closed traverse ABCDEA produced results shown in the table below.

Back station	Instrument station	Fore station	Clockwise horizontal angle
A	B	C	283° 31′ 40″
B	C	D	329° 06′ 50″
C	D	E	90° 47′ 20″
D	E	A	299° 43′ 00″
E	A	B	256° 50′ 20″

(a) Calculate the error of closure of the angles.
(b) Adjust the angles for complete closure.
(c) Calculate the corrected QUADRANT bearings of BC, CD, DE, and EA, given that the bearing of AB is S27° 35′ 20″E.

14. Tacheometry

The word tacheometry means 'speedy measurement'. It is derived from the Greek *tacheos* (fast) and *metron* (measurement) and is, in fact, a method of measuring distances without the use of a tape. The distances, both horizontal and vertical, are measured by using the optical properties of the telescope.

The accuracy attainable by tacheometric methods varies from about 1:500 to 1:10 000. It has the advantage that poor surface measuring conditions do not affect it and in many instances the accuracy is higher than that obtained by normal ground taping.

There are many systems of tacheometry but in the syllabuses of the various construction bodies the systems considered are those in which an ordinary theodolite can be used in conjunction with some form of staff.

In each system a small angle called the 'parallactic' angle is measured by theodolite to a short baseline defined on a staff held either horizontally or vertically. The distance between theodolite and staff is then some function of the parallactic angle. The angle may be fixed or variable, resulting in the following systems of tacheometry.

Parallactic angle	Staff position	Tacheometric system
Variable	Vertical	(1) Tangential
Fixed	Vertical	(2) Stadia
Variable	Horizontal	(3) Subtense bar
Fixed	Horizontal	(4) Optical wedge

Each system is now considered in turn.

Tangential tacheometry

Although it is dangerous to generalize, tangential tacheometry is probably the least accurate system. It is, however, most easily understood.

In the simplest case (Fig. 14.1) the plan length AB and difference in level AB are required:

Procedure

(a) The theodolite is set up at station A and the height of the instrument AT (1.410 m) above the station is noted.

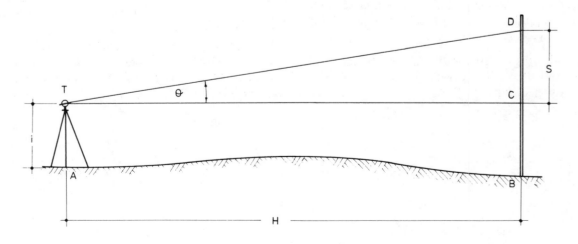

Figure 14.1

(b) The vertical circle is set to zero and a sight (C) taken to a staff held vertically on station B. The reading (1.022 m) is noted.

(c) The vertical circle clamp is released and a second point (D) is sighted on the staff. The staff reading (3.022 m) and vertical circle reading (+01° 30′ 00″) are noted.

(d) The difference between the two staff readings is $3.022 - 1.022 = 2.000$ m. This is called the staff intercept and is generally denoted by the letter S.

Basic formula

In triangle TCD, angle TCD is right angled, TC = horizontal distance H and DC = staff intercept S

$$\frac{S}{H} = \tan \theta$$

$$\text{therefore } H = \frac{S}{\tan \theta}$$

$$= \frac{2.000}{0.026\ 186}$$

$$= \underline{76.37 \text{ m}}$$

The reduced level of station A is 104.238 m above datum, therefore the reduced level of station B is

$$104.238 + AT - CB$$
$$= 104.238 + 1.410 - 1.022$$
$$= \underline{104.626 \text{ m}}$$

This method is not greatly employed but it serves to illustrate the principle of tangential tacheometry used in the following cases (Fig. 14.2).

The horizontal distance and difference in level between the points A and B are required in both cases.

Procedure

Vertical angles θ and α are measured to points C and D on the vertical staffs held at B. The instrument height i, staff readings C and D and vertical angles θ and α are all noted.

Basic formulae

In triangles DTE and CTE angle E is a right angle. ET = horizontal distance H and $(D - C)$ = staff intercept S.

$$DE = H \tan \alpha$$

$$\text{and } CE = H \tan \theta$$

$$\text{therefore } DE - CE = H \tan \alpha - H \tan \theta$$

$$\text{i.e., } S = H(\tan \alpha - \tan \theta)$$

$$\text{therefore } H = \frac{S}{(\tan \alpha - \tan \theta)}$$

In fig. 14.2(a), the height V between the theodolite trunnion axis and staff reading C is

$$V = H \tan \theta$$

The difference in level, ΔL, between the stations A and B is therefore

$$\Delta L = i + V - h$$

$$= I + H \tan \theta - h$$

while in Fig. 14.2(b)

$$\Delta L = i - V - h$$

$$= i - H \tan \alpha - h$$

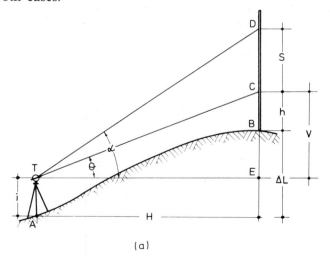

(a)

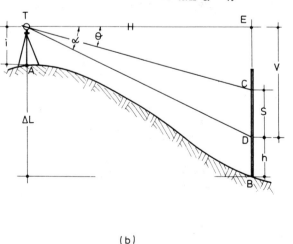

(b)

Figure 14.2

One further variation is possible. In Fig. 14.3, θ is an angle of elevation while α is an angle of depression.

In triangle TED

$$DE = H \tan \alpha$$

and in triangle TEC

$$CE = H \tan \theta$$

therefore $DE + CE = H \tan \alpha + H \tan \theta$

that is, $S = H(\tan \alpha + \tan \theta)$

therefore $H = \dfrac{S}{(\tan \alpha + \tan \theta)}$

Difference in level $\Delta L = i - V - h$

$$= i - H \tan \alpha - h$$

Example 1 The following data relate to Fig. 14.3.

Vertical
    angle $\theta = 00° 26' 30''$ Staff reading $C = 4.390$ m

Vertical
    angle $\alpha = 00° 51' 26''$ Staff reading $D = 1.890$ m

Instrument
    height = 1.350 m   Reduced level $A = 32.420$ m

Calculate the plan length AB and the reduced level of B.

Solution

$$S = 4.390 - 1.890$$

$$= 2.500 \text{ m}$$

$$H = \frac{S}{\tan \alpha + \tan \theta}$$

$$= \frac{2.500}{0.014\,965 + 0.007\,709}$$

$$= \frac{2.500}{0.022\,674}$$

$$= \underline{110.26 \text{ m}}$$

$$\Delta L = i - H \tan \alpha - h$$

$$= 1.350 - (110.26 \times 0.014\,965) - 1.890$$

$$= 1.350 - 1.650 - 1.890$$

$$= -2.190$$

$$\text{Red. lev. B} = 32.420 - 2.190$$

$$= \underline{30.230 \text{ m}}$$

Errors in tangential tacheometry

The sources of gross error are those which result from:
(a) Wrong reading of the staff.
(b) Wrong reading of vertical angle.
(c) Wrong booking.

In tacheometry the observer should not act as booker since there are so many observations to be recorded. With an experienced booker, booking errors are rare. The vertical angle is very seldom checked on both faces in tacheometry, consequently great care must be taken in observing and reading the vertical angles and staff readings.

Systematic errors are those which result from:

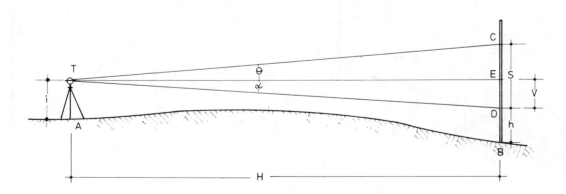

Figure 14.3

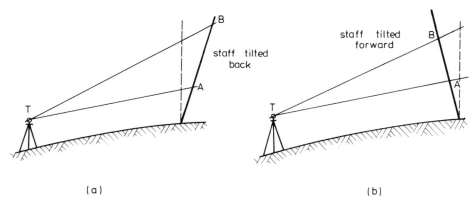

Figure 14.4

(a) Non-perpendicularity of the staff.

(b) Differential refraction.

(a) In Fig. 14.4(a) the staff is leaning backwards from the normal position. The intercept AB used in the calculation is therefore greater than it would be with the staff in its correct vertical position with the result that H is greater.

Similarly H is smaller in Fig. 14.4(b) though it could be greater if the error in verticality were sufficiently great.

Tacheometric staffs should always be fitted with handles and a good spirit level to ensure that the staff is vertical for every sight.

(b) Refraction has an important bearing on the accuracy of tacheometric observations. It should be recalled from Chapter 5 that rays of light are 'bent' when passing from one substance to another of different density. The density of the atmosphere changes fairly rapidly close to the earth with the result that the lower line of sight is refracted more than the upper in tangential tacheometry.

In order to minimize this effect the lower line of sight should be kept as high as possible. In practice this means that the lower staff reading should not be less than 1 metre.

Methods of reducing calculations

Many of the former field methods devised to reduce time in calculating results have been made redundant by the scientific calculator

Recalling the basic formulae:

$$\text{horz. dist. } H = S/(\tan \alpha \pm \tan \theta)$$

$$\text{and } \Delta \text{ level } = i \pm H \tan \theta - h$$

the only methods of reducing the calculations are as follows.

(a) The staff readings are observed to the whole metre graduations and the intercept is made either 2 m or 3 m.

The staff intercept generally adopted is 2 m. The lower staff reading must be kept above 1 m to minimize differential refraction effects, while the upper staff reading should be kept low to minimize the effects of a non-perpendicular staff. There is therefore little choice but to accept staff readings of 1.000 m and 3.000 m whereby $S = 2.000$ m.

In practice, targets are fitted to the staff at these heights to make sighting easier.

(b) The staff reading h, is made equal to the instrument height, i, whereupon the difference in level, ΔL, between the instrument station and staff station becomes

$$\Delta L = i + H \tan \theta - h$$

but $h = i$

$$\text{therefore } \underline{\Delta L = H \tan \theta}$$

This method can be combined with procedure (a) where $S = 2$ m. Targets are fitted to the staff at heights of i and $(i + 2)$ m.

The booking shown in Table 14.1 illustrates these field methods.

(c) A computer program is used.

The program has been included in Chapter 22, program 7, and shows how the fieldwork data given in Table 14.1 are computed and presented. It has already been stated that it is not the remit of this textbook to teach computing.

In order to enable the casual computer user to follow the program an accompanying explanation and simple user instructions are provided.

Booking the observations

In tacheometry, a sketch (Fig. 14.5) is an invaluable aid when reducing observations and should always be included in any table of booking. A suitable method of booking is given in Table 14.1. Columns 1–4 and 9 are filled in the field while columns 5–8 are completed later from the field observations.

Target station	Staff reading	Intercept S	Vert. angle	Tangent vert. angle	H	V	Red. lev. target	Horiz. angle
B	1.000		+4° 33′ 00″	0.079 579 8				00° 00′
B	3.000	2.000	+5° 06′ 00″	0.089 247 6	206.872	+16.463	89.06	
C	1.000		+6° 14′ 00″	0.109 223 4				23° 35′
C	1.945	0.945	+6° 48′ 00″	0.199 242 8	94.317	+10.302	82.90	
D	1.395		−0° 18′ 00″	−0.005 236 0				71° 48′
D	3.395	2.000	+0° 30′ 00″	0.008 726 9	143.237	−0.750	71.45	
Col. 1	2	3	4	5	6	7	8	9

Table 14.1

Calculations

A to B $H = \dfrac{2.000}{0.089\ 247\ 6 - 0.079\ 579\ 8} = 206.872$ m

$V = 206.872 \times 0.079\ 579\ 8 = 16.463$ m

$\Delta L = +1.395 + 16.463 - 1.000 = 16.858$ m

RL 'B' = $72.201 + 16.858 = 89.06$ m

A to C $H = \dfrac{0.945}{0.119\ 242\ 8 - 0.109\ 223\ 4} = 94.317$ m

$V = 94.317 \times 0.109\ 223\ 4 = 10.302$ m

$\Delta L = +1.395 + 10.302 - 1.000 = 10.697$ m

RL 'C' = $72.201 + 10.697 = \underline{82.90}$ m

A to D $H = \dfrac{2.000}{0.005\ 236 + 0.008\ 726\ 9} = 143.237$ m

$V = -143.237 \times 0.005\ 236\ 0 = -0.750$ m

$\Delta L = V$ (since $i = h$)

RL 'D' = $72.201 - 0.750 = \underline{71.451\ \text{m}}$

Stadia tacheometry

The stadia system makes use of similar triangles.
  In Fig. 14.6, the acute triangles ABC and AEF are similar

therefore $\dfrac{AC}{AF} = \dfrac{BC}{EF}$

therefore $AC = \dfrac{BC}{EF} \times AF$

Similarly $AC = \dfrac{BD}{EG} \times AF$

$= BD \times \dfrac{AF}{EG}$

In the figure, AF is 40 mm long while EG is 8 mm, that is, ratio AF:EG is 40:8 or 5:1. Since BD measures 20 mm AC equals (20 × 5) = 100 mm.
  The principle is used in the surveying telescope and is the basis of stadia tacheometry.

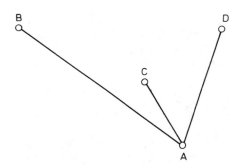

Figure 14.5

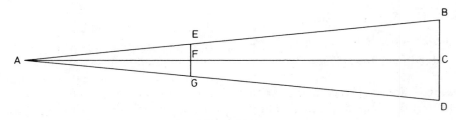

Figure 14.6

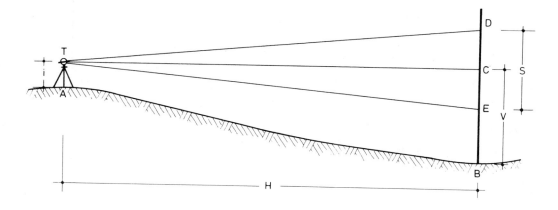

Figure 14.7

Procedure

Stadia tacheometry can be done using the level or the theodolite. In the simple case illustrated in Fig. 14.7 a level is being used. The horizontal distance H and difference in level between A and B are required.

The procedure is as follows:

(a) Set up the level over station A, level it accurately and record the instrument height i (1.300 m).

(b) Sight a vertical staff held on B and read the centre crosshair C as for normal levelling. Book the reading (2.340 m).

(c) Read the other crosshairs called the stadia lines, at D and E and note the readings (2.660 m and 2.020 m).

(d) The difference between readings D and E is the staff intercept, S.

$$S = 2.660 - 2.020$$

$$= \underline{0.640 \text{ m}}$$

(e) It will be shown that for a modern internal focusing telescope:

$$H = 100 \times S$$

$$= \underline{64.0 \text{ m}}$$

(f) The difference in level is by ordinary levelling methods:

$$(1.300 - 2.340) \text{ m} = \text{fall of } 1.040 \text{ m}$$

Basic formulae (derived from externally focusing telescope)

Chapter 5 dealt with the optical principles of instruments and should be revised at this point. Figure 14.8 is derived from Fig. 5.6. It shows the rays of light from a staff A_1B_1 passing through the object glass of a theodolite telescope to the diaphragm at AB.

The distance FO is the focal length, f, of the lens while the distances u and v are known as conjugate focal lengths.

The focal length, f, can be calculated from the lengths of the conjugate focal lengths since

$$\frac{1}{f} = \frac{1}{u} + \frac{1}{v}$$

Triangles ABO and A_1B_1O are similar.

Therefore
$$\frac{\text{Staff intercept } (S)}{\text{Image } (i)} = \frac{OC_1}{OC}$$

that is, $\dfrac{S}{i} = \dfrac{u}{v}$

Also
$$\frac{1}{f} = \frac{1}{u} + \frac{1}{v}$$

Multiply both sides by fu:

$$u = f + f\left(\frac{u}{v}\right)$$

Substitute $\dfrac{S}{i}$ for $\dfrac{u}{v}$:

$$u = f + \frac{fS}{i}$$

The distance $OC_1 = u$, is the horizontal distance from the object glass to staff, but the distance D from the centre of the instrument to the staff is required, therefore a constant c has to be added:

$$u + c = f + c + \frac{fS}{i}$$

Therefore
$$D = \left(\frac{f}{i}\right)S + (f + c)$$

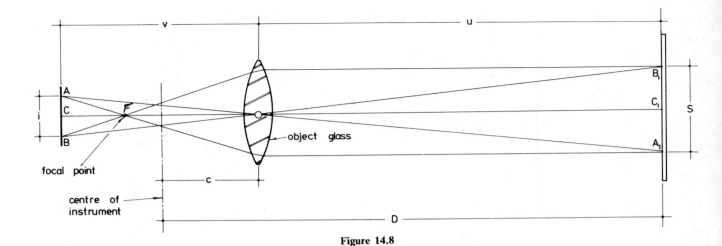

Figure 14.8

The ratio f/i is a constant by which the staff intercept S is multiplied while the quantity $(f + c)$ is a quantity which must be added for each observation. Calling f/i the multiplying constant, m, and $(f + c)$, the additive constant, k, the horizontal distance D becomes

$$D = mS + k$$

The values of m and k are supplied by the manufacturer of the instrument. Generally f/i is 100 to 1 (occasionally 50 to 1) while $(f + c)$ varies for each instrument depending on the focal length of the telescope. For a telescope 250 mm long, $(f + c)$ is 375 mm.

The simple theory outlined above is only possible for externally focusing telescopes since the focal length f is constant for that type only. In an internally focusing telescope the focal length f varies, owing to the action of the internal double concave lens (consult Fig. 5.8). As a result the multiplying constant m varies slightly but is compensated by the variable k. The discrepancy caused by treating both these variables as constant is, however, so small that it may be neglected.

The multiplying factor m is taken as being 100 while the additive constant is zero thus reducing the basic formula to

$$D = mS$$

It is of academic interest to note that in 1823 an Italian engineer named Porro succeeded in producing a tacheometric theodolite which did not have any additive constant $(f + c)$. An extra lens called an anallatic lens was introduced into the theodolite between the object glass and the vertical axis of the instrument. It had the effect of making the focal point f coincide with the vertical axis of the instrument and so obviated the necessity for adding the constant $(f + c)$.

Modern internally focusing telescopes are not anallatic but, as has already been explained, the error in treating them as being so is negligible.

Determination of constants by field tests

It is perfectly possible that the constants m and k may be unknown for any particular instrument, especially an older one. The constants can be determined by the following field test (Fig. 14.9).

(a) Choose a flat area of ground and set out pegs at distances of 50 m, 75 m, and 100 m from a station A.

(b) Set up the level or theodolite, level it accurately and in the case of the theodolite set the vertical circle to read zero.

(c) Read the three crosshairs on a vertical staff held in turn on each of the pegs. Note the three readings in each case.

(d) Let the readings be as in Table 14.2.

The constants are found from the simultaneous solution of the equations formed from the field observations.

Distance	Stadia readings			Staff
(m)	Top	Mid	Bottom	intercept
100	1.849	1.350	0.851	0.998
75	1.688	1.314	0.940	0.748
50	1.467	1.218	0.969	0.498

Table 14.2

General formula $\quad D = mS + k$

Therefore

$$100 = 0.998\,m + k \qquad (1)$$
$$75 = 0.748\,m + k \qquad (2)$$
$$50 = 0.498\,m + k \qquad (3)$$

Eq. (1) − Eq. (2): $\quad 25 = 0.25\,m \qquad (4)$

Eq. (2) − Eq. (3): $\quad 25 = 0.25\,m \qquad (5)$

therefore $\qquad m = 100$

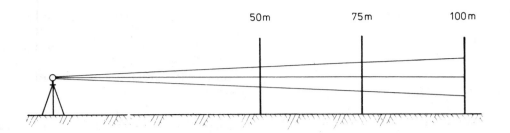

Figure 14.9

Substitute 100 for *m* in Eqs (1), (2), and (3).

In Eq. (1) $100 = 99.8 + k$

therefore $k = 0.20$ m

Similarly for Eqs (2) and (3).

Therefore multiplying constant = 100
and additive constant   = 0.20 m

Inclined sights

When the line of sight is inclined there are two possible cases.

(a) *Staff at right angles to line of sight*
The horizontal distance *D* and difference in level ΔL between the stations A and B are required (Fig. 14.10).

Field procedure   (i) Set up the theodolite at A and sight a staff on B. The staff must be held at right angles to the line of sight. This is usually accomplished by attaching a special sighting device to the staff at approximately the height of the instrument. The staff holder sights the theodolite through the device.
(ii) The observer reads all three stadia lines and the vertical angle to the mid staff reading.

Basic formulae (i) *Horizontal distance D*. Since the staff is at right angles to the line of sight

$$L = mS + k \text{ as before}$$

Triangle IMP is right angled at P.

$$\frac{IP}{IM} = \cos \theta$$

therefore $IP = IM \cos \theta$

i.e., $D_1 = L \cos \theta$

$$= (mS + k) \cos \theta$$

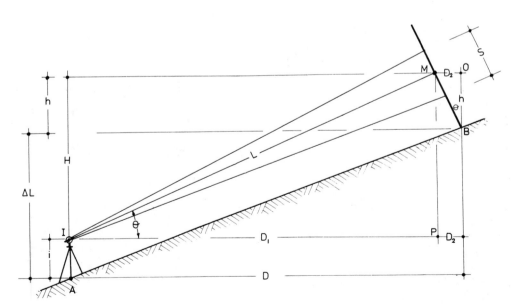

Figure 14.10

Triangle MOB is right angled at O and angle MBO = θ.

$$\frac{MO}{MB} = \sin \theta$$

therefore $MO = MB \sin \theta$

i.e., $D_2 = MB \sin \theta$

Since $D = D_1 + D_2$

$$D = (mS + k) \cos \theta + MB \sin \theta$$

(ii) Difference in level ΔL.

In IMP $\quad \dfrac{MP}{IM} = \sin \theta$

therefore $MP = IM \sin \theta$

i.e., $H = L \sin \theta$

$$= (mS + k) \sin \theta$$

In MOB $\quad \dfrac{OB}{MB} = \cos \theta$

therefore $OB = MB \cos \theta$

i.e., $h = MB \cos \theta$

Diff. in level, $\Delta L = i + H - h$

$$= i + (mS + k) \sin \theta - MB \cos \theta$$

The student should satisfy himself that when θ is an angle of depression:

$$D = (mS + k) \cos \theta - MB \sin \theta$$
$$\text{and } \Delta L = i - (mS + k) \sin \theta - MB \cos \theta$$

This system is used when the angle of elevation exceeds 30° though it may be used for any angle. Since angles of this magnitude are encountered less frequently than angles of lower magnitude the vertical staff stadia system is more common. This method is described later.

Example 2 Three points A, B, and C are collinear and lie on a steep hillside. A tacheometric survey was made of the three points using an instrument whose constants are $m = 100$ and $k = 0$. The staff was held normal to the line of sight at each pointing.

Given the following data calculate the gradient between the points A and C:

Inst. stn	Ht of inst	Target stn	Vertical angle	Top	Med	Bottom
B	1.350	A	+23° 30′	1.720	1.300	0.880
		C	−32° 24′	1.560	1.250	0.940

Solution

Line BA   Staff intercept $S = 1.720 - 0.880$
$$= 0.840 \text{ m}$$

Mid staff reading = 1.300 m

Horiz. dist. BA $= mS \cos \theta + MB \sin \theta$
$$= 100 \times 0.840 \times \cos 23° 30'$$
$$+ 1.300 \sin 23° 30'$$
$$= 77.55 \text{ m}$$

Diff. in level, $\Delta L = i + H - h$
$$= 1.350 + mS \sin \theta -$$
$$MB \cos \theta$$
$$= 1.350 + 84.0 \sin 23° 30'$$
$$- 1.300 \cos 23° 30'$$
$$= 33.65 \text{ m}$$

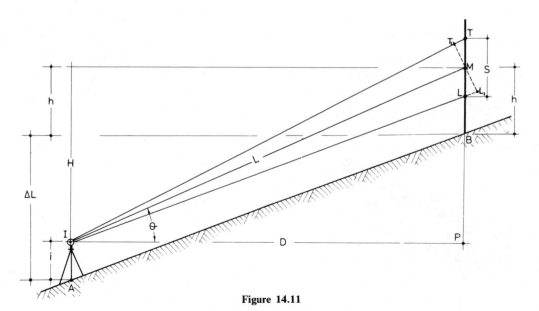

Figure 14.11

Line BC   Staff intercept $S = 1.560 - 0.940$
$$= 0.620 \text{ m}$$
Horiz. dist. BC $= mS \cos \theta - MB \sin \theta$
$$= 62.0 \cos 32° 24 - 1.250$$
$$\sin 32° 24'$$
$$= \underline{51.68 \text{ m}}$$
Diff. in level, $\Delta L = i - H - h$
$$= 1.350 - 62.0 \sin 32° 24'$$
$$- 1.250 \cos 32° 24'$$
$$= \underline{-32.93 \text{ m}}$$

Horizontal length AC $= (77.55 + 51.68) \text{ m}$
$$= 129.23 \text{ m}$$
Diff. in level, AC $= (33.65 + 32.93) \text{ m}$
$$= \underline{66.58 \text{ m}}$$

$$\text{Gradient AC} = 1 \text{ in } \frac{129.23}{66.58}$$
$$= \underline{1 \text{ in } 1.94}$$

(b) *Staff vertical*
In Fig. 14.11 the horizontal distance and difference in level between the points A and B are required. The field procedure is identical to the previous method except that the staff is held vertically by using the spirit level attached to the back

Basic formulae   Supposing for the moment that the staff is held at right angles to the line of sight. T_1L_1 is the staff intercept, therefore
$$L = (m \times T_1L_1 + k)$$
but the staff is vertical and the intercept is in fact TL = S.

Consider the triangle TT_1M.

Angle M $= \theta$ and angle T_1 is a right angle (almost)
$$\text{therefore } \frac{T_1M}{TM} = \cos \theta$$
$$\text{therefore } T_1M = TM \cos \theta$$

Also in the triangle LL_1M
$$L_1M = LM \cos \theta$$

Therefore $T_1L_1 = TL \cos \theta$

Now, since
$$L = m \times T_1L_1 + k$$
$$L = m \times TL \cos \theta + k)$$
$$= (mS \cos \theta + k)$$

In triangle IMP
$$\frac{IP}{IM} = \cos \theta$$

therefore IP = IM $\cos \theta$
$$\text{i.e., } D = L \cos \theta$$
$$= (mS \cos \theta + k) \cos \theta$$
$$= mS \cos^2 \theta + k \cos \theta$$

Also
$$\frac{MP}{IM} = \sin \theta$$

therefore MP = IM $\sin \theta$
$$\text{i.e., } H = L \sin \theta$$
$$= (mS \cos \theta + k) \sin \theta$$
$$= \underline{mS \cos \theta \sin \theta + k \sin \theta}$$

Alternatively, in triangle IMP
$$\frac{MP}{IP} = \tan \theta$$

therefore MP = IP $\tan \theta$
$$\text{i.e., } \underline{H = D \tan \theta}$$

Difference in level A to B $= \Delta L$
$$\Delta L = i + H - h$$

For angles of depression
$$\Delta L = i - H - h$$

Example 3 Stations M, N, and O for a right-angled triangle at station M (Fig. 14.12). A theodolite whose constants are $m = 100$ and $k = 0$ was used to determine the following tacheometric data

Instrument station	M
Height of instrument	1.410 m
Reduced level of stn	129.600 m

Target stn	Vertical angle	Stadia reading		
		Upper	Mid	Lower
N	−5° 40'	1.830	1.500	1.170
O	+2° 30'	2.810	2.610	2.410

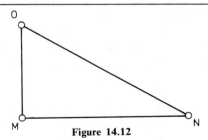

Figure 14.12

Calculate (a) horizontal lengths MN and MO
(b) horizontal length NO
(c) reduced levels of N and O.

Solution

Line MN

Staff intercept at N = 1.830 − 1.170
= 0.660 m

Horiz. length MN = $mS \cos^2 \theta + k \cos \theta$
= $66.0 \cos^2 5° 40'$
= 65.36 m

H = (Inst. ht M − Mid
    staff reading N) = $-mS \cos \theta \sin \theta + k \sin \theta$
= $-66.0 \cos 5° 40' \sin 5° 40'$
= −6.49 m

Reduced level of N = $129.600 + i - H - h$
= $129.600 + 1.410 - 6.490$
    -1.500
= 123.020 m

Line MO

Staff intercept at O = 2.810 − 2.410
= 0.400 m

Horiz. length   MO = $mS \cos^2 \theta + k \cos \theta$
= $40.0 \cos^2 2° 30'$
= 39.92 m

H = (Mid staff
reading O − Inst. ht M) = MO $\tan \theta$
= $39.92 \tan 2° 30'$
= +1.74 m

Reduced level of O   = $129.600 + i + H - h$
= $129.600 + 1.410 + 1.74$
    -2.610
= 130.140 m

Horiz. dist.   NO   = $\sqrt{(MN^2 + MO^2)}$
= $\sqrt{(65.36^2 + 39.92^2)}$
= 76.59 m

Errors in stadia tacheometry

The gross errors are the same as for the tangential system.

Wrong staff readings are more common in the stadia system simply because of the number of readings required and the fact that interpolation of the staff graduations has to be made each time.

The booker should be able to detect gross errors by comparing the difference between the top and mid staff readings with the difference between the mid and bottom readings. The differences should be the same; if not, the readings must be repeated.

Systematic errors are the same as for the tangential system, namely errors arising from non-perpendicular staffs and differential refraction.

Methods of simplifying calculations

The basic formulae for vertical staff stadia tacheometry as shown above are:

Horizontal distance $D = mS \cos^2 \theta$
Difference in level $\Delta L = i + D \tan \theta - h$
    or $\Delta L = i + (mS \cos \theta \sin \theta) - h$

All of these formulae produce fairly lengthy calculations and the following methods are designed to simplify the calculations.

(a) *Use of tacheometric tables*

Tacheometric tables, the work of D. T. F. Munsey MA, FRICS, have now been produced entirely in metric units (Table 14.3)

Values of m = 100 and k = 0 are used throughout the tables. The range of vertical angles is from 0° 10′ to 10° 00′ at 10′ intervals and from 10° 20′ to 20° 00′ at 20′ intervals. For each vertical angle the horizontal component $mS \cos^2 \theta$ and vertical component $mS \cos \theta \sin \theta$ are given for values of mS ranging from 10 m to 210 m at intervals of 1 metre.

In the tables, the distance mS is called the generating number G while the values $G \cos^2 \theta$ and $G \cos \theta \sin \theta$ are respectively called D for distance and H for height.

In Example 3, the calculations showed the horizontal distance MN to be 65.36 while the corresponding height was −6.49.

In Table 14.3, D and H for vertical angle 5° 40′ are underlined where G = 66, the values agreeing with those obtained above.

Where G is not an integral number the values of H and D have to be interpolated.

Example 4   Find the values of H and D corresponding to a staff intercept of 0.634 and vertical angle 5° 40′.

Answer

	G	D	H
From the tables	63	62.39	6.19
	64	63.38	6.29
By interpolation	63.4	62.79	6.23

In the field, the vertical angle must be set to some multiple of 10 minutes if under 10° 00′ or set to some multiple of 20 minutes if the angle exceeds this value.

(b) The calculation of differences in level is reduced, if, in the field the centre crosshair is set to read on the staff the value of the instrument height i. The formula for difference in level then becomes simply $\Delta L = D \tan \theta$ since h and i cancel each other.

Using this method, the tacheometric tables cannot be used since the vertical angle might not be a multiple of 10 minutes.

G	D	H	G	D	H	G	D	H	G	D	H
10	9.90	0.98	60	59.41	5.90	110	108.9	10.81	160	158.4	15.72
11	10.89	1.08	61	60.41	5.99	111	109.9	10.91	161	159.4	15.82
12	11.88	1.18	62	61.40	6.09	112	110.9	11.00	162	160.4	15.92
13	12.87	1.28	63	62.39	6.19	113	111.9	11.10	163	161.4	16.02
14	13.86	1.38	64	63.38	6.29	114	112.9	11.20	164	162.4	16.11
15	14.85	1.47	65	64.37	6.39	115	113.9	11.30	165	163.4	16.21
16	15.84	1.57	66	65.36	6.49	116	114.9	11.40	166	164.4	16.31
17	16.83	1.67	67	66.35	6.58	117	115.9	11.50	167	165.4	16.41
18	17.82	1.77	68	67.34	6.68	118	116.8	11.59	168	166.4	16.51
19	18.81	1.87	69	68.33	6.78	119	117.8	11.69	169	167.4	16.61
20	19.80	1.97	70	69.32	6.88	120	118.8	11.79	170	168.3	16.70
21	20.80	2.06	71	70.31	6.98	121	119.8	11.89	171	169.3	16.80
22	21.79	2.16	72	71.30	7.07	122	120.8	11.99	172	170.3	16.90
23	22.78	2.26	73	72.29	7.17	123	121.8	12.09	173	171.3	17.00
24	23.77	2.36	74	73.28	7.27	124	122.8	12.18	174	172.3	17.10
25	24.76	2.46	75	74.27	7.37	125	123.8	12.18	175	173.3	17.20
26	25.75	2.55	76	75.26	7.47	126	124.8	13.38	176	174.3	17.29
27	26.74	2.65	77	76.25	7.57	127	125.8	12.48	177	175.3	17.39
28	27.73	2.75	78	77.24	7.66	128	126.8	12.58	178	176.3	17.49
29	28.72	2.85	79	78.23	7.76	129	127.7	12.68	179	177.3	17.59
30	29.71	2.95	80	79.22	7.86	130	128.7	12.77	180	178.2	17.69
31	30.70	3.05	81	80.21	7.96	131	129.7	12.87	181	179.2	17.78
32	31.69	3.14	82	81.20	8.06	132	130.7	12.97	182	180.2	17.88
33	32.68	3.24	83	82.19	8.16	133	131.7	13.07	183	181.2	17.98
34	33.67	3.34	84	83.18	8.25	134	132.7	13.17	184	182.2	18.08
35	34.66	3.44	85	84.17	8.35	135	133.7	13.26	185	183.2	18.18
36	35.65	3.54	86	85.16	8.45	136	134.7	13.36	186	184.2	18.28
37	36.64	3.64	87	86.15	8.55	137	135.7	13.46	187	185.2	18.37
38	37.63	3.73	88	87.14	8.65	138	136.7	13.56	188	186.2	18.47
39	38.62	3.83	89	88.13	8.74	139	137.6	13.66	189	187.2	18.57
40	39.61	3.93	90	89.12	8.84	140	138.6	13.76	190	188.1	18.67
41	40.60	4.03	91	90.11	8.94	141	139.6	13.85	191	189.1	18.77
42	41.59	4.13	92	91.10	9.04	142	140.6	13.95	192	190.1	18.87
43	42.58	4.23	93	92.09	9.14	143	141.6	14.05	193	191.1	18.96
44	43.57	4.32	94	93.08	9.24	144	142.6	14.15	194	192.1	19.06
45	44.56	4.42	95	94.07	9.33	145	143.6	14.25	195	193.1	19.16
46	45.55	4.52	96	95.06	9.43	146	144.6	14.35	196	194.1	19.26
47	46.54	4.62	97	96.05	9.53	147	145.6	14.44	197	195.1	19.36
48	47.53	4.72	98	97.04	9.63	148	146.6	14.54	198	196.1	19.46
49	48.52	4.81	99	98.03	9.73	149	147.5	14.64	199	197.1	19.55
50	49.51	4.91	100	99.02	9.83	150	148.5	14.74	200	198.0	19.65
51	50.50	5.01	101	100.0	9.92	151	149.5	14.84	201	199.0	19.75
52	51.49	5.11	102	101.0	10.02	152	150.5	14.94	202	200.0	19.85
53	52.48	5.21	103	102.0	10.12	153	151.5	15.03	203	201.0	19.95
54	53.47	5.31	104	103.0	10.22	154	152.5	15.13	204	202.0	20.04
55	54.46	5.40	105	104.0	10.32	155	153.5	15.23	205	203.0	20.14
56	55.45	5.50	106	105.0	10.42	156	154.5	15.33	206	204.0	20.24
57	56.44	5.60	107	106.0	10.51	157	155.5	15.43	207	205.0	20.34
58	57.43	5.70	108	106.9	10.61	158	156.5	15.52	208	206.0	20.44
59	58.42	5.80	109	107.9	10.71	159	157.4	15.62	209	207.0	20.54
60	59.41	5.90	110	108.9	10.81	160	158.4	15.72	210	208.0	20.63

Table 14.3

(c) The calculation of the staff intercept S is simplified if the lower crosshair is set to a round value of 0.1 metre on the staff. The other two hairs are read in the usual manner.

This is probably the least useful simplification. Since tables cannot be used the difference in level, ΔL, has to be worked out in full since h will probably not equal i.

(d) A compromise field method is sometimes used which does not introduce any appreciable error.

The vertical angle is set to some multiple of 10 minutes and the centre hair reading is noted. The vertical slow motion screw is touched to bring the lower crosshair onto some round 0.1 m value and the upper and lower crosshairs read. S is then easily found and the tacheometric tables can be used.

The staff intercept S is unchecked by this method since the centre crosshair is not the mean of the upper and lower readings.

Booking the observations

Table 14.4 is only one of a number of tables suitable for booking tacheometric observations. This table is used for vertical staff tacheometry and illustrates the first three simplification methods outlined above. Columns 1–6 are completed in the field and 7–9 are filled in later

Calculations

Line AB
From tables $D = 43.57$ m
(when G is 44) $H = $ 4.32 m
             $\Delta L = $ $1.390 + 4.32 - 1.820 = +3.89$ m

Line AC

$$D = 56 \cos^2 2° 26' \qquad = 55.90 \text{ m}$$
$$H = 56 \cos 2° 26' \sin 2° 26' \quad = +2.38 \text{ m}$$
$$\Delta L = 1.390 + H - 1.390 \qquad = +2.38 \text{ m}$$

Line AD

$$D = 38 \cos^2 8° 10' \qquad = 37.23$$
$$H = -37.23 \tan 8° 10' \qquad = -5.34$$
$$\Delta L = 1.390 - 5.340 - 1.190 \quad = -5.14$$

Use of tangential and stadia tacheometry

The maximum accuracy obtainable by these methods is 1/1000. Consequently the uses are limited to obtaining minor details and spot heights for contouring.

Figure 14.13 shows the route of a proposed roadway along which contours are required. Since several instrument stations are necessary the survey is laid out in the form of an open traverse, wherein the horizontal clockwise angles between stations RO and D are measured. The distances are measured tacheometrically in both directions and the traverse calculated, plotted, and adjusted as previously described.

At each station several rays are observed and oriented to the RO with a view to determining the position and spot level of a number of points. The order of observing is shown by the numbers in the figure.

The spot levels are plotted onto the traverse plan generally by protractor and scale, and contours are interpolated from them.

The tangential and stadia methods are generally adopted when the terrain is unsuitable for linear measuring and when the result is required to a fairly low degree of accuracy. The methods also have the advantage of being speedy and require the minimum amount of equipment and least number of personnel.

The following example illustrates a further use of tacheometry and is fairly typical of examination questions on this subject.

	Instrument station A Height of Instrument 1.390					Reduced level 116.210			
			Stadia						
Target stn	Horiz. circle	Vert. angle	Top bottom	'S'	Mid	D	H	ΔL	Red. lev. target
RO	00° 00′								
B	12° 30′	+5° 40′	2.040 1.600	0.440	1.820	43.57	4.32	+3.89	120.100
C	34° 15′	+2° 26′	1.670 1.110	0.560	1.390	55.90	+2.38	+2.38	118.590
D	63° 26′	−8° 10′	1.380 1.000	0.380	1.190	37.23	−5.34	−5.14	111.070
1	2	3	4	5	6	7	8	9	10

Table 14.4

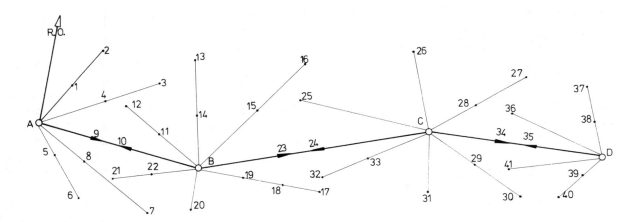

Figure 14.13

Example 5 The readings in Table 14.5 were taken along part of the boundary of a site using a theodolite whose constants were $m = 100$ and $k = 0$.

Calculate the plan length of the perimeter ABCDA and the reduced levels of each station given that A is 135.200 AOD.

Figure 14.14 illustrates the problem.

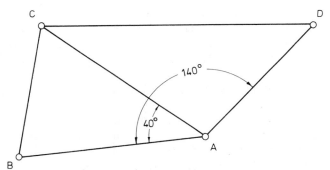

Figure 14.14

Line AB

$$S = 1.790 - 1.050$$
$$= 0.740$$

$$D = mS \cos^2 \theta \qquad H = D \tan \theta$$
$$= 74 \cos^2 7° 10' \qquad = 72.85 \tan 7° 10'$$
$$= 72.85 \text{ m} \qquad = +9.16 \text{ m}$$

Reduced level B $= 135.200 + 1.420 + 9.160$
$$-1.420$$
$$= 144.360 \text{ m AOD}$$

Line AC
$$S = 1.888 - 0.952$$
$$= 0.936$$
$$D = mS \cos^2 \theta \qquad H = D \tan \theta$$
$$= 93.6 \cos^2 3° 20' \qquad = 93.29 \tan 3° 20'$$
$$= 93.29 \text{ m} \qquad = +5.43 \text{ m}$$

Reduced level C $= 135.200 + 1.420 + 5.430$
$$- 1.420$$
$$= \underline{140.630 \text{ m AOD}}$$

Line AD
$$S = 2.860 - 2.000$$
$$= 0.860$$
$$D = mS \cos^2 \theta \qquad H = -D \tan \theta$$
$$= 86 \cos^2 2° 40' \qquad = -85.81 \tan 2° 40'$$
$$= 85.81 \text{ m} \qquad = -4.00 \text{ m}$$

Reduced level D $= 135.200 + 1.420 - 4.000$
$$- 2.430$$
$$= \underline{130.190 \text{ m AOD}}$$

Inst. stn	Inst. height	Target stn	Stadia			Whole circle bearing	Vertical angle
			Top	Mid	Bottom		
A	1.420	B	1.790	1.420	1.050	00° 00′	+7° 10′
		C	1.888	1.420	0.952	40° 00′	+3° 20′
		D	2.860	2.430	2.000	140° 00′	−2° 40′

Table 14.5

In triangle ABC

$$BC^2 = AB^2 + AC^2 - 2.AB.AC \cos 40°$$
$$= 72.85^2 + 93.29^2 - 2 \times 72.85 \times 93.29$$
$$\times 0.76604$$
$$= 3597.86 \text{ m}$$

therefore $BC = \sqrt{3597.86}$
$$= \underline{59.98 \text{ m}}$$

In triangle ACD

$$CD^2 = AC^2 + AD^2 - 2.AC.AD \cos 100°$$
$$= AC^2 + AD^2 + 2.AC.AD \cos 80°$$
$$= 93.29^2 + 85.81^2 + 2 \times 93.29 \times 85.81$$
$$\times 0.173\,65$$
$$= 18\,846.59$$

therefore $CD = \sqrt{18\,846.59}$
$$= 137.28 \text{ m}$$

Total perimeter ABCDA

$$= (72.85 + 59.98 + 137.28 + 85.81) \text{ m}$$
$$= \underline{355.92 \text{ m}}$$

Horizontal staff tacheometry

It is known that one of the principal errors in vertical staff tacheometry is that caused by differential refraction of the rays from the staff.

The error can be eliminated by simply holding the staff horizontally. The rays then pass through uniform atmospheric conditions.

The following methods are used.

1. Subtense bar (Variable parallactic angle)

The equipment and methods are primarily designed for obtaining accurate horizontal distances. It is possible though uncommon to obtain differences in level.

The precision of the distance measurements depends to a great extent on the type of theodolite used. It is recommended that a 1-second instrument be employed though distances can be obtained to a lower degree of precision using a 20-second instrument.

The subtense bar (Fig. 14.15) consists of two hollow aluminium alloy hinged arms, which fold together and fit compactly into a wooden box. Fitting within the arms are two 'Invar' steel rods to which targets are attached. When opened out, the 'Invar' rods form a bar exactly 2 metres long with a target at both ends. A sighting telescope is provided on one of the arms with its line of sight at right angles to the bar. The bar also carries a spirit level for setting the bar horizontally.

A third target can be fitted to the centre of the bar such that it is directly over the ground station. It can therefore be used as a normal traverse target for measuring horizontal angles.

The bar is fitted with a conventional three-screw levelling base which is mounted on a normal theodolite tripod. Alternatively, and this is more common, the levelling base is detachable so that subtense bar and theodolite are interchangeable. They can therefore be used with the three-tripod traverse system.

Procedure
The horizontal distance AB is required in Fig. 14.16.
(a) Tripods are set at stations A and B and accurately centred and levelled using an optical plummet.
(b) The theodolite is set at A and the subtense bar at B.
(c) The parallactic angle CTD is measured by repetition or reiteration several times. The number of times depends upon the accuracy to which the distance is required and upon the mean square error of the measured angle.

In general, it is true to say that the more often the angle is measured the smaller will be the mean square error and the more accurate will be the distance.

It is important to note that the measured angle is the horizontal angle regardless of whether the bar is

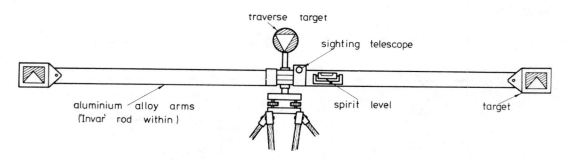

traverse target
sighting telescope
aluminium alloy arms
('Invar' rod within)
spirit level
target

Figure 14.15

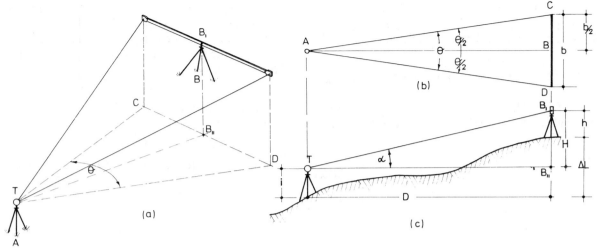

Figure 14.16

above or below the collimation level of the theodolite.

Basic formula

In triangle CAB, (Fig.14.16 (b)), CB is half the bar length = $b/2$ = 1 metre. Angle ABC is a right angle and angle CAB is half the parallactic angle = $\theta/2$

$$\text{therefore } \frac{AB}{CB} = \cot \theta/2$$

$$\text{and } AB = CB \cot \theta/2$$

i.e., horiz. dist. $AB = b/2 \cot \theta/2$ metres

$$\text{But } b/2 = 1 \text{ metre}$$

$$\text{so } \underline{AB = \cot \theta/2 \text{ metres}}$$

For example, if $\theta = 03° 40' 50''$

$$AB = (\cot 01° 50' 25'') \text{ metres}$$

$$= \underline{31.124 \text{ m}}$$

The difference in level, ΔL, can be calculated by measuring the vertical angle to the centre target and measuring the height of the target above the ground station by tape (Fig. 14.16(c)).

Then, in triangle TB_1B_{11}

$$B_1B_{11} = H = D \tan \alpha$$

therefore difference in level $\Delta L = i + H - h$
If α is an angle of depression

$$\underline{\Delta L = i - H - h}$$

Accuracy of subtense measurements

The accuracy of the horizontal length is dependent upon three factors:

(a) *The length of the subtense bar* It should be obvious that if the bar changes length the horizontal distance D will change. Since the bar is made of 'Invar' steel, temperature has virtually no effect on the length of the bar and it can be considered to be constant at 2 metres.

(b) *Orientation of the bar* If the bar is not exactly at right angles to the line of sight the horizontal distance D will again be in error.

It can be shown that the bar would have to be out of alignment by $1° 17'$ for the error in D to exceed $1/10\,000$ of the length.

The sighting device is capable of orienting the bar far more accurately than $1° 17'$ and the effect of mis-orientation can be ignored.

(c) *Precision of parallactic angle* If the distance is to be measured with high accuracy (1 in 10 000), the parallactic angle must be measured with an accuracy of $\pm 1''$. To ensure the required degree of precision the angle must be measured ten times. For a given distance the accuracy in D is directly proportional to the accuracy in measurement of the angle.

Length of sights

The accuracy of a main traverse in construction surveying is required to be of the order of 1 in 10 000— certainly not less than 1 in 5000.

Assuming that the parallactic angle can be measured with an accuracy of $\pm 1''$, Table 14.6 shows the maximum length of sight which can be measured to produce the accuracies shown.

Accuracy	Length of sight (m)
1/2500	164.8
1/5000	82.4
1/10 000	41.2

Table 14.6

If the length of sight must be increased the auxiliary base system must be used. This is by no means the only auxiliary method but is the simplest.

Auxiliary base method
Figure 14.17 shows the required horizontal length AB.
(a) The theodolite is set at B, aligned to A and a right angle ABC is turned off for a distance BC.

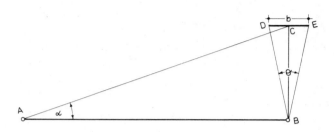

Figure 14.17

Triangles DBE and ABC are fairly similar and the ratio $b/CB = CB/AB$ is approximately correct. Therefore $CB = \sqrt{(AB.b)}$.
(b) The distance CB is obtained directly by measuring the parallactic angle:

$$BC = \cot \theta/2 \text{ (as before)}$$

(c) The theodolite is removed to A and angle α is measured by normal repetitive or reiteration methods.
(d) In triangle ABC

$$\frac{AB}{BC} = \cot \alpha$$

therefore $AB = BC \cot \alpha$

$$= \cot \theta/2 \cot \alpha$$

(e) With an error in the parallactic angle of $\pm1''$ the maximum length of sight is 410 metres if an accuracy of 1/10 000 is to be achieved.

Example 6 Figure 14.18 shows two lines of a subtense traverse AB and BC.

Given that the bearing AB is due North and A is origin, calculate the coordinates of stations B and C.

Solution

(a) Base at A $= \cot \dfrac{05° \ 50' \ 20''}{2}$

$$= \cot 02° \ 55' \ 10''$$

(b) $\qquad AB = \cot 02° \ 55' \ 10'' \times \cot 05° \ 40' \ 17''$

$$= \underline{197.454 \text{ m}}$$

(c) $\qquad BC = \cot \dfrac{03° \ 02' \ 10''}{2}$

$$= \cot 01° \ 31' \ 05''$$

$$= \underline{37.734 \text{ m}}$$

(d) Coordinates of B = 00.00 E, 197.454 N

(e) $\qquad$ Bearing BC = N 30° 45′ 00″ E

Part. dep. BC = 37.734 sin 30° 45′ 00″
$$= 19.293 \text{ m}$$

Part. lat. BC = 37.734 cos 30° 45′ 00″
$$= 32.429 \text{ m N}$$

(f) Coordinates of C = 00.00 + 19.293
$$= 19.293 \text{ m E}$$
$$= 197.454 + 32.429$$
$$= 229.883 \text{ m N}$$

2. Optical wedge system

Once again the equipment is designed for obtaining distances rather than differences in level, though the latter is possible. The distance is, however, the slant distance which is a slight disadvantage. Accuracies of the order of 1/5000 are readily attainable.

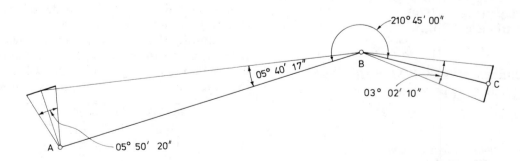

Figure 14.18

Equipment

As the name suggests the optical wedge is a wedge-shaped glass prism which fits over the objective end of the telescope. Only one-third of the objective is covered by the prism as in Fig. 14.19 (top).

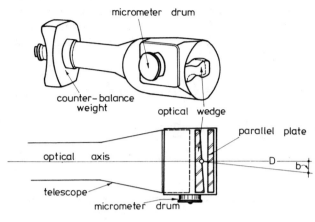

Figure 14.19

Rays passing through the uncovered part of the object glass are free to travel to the diaphragm in the normal manner. Rays which pass through the prism are deflected. The prism is manufactured in such a manner that the displacement b is 1/100 of the distance D, therefore if two objects situated 100 metres from the prism are 1 metre apart, they will appear to coincide when viewed through the telescope.

A special staff is used in conjunction with the prism and is shown in Fig. 14.20(a). The stand pipe is centred over the ground station and the sighting device attached to the staff is used to orientate the bar at right angles to the line of sight.

The staff illustrated is manufactured by Hilger & Watts. It is graduated every 2 cm from 20 cm to 150 cm and is figured every 10 cm. A vernier is positioned above the main scale and covers the first 20 cm of the latter.

Procedure

(a) The staff is set above the survey mark and raised or lowered until it is at the same height as the instrument.

(b) The optical wedge is attached to the objective end of the theodolite and a counter-weight placed on the eyepiece end to maintain equilibrium.

(c) The staff is sighted and the telescope is moved slowly vertically until the two images of the staff appear to meet along a common straight dividing line (Fig. 14.20(b)). The vernier will have shifted optically by 1/100 of the slant distance from theodolite to staff.

(d) It will often happen that no vernier division will coincide exactly with a main scale division. Before the vernier can be read, however, one of the divisions must be made to coincide. This is achieved by placing a parallel plate micrometer in front of the wedge.

The micrometer is geared to a drum which records movement of the parallel plate. The drum is divided into twenty parts each of which represents 0.01 cm.

(e) The distance recorded in Fig. 14.20(b) is therefore

Staff reading	42.0 cm
Vernier reading	1.2 cm
Micrometer reading	0.04 cm

therefore slope distance = 43.24 × 100 cm
= 43.24 m

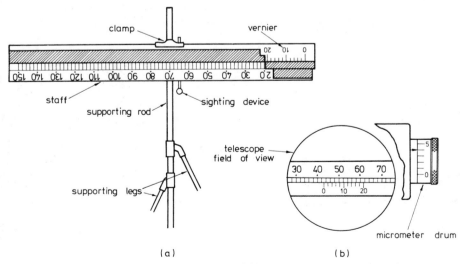

(a) (b)

Figure 14.20

(f) The vertical angle is read and used to calculate the plan distance and difference in level between the stations.

Basic formula

Plan distance D = slope dist. × cos vert. angle
Diff. in height H = slope dist. × sin vert. angle
Diff. in level = $i \pm H - h$

Self-reducing tacheometers

It should be understood from the foregoing descriptions that the end results of tacheometry are the horizontal and vertical dimensions from the instrument station to the staff station. A fair measure of calculation is involved in all of the methods and since the calculations are lengthy and potential sources of error, clearly it would be beneficial to eliminate or minimize them. In order to do so the instrument would have to be capable of directly producing the end results. Many such instruments have been developed by the various manufacturers, one of the simplest being the diagram circle tacheometer RDS produced by Wild Heerbrugg Ltd of Switzerland.

Diagram circle tacheometer

In Fig. 14.21 the horizontal distance AB is $AB = mS = 100S$ for a modern instrument where S is the stadia intercept. This formula was proved on page 187. The equivalent horizontal distance for D_{AC} is

$$D_{AC} = \left(\frac{f}{i}\right) S \cos^2 \theta$$

where θ is the vertical angle. The difference in elevation H_{AC} is

$$H_{AC} = \left(\frac{f}{i}\right) S \cos \theta \sin \theta$$

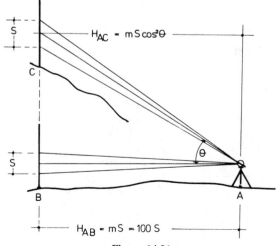

Figure 14.21

1. *Horizontal distances*

In order to eliminate the calculation $(f/i)\, S \cos^2 \theta$, the instrument has to be capable of automatically reducing this expression to $100S$.

For automatic reduction an equivalent staff intercept E has to be substituted for S. Thus E must vary with the angle of elevation θ so that, when it is multiplied by (f/i), it will produce the horizontal distance directly. This can be accomplished by introducing a variable stadia interval v to replace the fixed stadia interval i.

From above $D_{AC} = \left(\dfrac{f}{i}\right) S \cos^2 \theta = 100S$ (1)

This expression has to become

$$D_{AC} = \left(\frac{f}{v}\right) E \cos^2 \theta = 100E \qquad (2)$$

In order to do so v must be made to equal $i \cos^2 \theta$.

When substituted in Eq. (2) above:

$$D = \left(\frac{f}{i \cos^2 \theta}\right) E \cos^2 \theta \qquad (3)$$

$$= \left(\frac{f}{i}\right) E$$

$$= 100E$$

In practice then, the conventional constant stadia interval has to be physically altered to produce an ever-changing stadia interval which will vary with the magnitude of the vertical angle being observed. This is achieved by etching the curve $i \cos^2 \theta$ on a circle called the diagram circle which is mounted on the theodolite trunnion axis, at the side of the telescope (Fig. 14.22). This curve is called the distance

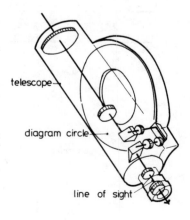

Figure 14.22

curve and replaces the top stadia wire. The bottom stadia curve is called the datum or base curve. Prisms are inserted in the optical train from the object glass to the eyepiece so that, as the telescope is inclined, different parts of the diagram circle come into view in the eyepiece (Fig. 14.22). The observer then sees the staff image with the distance and base curves superimposed. Because the distance curve has a value of $i \cos^2 \theta$, the horizontal distance D is given directly as the intercept E multiplied by 100. In Fig. 14.23 the horizontal distance D is

$$
\begin{aligned}
D &= E \times 100 \\
&= 0.355 \times 100 \\
&= 35.5 \text{ m}
\end{aligned}
$$

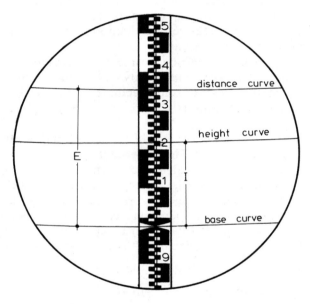

Figure 14.23

2. Differences in height

Similar theory exists for the derivation of differences in height between instrument and staff stations. On page 187 it was shown that the difference in height H_{AC} is

$$
H_{AC} = \left(\frac{f}{i} \right) S \cos \theta \sin \theta
$$

For automatic reduction, a second equivalent staff intercept I has to be substituted for S, and a second variable stadia interval v has to replace the fixed stadia interval i.

Although the mathematics are complex, it can be shown that the variable stadia interval v has to be of the form:

$$
v = \frac{\tfrac{1}{2} f \sin 2\theta}{f/i \pm \sin^2 \theta}
$$

For a simpler understanding it can be taken as $v = \frac{1}{2} \sin 2\theta$.

This second equation produces a curve called the height curve, when plotted against increasing values of θ. This height curve is etched on the diagram circle and is observed through the telescope eyepiece simultaneously with the horizontal distance curve (Fig. 14.23). The difference in height can then be read directly as an intercept between the base and height curves.

Plotting the reduction curves

Figure 14.24(a) shows the datum curve, distance curve and height curve plotted against vertical angles varying between 0° and 45°. The figure shows that the distance curve is always the topmost curve. In any practical situation the horizontal distance and difference in height will be identical for any slope length at an angle of 45° elevation; hence the curves coincide at this point.

Figure 14.24(b) shows the curves plotted in polar form as they appear on the diagram circle of the

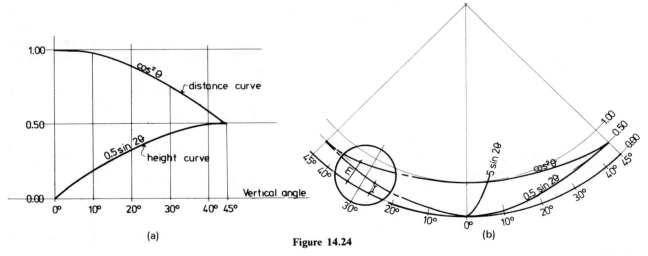

Figure 14.24

theodolite. The view obtained through the eyepiece at elevation 30° is superimposed on the figure. The horizontal distance intercept E and the difference in height intercept I are easily read against the staff at this angle.

At low angles of inclination, for example, between 1° and 4°, the base curve and height curve are very close together, and this would inevitably cause errors in the determination of heights. In order to overcome this problem, the values of the curve ½ sin 2θ are magnified and replotted.

Figure 14.24(b) shows the height curve magnified by a factor of 10 between angles of elevation 0° and 4°. The curve is therefore plotted as 5 sin 2θ. The view through the eyepiece at elevation 2° is shown in Fig. 14.25 where the height intercept I is easily read against a staff. Since the height curve is

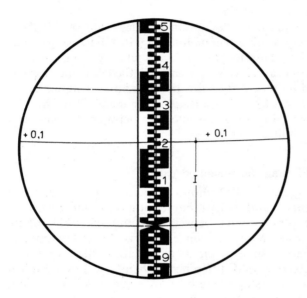

Figure 14.25

magnified by factor 10, the resulting intercept must be divided by 10 or, more easily, multiplied by 0.1. The difference in height H is therefore

$$H = (\text{intercept } I \times 100) \div 10$$

i.e.,
$$H = (\text{intercept } I \times 100) \times 0.1$$
$$= 2.2 \text{ m} \times 0.1$$
$$= 0.22 \text{ m}$$

As the resolution of the intercept values increases with an increase in inclination, the magnification of the height curve can be correspondingly decreased. Between angles of inclination 4° and 9°, the magnification factor is 5 and the curve is plotted as 2.5 sin 2θ.

Hence
$$H = (\text{intercept } I \times 100) \div 5$$
$$= (\text{intercept } I \times 100) \times 0.2$$

Similarly between angles of inclination 9° and 22°, the magnification factor is 2 and the curve is plotted as sin 2θ.

Hence
$$H = (\text{intercept} \times 100) \div 2$$
$$= (\text{intercept} \times 100) \times 0.5$$

Field procedure

(a) The tacheometer is set on face left over the survey station in the same manner as a conventional theodolite. The instrument height is measured above the survey point.

(b) The staff is held vertically on the detail point or a specially produced short staff is set on its tripod over the point and is raised or lowered on its centring rod until the zero graduation is at the same height as the tacheometer.

(c) The telescope is tilted using the slow motion screw until the base curve reads zero on the special short staff or is set to some decimetre value on the ordinary staff. This reading is noted in the field book.

(d) The distance curve is read against the staff and the reading is noted. The horizontal distance intercept E can then be calculated when required.

(e) The height curve is read against the staff; the reading is noted together with the sign and magnitude of the multiplication factor. The difference in height intercept I can then be calculated.

(f) The horizontal distance D is calculated as

$$D = E \times 100$$

while the difference in height H is

$$H = (I \times 100) \times \text{multiplication factor}$$

Accuracy of the method

The accuracy of the method clearly depends upon how accurately the vertical staff can be read. On any staff graduated in centimetres, the best estimation must be 1 mm; therefore the error in horizontal distance must be 100 mm. Over a maximum sighting range of 100 m, the order of accuracy is therefore 1:1000.

The accuracy of height determination depends upon the height curve used and varies from about 1/5000 with the × 0.1 curve to about 1/500 with the × 1.0 curve.

It is important in this method of tacheometry that sights should never be taken within one metre of the base of the staff, since the effects of differential refraction are considerable close to the ground.

Application of self-reduction methods

Since the method is economical in manpower and

is reasonably accurate, it is used on surveys where a great number of points of detail are to be surveyed. The principal application in construction surveys is in the determination of volumes of earthworks. The instrument has been used with great success in the surveys of quarries.

Test questions

1 The following observations were made using a theodolite where the staff was held vertically.
The instrument constants were 100 and zero.

Inst. stn	Staff stn	Horiz. circle reading	Vert. angle	Staff readings Top	Mid	Bottom
A	B	350° 10′	+2° 40′	4.200	3.000	1.800
	C	40° 10′	+4° 00′	3.760	2.500	1.240
	D	78° 10′	+5° 20′	3.050	2.000	0.950

Calculate the area in hectares enclosed by stations ABC and D.

2 Observations from a station A to another point B were made to a vertical staff as follows:

Inst. stn	Staff	Vert. angle	Staff reading
A	B	+4° 30′	1.000 m
	B	+5° 18′	3.000 m

At a second station C, the staff could not be held vertically because of the overhanging branches of a tree. Observations were made to a horizontal staff at C held at right angles to the line AC with the following results.

Inst. stn	Staff	Parallactic angle	Staff intercept
A	C	04° 10′	3.000 m

The horizontal angle BAC was measured as 95° 10′. Calculate the horizontal length BC.

3 The following tacheometric observations were taken along the centre line of a proposed roadway at approximately 20-metre intervals.
Instrument stn 'D' (approx. 60 m chainage)
Instrument height 1.45 m
Reduced level of station = 234.21 m AOD.

Target stn	Vert. angle	Stadia Top	Mid	Bottom	Remarks
A	−4° 20′	1.750	1.450	1.140	Chainage 0 m
B	−4° 20′	1.400	1.200	1.000	Approx. 20 m
C	−4° 20′	1.210	1.105	1.000	Approx. 40 m
E	−3° 00′	1.200	1.100	1.000	Approx. 80 m
F	−3° 00′	1.650	1.450	1.250	Approx. 100 m

Instrumental constants $m = 100$, $k = 0$.

The new roadway is to begin at point A and rise at 1 in 50 towards B. Calculate for each station:
(a) the reduced level
(b) the true chainage
(c) the proposed roadway level
(d) the depth of cutting required.

4 Table 14.7 shows the readings obtained on a tacheometric survey by a theodolite set up at A and of reduced level 99.0 m AOD.
Calculate:
(a) The horizontal distances, A–B and A–C.
(b) The reduced levels of stations B and C.

5 The results of a tacheometric survey along the line of a proposed straight roadway ABCD are shown in Table 14.8. The theodolite used had a multiplying constant of 100 and an additive constant of 0.
Calculate:
(a) The plan length of AD.
(b) the gradient from A to D.

Target station	Horizontal circle reading	Face	Vertical angle	Stadia hairs Top Mid Base	Height of target	Horizontal distance	Difference in height	Reduced level of target station	Remarks
B		L	+7° 10′ 00″	2.139 2.007 1.875					Staff vertical
C	78° 52′ 00″	L	+9° 32′ 20″	3.506					Staff vertical
C	78° 52′ 00″	L	+8° 05′ 40″	2.917					Staff vertical

(SCOTVEC, OND Building)

Table 14.7

Instrument station	Instrument height (m)	Staff station	Vertical angle	Top	Mid	Bottom
A	1.492	B	5° 27′	1.913	1.500	1.087
B	1.300	C	−4° 42′	2.319	2.000	1.681
C	1.410	D	4° 20′		4.000	
		D	2° 12′		1.000	

(SCOTVEC, OND Building)

Table 14.8

6 Worksheet (Table 14.9) shows the tacheometric readings obtained by different methods, using 6″ (second) optical scale theodolite, with zero additive constant and a multiplying constant of 100.

(a) Calculate the horizontal length AB and reduced level for point B.
(b) Explain three factors which could have contributed to the differences in the results.

Station point	Reduced level of station	Height of instrument	Staff point	Vertical angle	Stadia readings	Horizontal length	Reduced level of point	Remarks
A	100.0	1.49	B	+5° 51′ 42″	1.076 1.000 0.923			Staff vertical
A	100.0	1.49	B	+5° 51′ 42″	1.000			Mid hair readings
				+9° 34′ 06″	2.000			Staff vertical

(OND Building)

Table 14.9

15. Mensuration

(a) Areas

On even the smallest site, calculations have to be made of a wide variety of areas and volumes, for example, the area of the site itself, the volume of earthworks, cuttings, embankments, etc.

Many of the figures encountered can be calculated by the direct application of the accepted mensuration formulae but very often the figures are irregular in shape.

Regardless of the shape of the area, relevant data have to be presented in some way in order to make the calculations:

(a) The data are gathered in the field by some form of survey. The area is then calculated directly from these notes.

(b) The data are converted into coordinates or a plotted plan from which the area is computed.
(c) The data already exist in the form of a map or plan, e.g., Ordnance Survey plans.

Regular areas

Figure 15.1 shows the common regular figures and the formulae required to calculate their areas. In modern construction practice, many developments include shapes comprised of several of the common geometrical figures.

Example 1 The central piazza of a town centre development is drawn to scale 1.500 in Fig. 15.2. The

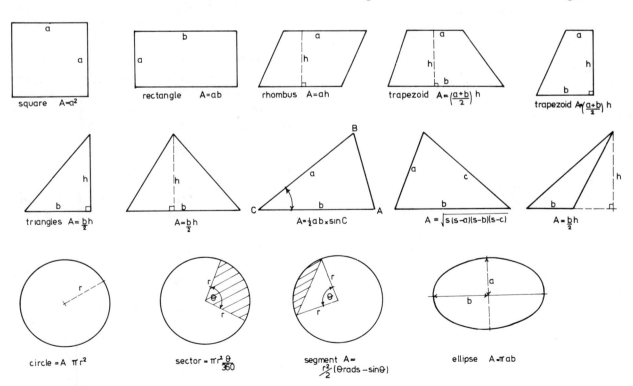

square $A = a^2$

rectangle $A = ab$

rhombus $A = ah$

trapezoid $A = \left(\dfrac{a+b}{2}\right) h$

trapezoid $A = \left(\dfrac{a+b}{2}\right) h$

triangles $A = \dfrac{bh}{2}$

$A = \dfrac{bh}{2}$

$A = \frac{1}{2} ab \times \sin C$

$A = \sqrt{s(s-a)(s-b)(s-c)}$

$A = \dfrac{bh}{2}$

circle $= A = \pi r^2$

sector $= \pi r^2 \dfrac{\theta}{360}$

segment $A = \dfrac{r^2}{2}(\theta \text{ rads} - \sin\theta)$

ellipse $A = \pi ab$

Figure 15.1

piazza has been split into six figures on the plan in order to calculate its area. Using the relevant surveyed dimensions, calculate the total area of the piazza in m².

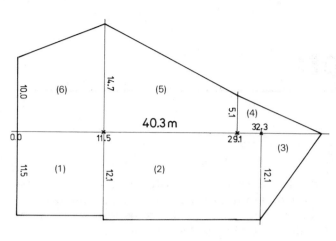

Figure 15.2

Solution
Area
(1) Square	11.5 × 11.5	= 132.25
(2) Rectangle	12.1 × 20.8	= 251.68
(3) Triangle	0.5 × 12.1 × 8.0	= 48.40
(4) Triangle	0.5 × 5.1 × 11.2	= 28.56
(5) Trapezoid	0.5 × (5.1 + 14.7) × 17.6	= 174.24
(6) Trapezoid	0.5 × (14.7 + 10.0) × 11.5	= 142.03
	Total	= 777.16 m²

Example 2 Figure 15.3 shows the surveyed dimensions of the concourse of a shopping centre which is to be floored with terrazzo tiles. The central area is to be a water feature with a fountain.

Calculate the total area to be tiled.

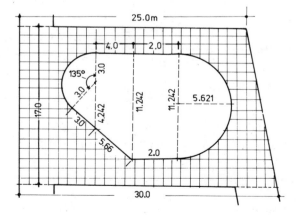

Figure 15.3

Solution
Overall area (trapezoid)
Area = 0.5 (25.0 + 30.0) × 17.0 = 27.5 × 17 = 467.50²

Water feature
1. Semi-circle	= π × 5.621² × 0.5	= 49.63
2. Rectangle	= 11.242 × 2.0	= 22.48
3. Trapezoid	= 0.5 (11.242 + 7.242) × 4.0	= 36.97
4. Triangle	= 0.5 (3 × 3)	= 4.50
5. Sector	= π × 3² × (135 ÷ 360)	= 10.60
	Total	= 124.18 m²

Area to be tiled = 467.50 − 124.18

= 343.32 m²

Irregular areas

Figures 15.4 and 15.5 are examples of the irregular areas which require calculation in a construction project. In Fig. 15.4, PQR is the area of a proposed factory development taken from an OS map.

PR is a new boundary fence denoting the northern limit of the site. A linear survey was made to determine the area of the site. The relevant field notes are used in Example 3.

In Fig. 15.5 a new roadway is to be constructed in a cutting. ABCD is the surface area of topsoil; ABE and CDF are the side slope areas which will be seeded with grass after the excavation is completed. These three areas are calculated in Example 4.

1. Areas from field notes

In a linear survey the area is divided into triangles, and the lengths of the three sides of each triangle are measured. The area contained within any one triangle ABC is found from the formula

$$\text{Area} = \sqrt{[s(s - a)(s - b)(s - c)]}$$

where s is the semi-perimeter.

Example 3 Figure 15.4 shows a simple chain survey composed of one triangle PQR, of which the lengths of the sides are

PQ = 60.0 m
QR = 104.6 m
RP = 70.0 m

The area within PQR is found thus:
(a)   In triangle PQ = r = 60.0 m
               PQR   QR = p = 104.6 m
                        RP = q = 70.0 m
Perimeter of PQR = 234.6 m

therefore semi-perimeter s = 117.3 m.

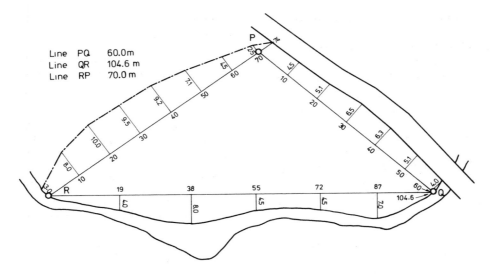

Figure 15.4

(b)

$$s - r = 57.3$$
$$s - p = 12.7$$
$$s - q = 47.3$$

Check = 117.3 = s

(c) Area of triangle PQR = $\sqrt{[s(s-r)(s-p)(s-q)]}$
$$= \sqrt{(117.3 \times 57.3 \times 12.7 \times 47.3)}$$
$$= 2009.3 \text{ m}^2$$

The boundaries of the chain survey are established by measuring offsets from the chain lines.

In Fig. 15.4 the area between the survey line and stream is composed of a series of trapezoids and triangles. The area of each figure must be computed separately as follows.
On line RQ:

Area of triangle (1) = $\frac{1}{2} \times 19 \times 4$ = 38.0
trapezoid (2) = $\frac{1}{2}(4 + 8) \times (38 - 19)$ = 114.0
trapezoid (3) = $\frac{1}{2}(8 + 4.5) \times (55 - 38)$ = 106.25
rectangle (4) = $4.5 \times (72 - 55)$ = 76.5
trapezoid (5) = $\frac{1}{2}(4.5 + 7) \times (87 - 72)$ = 86.25
triangle (6) = $\frac{1}{2}(104.6 - 87) \times 7$ = 61.6

= 482.6 m²

The area between survey line PQ and the road is again composed of a series of separate figures. It must be noticed however, that the offsets are at regular intervals of 10 metres in this instance.

Calling each offset y the area between *any* two offsets is calculated thus:
Area between chainage 20 m and chainage 30 m
$$= \frac{1}{2}(y_{20} + y_{30}) \times 10$$
Therefore total area

$$= \frac{1}{2}(y_0 + y_{10}) \times 10 + \frac{1}{2}(y_{10} + y_{20}) \times 10 + \frac{1}{2}(y_{20} + y_{30})$$
$$\times 10 + \ldots + \frac{1}{2}(y_{50} + y_{60}) \times 10$$
$$= \frac{1}{2} \times 10(y_0 + y_{10} + y_{10} + y_{20} + y_{20} + y_{30} + \ldots + y_{50} + y_{60})$$
$$= \frac{1}{2} \times 10(y_0 + y_{60} + 2y_{10} + 2y_{20} + 2y_{30} + 2y_{40} + 2y_{50})$$
$$= 10\left(\frac{y_0 + y_{60}}{2} + y_{10} + y_{20} + y_{40} + y_{50}\right)$$

This is the trapezoidal rule and is usually expressed thus:
Area = Strip width × (Average of first and last offsets + sum of others)
(d) In Fig. 15.4 the area is as follows:

$$\text{Area} = 10\left(\frac{4 + 4}{2} + 4.5 + 5.1 + 6.5 + 6.3 + 5.1)\right)$$

$$= 315.0 \text{ m}^2$$

The area can be found slightly more accurately by Simpson's rule. A knowledge of the integral calculus is required to prove the rule but it can be shown to be

Area = ⅓ strip width (first + last offsets + twice sum of odd offsets + four times sum of even offsets).

Note: (i) There must be an ODD number of offsets.

(ii) The offsets must be at regular intervals.
Using Simpson's rule the area between the line PQ and the road is as follows:

$$\text{Area} = 10/30[y_0 + y_{60} + 2(y_{20} + y_{40})$$
$$+ 4(y_{10} + y_{30} + y_{50})]$$
$$= 10/3[4 + 4 + 2(5.1 + 6.3) + 4(4.5 + 6.5 + 5.1)]$$
$$= 10/3[8 + 2(11.4) + 4(16.1)]$$
$$317.3 \text{ m}^2$$

(e) Finally the area between line RP and the wood is calculated.

Question: Which, if any, of the two rules can be used to calculate the area?

Answer: Trapezoidal rule—there is an even number of offsets between R and P at regular 10-metre spacing. The area between chainages 70 m and 74 m is calculated separately.

Area between line RP and wood is as follows:

$$\text{Area} = 10\left(\frac{3 + 2.5}{2} + 8 + 10 + 9.5 + 9.2 + 7.1 + 4.5\right)$$

$$= 510.5 + 5.0$$

$$= \underline{515.5 \text{ m}^2}$$

Total area of chain survey = 2009.3 + 482.6
+ 317.3 + 515.5
= $\underline{3324.7 \text{ m}^2}$

Example 3 illustrates the method of dealing with irregularly shaped areas. In general terms, ordinates, i.e., the offsets in Example 3, are measured at right angles to a base line and Simpson's or Trapezoidal rules applied. In Example 4 the base line is the centre line of the cutting and the ordinates are the widths of the cross-sections.

Example 4 In Fig. 15.5 the survey of a proposed cutting shows that the depths at 20 m intervals are 0.0 m, 0.9 m, 1.5 m, 3.2 m, and 3.3 m. Given that the roadway is to be 5 m wide and that the cutting has 45° side slopes, calculate:
(a) The plan surface area of the excavation, ABCD.
(b) The actual area of the side slopes, ABE and CDF.

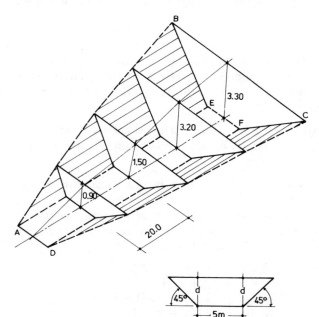

Figure 15.5

Solution
(a) *Area ABCD*

Chainage	Depth (d)	Top width [= (5.0 m + 2d)]
1 0	0.0	5.0
2 20	0.9	6.8
3 40	1.5	8.0
4 60	3.2	11.4
5 80	3.3	11.6

Using Simpson's rule:

$$\text{Area} = \frac{20}{3}[5.0 + 11.6 + 2(8.0) + 4(6.8 + 11.4)]\text{m}^2$$

$$= \frac{20}{3} \times 105.4$$

$$= \underline{702.7 \text{ m}^2}$$

(b) *Plan areas ABE and CDF*

area ABE + area CDF = area ABCD − area ADEF

$$= 702.7 - (5 \times 80)$$

$$= \underline{302.7 \text{ m}^2}$$

therefore area ABE = area CDF = 151.35 m²

Actual side area = plan area ÷ cos 45°

$$= 151.35 \div 0.7071$$

$$= \underline{214.04 \text{ m}^2}$$

Alternatively, the side slope areas may be computed independently. At each chainage point the side width = depth since the slopes have 45° gradients.
Using Simpson's rule:

$$\text{Side area} = \frac{20}{3}[0.0 + 3.3 + 2(1.5) + 4(0.9 + 3.2)]$$

$$= \frac{20}{3} \times 22.7$$

$$= \underline{151.33 \text{ m}^2} \text{ (plan)}$$

Using Trapezoidal rule:

$$\text{Side area} = 20\left[\frac{0.0 + 3.3}{2} + (0.9 + 1.5 + 3.2)\right]$$

$$= \underline{145.0 \text{ m}^2} \text{ plan}$$

Actual area = 151.33 ÷ cos 45° or 145.0 ÷ cos 45°

$$= \underline{214.02 \text{ m}^2} \qquad \text{or } \underline{205.1 \text{ m}^2}$$

2. Areas from coordinates

The calculation of areas is not made directly from the actual field notes but from the coordinate calculations made from the notes.

Method of double longitudes

The meanings of the terms partial departure, partial latitude, total departure, and total latitude have already been made clear and should be revised at this point.

One further definition is required, namely: The longitude of any line is the total departure of the mid-point of the line. In other words the longitude of any line is the mean of the total departures of the stations at the ends of the line. In Fig.15.5:

longitude of line AB = ½(total dep. A
+ total dep. B)

therefore double longitude AB = (total dep. A
+ total dep. B)

Example 5 Figure 15.6(a) is an example of a co-ordinated figure where the coordinates of the stations are:

	Total coords	
Station	Departure (m)	Latitude (m)
A	30.0	60.0
B	90.0	100.0
C	120.0	20.0

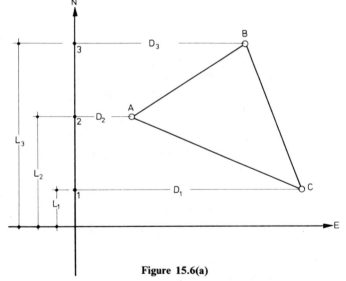

Figure 15.6(a)

Area of triangle ABC =
Area of trapezium 3BC1 − Area of trapezium 3BA2 − Area of trapezium 2AC1, i.e.,

$$\text{triangle ABC} = \frac{(D_3 + D_1)}{2} \times (L_3 - L_1)$$
$$- \frac{(D_3 + D_2)}{2} \times (L_3 - L_2)$$
$$- \frac{(D_2 + D_1)}{2} \times (L_2 - L_1)$$

therefore 2 × triangle ABC
$$= (D_3 + D_1) \times (L_3 - L_1) - (D_3 + D_2) \times (L_3 - L_2)$$
$$- (D_2 + D_1) \times (L_2 - L_1)$$
$$= (210 \times 80) \quad - (120 \times 40) \quad - (150 \times 40)$$
or $\doteq (+210 \times +80) + (+120 \times -40) + (+150 \times -40)$

The values of +210, +120, and +150 are the double longitudes of lines, CB, BA, and AC respectively while the values +80, −40, and −40 are the partial latitudes of the same lines when moving around the figure in an anti-clockwise direction.

Therefore 2 × triangle ABC = 16 800 − 4800 − 6000
= 6000 m²
Triangle ABC   = 3000 m²

In order to find the area of any polygon, the following is the sequence of operations.
(a) Find double longitude and partial latitude of each line.
(b) Multiply double longitude by partial latitude.
(c) Add these products algebraically.
(d) Halve the sum.

The calculations are more neatly set out in tabular form as in Table 15.1 where the area of triangle ABC is calculated.

It should be noted that a negative result is perfectly possible. For example, if the coordinates in Example 2 were written in a clockwise direction the area would have been − 3000 m². In such cases the negative sign is ignored.

Example 6 The following coordinates refer to a closed theodolite traverse ABCDEA.

	Total coordinates			Double	Partial	Double areas (m²)	
Stn	Dep. (m)	Lat. (m)	Line	longitude	latitude	+	−
C	+120.0	+20.0					
B	+90.0	+100.0	CB	+210.0	+80.0	16 800	
A	+30.0	+60.0	BA	+120.0	−40.0		4800
C	+120.0	+20.0	AC	+150.0	−40.0		6000
						16 800	10800
						−10 800	
					÷ 2 =	6 000	
						3 000 m²	

Table 15.1

	Total coordinates			Double	Partial	Double areas (m²)	
Stn	Dep. (m)	Lat. (m)	Line	longitude	latitude	+	−
A	+51.0	−150.2					
B	+300.1	−24.6	AB	+351.1	+125.6	44 098.16	
C	+220.1	+151.3	BC	+520.2	+175.9	91 503.18	
D	−50.0	+175.0	CD	+170.1	+23.7	4 031.37	
E	−125.2	−51.1	DE	−175.2	−226.1	39 612.72	
A	+51.0	−150.2	EA	−74.2	−99.1	7 353.22	

$$\text{Area} = \frac{186\,598.65}{2} = 93\,299.325 \text{ m}^2$$
$$= 9.3299 \text{ hectares}$$

Table 15.2

Station	Total dep (m)	Total lat. (m)
A	+ 51.0	−150.2
B	+300.1	− 24.6
C	+220.1	+151.3
D	− 50.0	+175.0
E	−125.2	− 51.1

Calculate the area in hectares enclosed by the stations.

Solution
See Table 15.2.

Method of total coordinates products
Figure 15.6(b) shows the same triangular area as Fig. 15.6(a) where the coordinates of the stations are:

	Total coordinates	
Station	Departure (m)	Latitude (m)
A	30.0	60.0
B	90.0	100.0
C	120.0	20.0

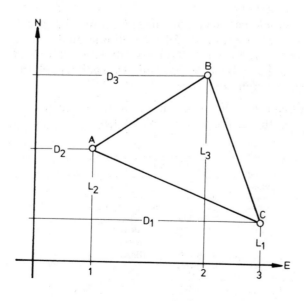

Figure 15.6(b)

Area of triangle ABC = Area of trapezium 1AB2 + Area of trapezium 2BC3 − Area of trapezium 1AC3, i.e.

$$\text{triangle ABC} = \left(\frac{L_2 + L_3}{2}\right) \times (D_3 - D_2)$$

$$+ \left(\frac{L_3 + L_1}{2}\right) \times (D_1 - D_3)$$

$$- \left(\frac{L_2 + L_1}{2}\right) \times (D_1 - D_2)$$

Therefore 2 × area of triangle ABC
$$= (L_2 + L_3) \times (D_3 - D_2) + (L_3 + L_1)$$
$$\times (D_1 - D_3) - (L_2 + L_1) \times (D_1 - D_2)$$
$$= L_2D_3 - L_2D_2 + L_3D_3 - L_3D_2 + L_3D_1$$
$$- L_3D_3 + L_1D_1 - L_1D_3 - L_2D_1 + L_2D_2$$
$$- L_1D_1 + L_1D_2$$

Six of these twelve terms cancel, therefore, when the remaining six terms are rearranged, the double area of the triangle ABC is

$$2ABC = L_1D_2 + L_2D_3 + L_3D_1 - L_1D_3 - L_2D_1 - L_3D_2$$
$$= (D_1L_3 + D_2L_1 + D_3L_2)$$
$$- (D_1L_2 + D_2L_3 + D_3L_1)$$

= sum (departure of station

   × latitude of preceding station)

− sum (departure of station

   × latitude of following station)

The final result is easily remembered by writing the array in the following manner:

$$2 \times \text{area of figure} = \begin{matrix} D_1 & L_1 \\ D_2 & L_2 \\ D_3 & L_3 \\ D_1 & L_1 \end{matrix}$$

The following is the sequence of operations in the calculation of the area of any polygon:
(a) Write the array of departures and latitudes as shown above. NOTE that D_1 and L_1 are repeated at the bottom of the columns.
(b) Multiply the departure of each station by the latitude of the preceding station and find the sum.
(c) Multiply the departure of each station by the latitude of the following station and find the sum.
(d) Find the algebraic difference between (b) and (c) above.
(e) Halve this figure to give the area of the polygon. Again the results are always set out in tabular form as in Table 15.3.

| | Table coordinates | | Areas | |
Stn	Dep. (m) (D)	Lat. (m) (L)	D_2L_1 +	D_1L_2 etc. −
A	+30.0	+60.0		
B	+90.0	+100.0	+5 400	+3 000
C	+120.0	+20.0	+12 000	+1 800
A	+30.0	+60.0	+ 600	+7 200
			+18 000	+12 000
			6 000	
			÷ 2 = 3000 m²	

Table 15.3

Example 7 The coordinates listed below refer to a closed traverse PQRS:

Station	Easting (m)	Northing (m)
P	+35.2	+46.1
Q	+162.9	+151.0
R	+14.9	+218.6
S	−69.2	−25.2

Calculate the area in hecatres enclosed by the stations.

Solution See Table 15.4.

| | Easting | Northing | Area | |
Station	(D)	(L)	D_2L_1(+ve)	D_1L_2(−ve)
P	+35.2	+46.1		
Q	+162.9	+151.0	7 509.69	5 315.20
R	+14.9	+218.6	2 249.90	35 609.94
S	−69.2	−25.2	−15 127.12	−375.48
P	+35.2	+46.1	− 887.04	−3 190.12
			−6 254.57	37 359.54
			−37 359.54	
			−43 614.11	
			÷ 2 = 21 807.06 m²	

Table 15.4

3. Measuring areas from plans

Several methods are available for calculating the area of a figure from a survey plot:
(a) By Simpson's or Trapezoidal rules. The area is divided into a series of equidistant strips, the ordinates are measured and the rules applied.
(b) Graphically—a piece of transparent graph paper is laid over the area, the squares are counted and the area calculated.
(c) Mechanically—by planimeter.

Planimeter

The area of any irregular figure may be found from a plan, by using a mechanical device for measuring areas known as a planimeter.

Two kinds of planimeter are available:
1. A fixed index model.
2. A sliding bar model.

The construction of both models is essentially the same (Fig. 15.7) consisting of:

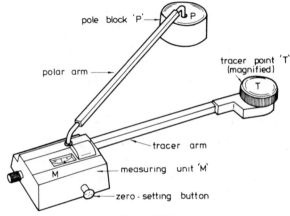

Figure 15.7

1. An arm of fixed length, known as the polar arm. The polar arm rests within the pole block P, which, in turn, rests upon the plan in a stationary position.
2. A tracer arm carrying a tracer point, T, which can be moved in any direction across the plan.
3. Attached to both of these arms is the measuring unit, M, which is, in effect, a rolling wheel. As the tracer point moves, the rolling wheel rotates. The wheel is divided into ten units each of which is sub-divided into ten parts. The drum therefore reads directly to hundredths of a revolution and a vernier reading against the drum allows thousandths of a revolution to be measured. A horizontal counting wheel is directly geared to the rolling wheel and records the number of complete revolutions. The counting wheel can be made to read zero by simply pressing the zero-setting button.

Use of planimeter
1. Fixed index model
If only a small area is to be measured the following procedure is carried out:

(a) Position the pole block outside the area.

(b) Set the tracer point over a well-defined mark such as the intersection of two fences.

(c) Press the zero-setting button. The instrument of course reads 0.000 revolutions.

(d) Move the tracer point carefully around the boundary of the area being measured and return to the starting point.

(e) Note the reading; let it be 3.250 revolutions.

(f) Repeat all of the operations twice more and obtain a mean value of the number of revolutions of the wheel.

This particular model of the Stanley Allbrit planimeter reads the area in square centimetres. Each revolution of the measuring wheel is equivalent to 100 cm² of area.

In the example the area is therefore

$$(100 \times 3.250) \text{ cm}^2$$
$$= 325.0 \text{ cm}^2$$

If the scale of the plan is full size, the actual area measured is 325 cm². If as is likely the scale is much smaller, for example 1:500, the actual area must be obtained by calculation.

$$\text{On } 1:500 \text{ scale } 1 \text{ cm} = 500 \text{ cm}$$

$$\text{therefore } 1 \text{ cm}^2 = (500 \times 500) \text{ cm}^2$$

$$= \left(\frac{500 \times 500}{100 \times 100} \right) \text{ m}^2$$

$$= 25 \text{ m}^2$$

$$\text{therefore } 325 \text{ cm}^2 = (25 \times 325) \text{ m}^2$$

$$= 8125 \text{ m}^2$$

If a large area is to be measured, account must be taken of the instrument's 'zero circle'. Every planimeter has a zero circle. When the polar arm and tracer arm form an angle of, say 90°, and the angle is maintained as the tracer point is moved round in a circle the drum will not revolve and the area of the circle swept out on the plan by the polar arm will be zero. The manufacturer supplies the actual area of the zero circle in the form of a constant which is added to the number of revolutions counted on the drum.

If the enclosure in Fig.15.8 is on a scale of 1:2500 and its area is required, the pole block is placed within the enclosure, and the tracer point is moved around the boundary as before. The average number of revolutions after following the boundary three times is perhaps 5.290.

$$\text{Number of revs} = 5.290$$

$$\text{Add zero circle constant} = 22.300$$

$$\text{Total revs} = 27.590$$

$$\text{Total cm}^2 = 27.59 \times 100$$

$$= 2759 \text{ cm}^2$$

$$\text{On } 1:2500 \text{ scale, } 1 \text{ cm}^2 = \left(\frac{2500 \times 2500}{100 \times 100} \right) \text{ m}^2$$

$$= 625 \text{ m}^2$$

$$\text{Therefore total area of enclosure} = 2759 \times 625$$

$$= 1\,724\,375 \text{ m}^2$$

$$= 172.438 \text{ hectares}$$

Great care must be taken when using the planimeter with the pole block inside the area. It is perfectly possible that the zero circle is larger in area than is the parcel of land being measured (Fig. 15.9). In such a case the rolling drum actually moves backwards and the second reading is subtracted from the first to obtain the area in square centimetres. If the first reading is called 10.000 instead of 0.000 the subtraction is simple; for example:

$$\text{1st reading} = 10.000$$
$$\text{2nd reading} = 7.535$$
$$\text{No. of revs} = 2.465$$

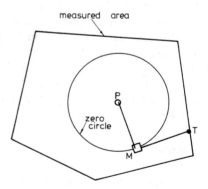

Figure 15.8

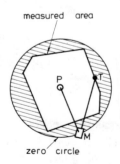

Figure 15.9

This area is, however, the area of the shaded portion of Fig. 15.5 and the true area is found by subtracting the number of revolutions from the zero circle constant.

$$\text{No. of revs} = 22.300 - 2.465$$
$$= 19.835$$
$$= \underline{1983.5 \text{ cm}^2}$$

If the scale of the plan is 1:100:

$$\text{true area} = \underline{1983.5 \text{ m}^2}$$

2. Sliding bar model

On the Allbrit sliding bar planimeter the tracer arm is able to slide through the measuring unit, Fig. 15.10, and can be clamped at any position. The arm is graduated and any setting can be made against the index mark, which is attached to the measuring unit. A table on the base of the instrument gives the graduation reading for various scales in common use.

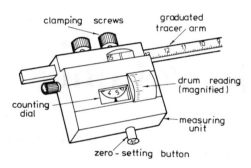

Figure 15.10

The number of revolutions is obtained by the methods previously described and the true area is calculated by multiplying by the appropriate conversion factor supplied by the manufacturer for the scale being used.

For example, when the graduation setting for a scale 1:2500 is 9.92, one revolution of the wheel corresponds to a true area of 4 hectares.

If the pole block was positioned inside the enclosure and the mean readings for four measurements were:

1st reading: 0.000   0.000   0.000   0.000
2nd reading: 4.320   4.320   4.321   4.323

The true area is found as follows:

Mean of four measurements =   4.321
Constant for zero circle =   25.730
$$\overline{ 30.051}$$
True area =   30.051 × 4 hectares
$$= \underline{120.204}$$

Example 8   Figure 15.11 shows an irregular area drawn on a plan to a scale of 1:500.

Calculate the area of the top of the embankment by the following methods:
(a) Counting squares.
(b) Simpson's and Trapezoidal rules.
(c) Planimeter.

Solution

(a) The graph paper shown superimposed on the area has squares of 5 mm side, therefore each

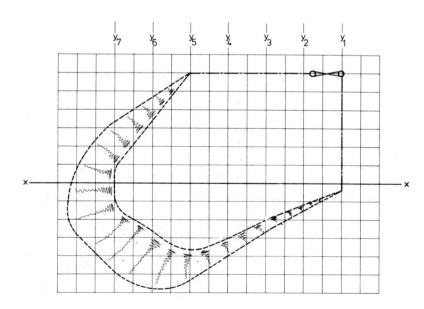

Figure 15.11

square represents a ground area of $(5 \times 500 \times 5 \times 500)$ mm²
$= 25 \times 0.25$ m²
$= 6.25$ m²

$$\text{Area} = (6.25 \times \text{No. of squares}) \text{ m}^2$$
$$= 6.25 \times 89$$
$$= \underline{556.25 \text{ m}^2}$$

(b) Consider the line marked xx as a baseline and every second line of the graph paper as an ordinate y, thereby producing seven in total (y_1 to y_7). The lengths of the respective ordinates are, by scaling 16 m, 18.3 m, 20 m, 22.5 m, 23.8 m, 15.3 m, and 0 m, and the spacing of the ordinates is 5 m along the base line.

By Simpson's rule:

$\text{Area} = \frac{5}{3}[16 + 0 + 2(20 + 23.8) + 4(18.3 + 22.5 + 15.3)]$
$= \underline{546.67 \text{ m}^2}$

By Trapezoidal rule:

$$\text{Area} = 5\left[\frac{(16 + 0)}{2} + 18.3 + 20 + 22.5 + 23.8 + 15.3\right]$$

$$= \underline{539.50 \text{ m}^2}$$

(c) By planimeter (fixed index):

No. of revolutions = 0.2158
At 1:500 1 rev = 2500 m²
therefore area = $\underline{539.5 \text{ m}^2}$

Example 9 Figure 15.12 shows an irregular area of ground lying between a straight kerb and a curved fence drawn to scale 1.500.

Calculate the area by
(a) Simpson's and Trapezoidal rules.
(b) Counting squares.
(c) Planimeter.

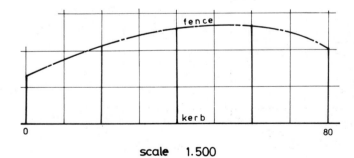

scale 1.500

Figure 15.12

Solution
(a) Using Simpson's rule:

$$\text{Area} = \frac{10}{3}[6.5 + 10.0 + 2(12.8) + 4(10.5 + 13.2)] \text{ m}^2$$

$$= \frac{10}{3}[136.9]$$

$$= \underline{456.3 \text{ m}^2}$$

Using Trapezoidal rule:

$$\text{Area} = 10\left[\frac{6.5 + 10.0}{2} + (10.5 + 12.8 + 13.2)\right]$$

$$= \underline{447.5 \text{ m}^2}$$

(b) By counting squares (Fig. 15.7):

Ground area = 18×25 m² = $\underline{450 \text{ m}^2}$

(c) By planimeter:

$$\text{Area} = 25 \times 18.1$$

$$= \underline{452.5 \text{ m}^2}$$

(b) Volumes

On almost every construction site, some form of cutting or embankment is necessary to accommodate roads, buildings etc.

In general the earthworks fall into one of two categories:
1. Long narrow earthworks of varying depths—roadway cuttings and embankments.
2. Wide flat earthworks—reservoirs, sports pitches, car parks, etc.

1. Cuttings and embankments (with vertical sides)

In Chapter 7, Fig. 7.2, partly reproduced below to a smaller scale as Fig. 15.13, the longitudinal and cross-sectional areas of a proposed roadway and sewer are drawn to scale. The reader should revise that particular section in Chapter 7 in order to appreciate the sources of data used below.

In Fig. 15.13 the sewer track has vertical sides; the depth varies along the length of the trench; the width is constant at 0.8 m and there is no ground slope across the section.

The volume is calculated by either:
(a) Computing the side area of the trench ABCD by Simpson's rule or Trapezoidal rule and multiplying the area by width 0.8 m, *OR*

(b) Computing the cross-sectional area of the trench at each chainage point and entering the values into Simpson's rule to produce the volume directly.

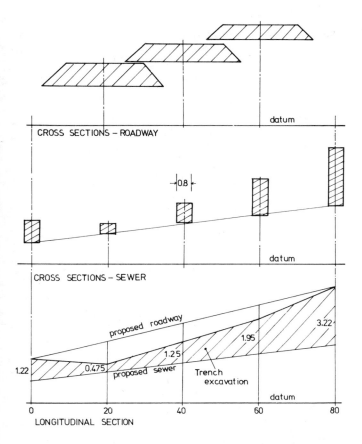

CROSS SECTIONS – ROADWAY

CROSS SECTIONS – SEWER

datum

datum

0.8

datum

proposed roadway

3.22

1.95

1.25

1.22

0.475

proposed sewer

Trench excavation

0   20   40   60   80

LONGITUDINAL SECTION

Figure 15.13

Example 10 In Fig. 15.13 the depths of the trench at 20 m intervals are 1.220 m, 0.475 m, 1.250 m, 1.950 m, and 3.220 m. Given that the trench is 0.8 m wide and has vertical sides, calculate the volume of material to be removed in the excavation.

Solution

(a) Area of side of trench = $\dfrac{20}{3}$[1.220 + 3.220 +

2(1.250) + 4(0.475 + 1.950)] m²

$= \dfrac{20}{3}$ (16.64) m²

$= \underline{110.93 \text{ m}^2}$

Volume of trench = 110.93 × 0.8

$= \underline{88.74 \text{ m}^3}$

(b) Alternatively, the cross-sectional areas at each chainage point are calculated (Area = depth × width).

Chainage	Depth (d)	Area (d × 0.8)
0	1.220	0.976
20	0.475	0.380
40	1.250	1.000
60	1.950	1.560
80	3.220	2.576

Volume = $\dfrac{20}{3}$ [0.976 + 2.576 + 2(1.000) +

4(0.380 + 1.560)] m³

$= \dfrac{20}{3}$ (13.312)

$= \underline{88.75 \text{ m}^3}$

Example 11 The ground levels at 30 m intervals along the centre line of a proposed trench are as follows:

Chainage (m)   0   30   60   90   120
Ground level (m) 36.20 35.50 35.60 35.75 34.20

The trench falls at gradient 1 in 40 from an invert level of 35.00 m at chainage 0 m.
Calculate:
(a) The invert level of the trench at each chainage point.
(b) The depth of excavation at the chainage points.
(c) The volume of material to be excavated given that the trench is 1.2 m wide and has vertical sides.

Solution

Chainage	0	30	60	90	120
(a) Invert level	35.00	34.25	33.50	32.75	32.00
(b) Depth		1.20	1.25	2.10	3.00	2.20

(c) Volume of excavation = $\underline{295.2 \text{ m}^3}$

2. Cuttings and embankments (with sloping sides)

In this category the field work connected with the determination of the volume generally consists of obtaining cross-sections at intervals along a centre line. The areas of the cross-sections are calculated and by using either Simpson's volume rule or the prismoidal formula the volume of material to be removed or deposited can be calculated.

Calculation of cross-sectional areas

In Fig. 15.13, partly reproduced from Fig. 7.2, the longitudinal and cross-sectional areas of a proposed roadway are shown. At each chainage point, only the centre line level was known. Each cross-section was produced by assuming that the ground across the section was horizontal. By adding the roadway formation level and side slopes, a trapezoidal cross-section was produced.

There are three distinct types of cross-section:
(a) A one-level section produced as above.
(b) A three-level section produced when the ground across the section is not horizontal.
(c) A cross fall section produced when the ground across the section has a regular gradient.

In all earthworks where the side slopes of the cross-sections are not vertical, the volume of excavation must be calculated by computing the area of the cross-sections and entering them into Simpson's rule or the prismoidal rule.

(a) *One-level section* In Fig. 15.14 the cross-section is trapezium shaped. Dimensions w and c are known as explained above, and D is the only unknown. Since the sides slope at 1 to s:

$$D = cs + w + cs$$
$$= 2cs + w$$

$$\text{Area of trapezium} = \left(\frac{D + w}{2}\right) \times c$$

$$= \left(\frac{2cs + w + w}{2}\right) \times c$$

$$= (cs + w)c$$

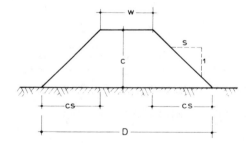

Figure 15.14

Example 12 The reduced ground level and formation level of an embankment at 0 m, 30 m, and 60 m chainages are shown below:

Chainage (m)	0	30	60
RL (m)	35.10	36.20	35.80
FL (m)	38.20	38.40	38.60

Given that the formation width of the top of the embankment is 6.00 m; that the transverse ground slope is horizontal, and that the embankment sides slope at 1 unit vertically to 2 units horizontally, calculate the cross-sectional areas at the various chainages.

Solution (Fig.15.14) Chainage 0 m

$$c = 38.20 - 35.10$$
$$= 3.10 \text{ m}$$
$$w = 6.00 \text{ m (given)}$$
$$D = 2cs + w$$
$$= (2 \times 3.10 \times 2) + 6.00$$
$$= 12.40 + 6.00$$
$$= 18.40 \text{ m}$$
$$\text{Area} = \left(\frac{D + w}{2}\right) \times c$$
$$= \left(\frac{18.40 + 6.00}{2}\right) \times 3.10$$
$$= 12.2 \times 3.10$$
$$= 37.82 \text{ m}^2$$

Alternatively, area $= (cs + w) \times c$

$$= (3.10 \times 2 + 6.00) \times 3.10$$
$$= 12.20 \times 3.10$$
$$= 37.82 \text{ m}^2$$

When several cross-sectional areas have to be calculated a tabular solution is best:

Chainage	RL	FL	c	cs	$(cs + w)$	$(cs + w)c$
0	35.10	38.20	3.10	6.20	12.20	37.82 m²
30	36.20	38.40	2.20	4.40	10.40	22.88 m²
60	35.80	38.60	2.80	5.60	11.60	32.48 m²

Example 13 In Fig. 15.5 the depths of cutting required to form the cutting at 20 m intervals are 0.00 m, 0.90 m, 1.50 m, 3.20 m, and 3.30 m. Given that the roadway is 5.0 m wide and the sides slope at gradient 1 in 1, calculate the cross-sectional areas of the cutting.

Solution

Chainage	Depth (c)	$(cs + w)$	$(cs + w)c$
0	0.00	5.00	0.00
20	0.90	5.90	5.31
40	1.50	6.50	9.75
60	3.20	8.20	26.24
80	3.30	8.30	27.39

In practice there is usually a large number of cross-sections in any earthworks. The volume is most easily calculated by using a simple computer program or a spreadsheet.

In Chapter 11, programs were used to calculate coordinates and the basic concept of looping was introduced. A loop is a repetitive calculation and since every cross-section of an earthwork is a repeat of any other, the computer is ideal for solving earthworks problems.

The simplest program involving looping, for any number of cross-sections, is given with an explanation in Chapter 22. The program is used to compute Example 12.

(b) *Three-level section* Figure 15.15, reproduced from Fig. 8.26 (page 326), shows a cross-section of an embankment drawn from contour line values. By adding the roadway formation level and side slopes a three-level section was produced. The three levels at the base of the side slopes and on the centre line are clearly shown.

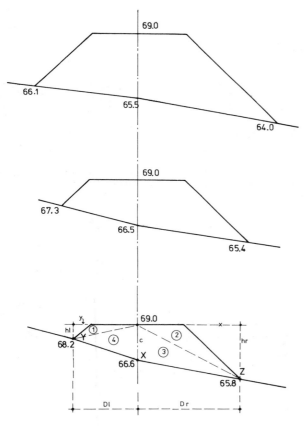

Figure 15.15

Since the figure is not a regular geometrical figure, it has to be split into triangles where the base and altitude can be readily calculated.

The differences between formation level and reduced levels X, Y, and Z produce the dimensions c, hl, and hr.

Since the sides slope at 1 to s.
$$x = (hr \times s) \text{ and } y = (hl \times s)$$
and
$$Dr = x + w/2 = (hr \times s) + w/2$$
$$Dl = y + w/2 = (hl \times s) + w/2$$

In all four triangles, area = ½ base × altitude

Area triangle 1 = $\frac{1}{2}(w/2 \times hl) = (w/4 \times hl)$

Area triangle 2 = $\frac{1}{2}(w/2 \times hr) = (w/4 \times hr)$

Area triangle 3 = $\frac{1}{2}(c \times Dr) = c/2 \times Dr$

Area triangle 4 = $\frac{1}{2}(c \times Dl) = c/2 \times Dl$

Therefore total area

$$= (w/4 \times hl) + (w/4 \times hr) + (c/2 \times Dr) + (c/2 \times Dl)$$
$$= w/4(hl + hr) + c/2(Dr + Dl)$$

Example 14 In Fig. 15.15, the width of the roadway is 5.0 m and the sides slope at gradient 1 to 1. Calculate the bottom cross-sectional area of the embankment.

Solution

Cross-section (bottom):

At Y, $hl = 69.0 - 68.2 = 0.80$ m

At X, $c = 69.0 - 66.6 = 2.40$ m

At Z, $hr = 69.0 - 65.8 = 3.20$ m

Since the sides slope at 1 to 1 and $w/2 = 2.50$ m:

$$Dl = (0.80 + 2.50) = 3.30 \text{ m}$$

$$Dr = (3.20 + 2.50) = 5.70 \text{ m}$$

Area of triangle 1 = $(w/4 \times hl) = 1.25 \times 0.80 = $ 1.000
Area of triangle 2 = $(w/4 \times hr) = 1.25 \times 3.20 = $ 4.000
Area of triangle 3 = $\frac{1}{2}(c \times Dr) = 1.20 \times 3.30 = $ 3.960
Area of triangle 4 = $\frac{1}{2}(c \times Dl) = 1.20 \times 5.70 = $ 6.840
Total area = 15.800 m²

Alternatively, total area

$$= w/4(hl + hr) + c/2(Dr + Dl)$$
$$= 1.25(0.80 + 3.20) + 1.20(3.30 + 5.70) \text{ m}^2$$
$$= 15.80 \text{ m}^2$$

Example 15 In Fig. 15.15 calculate the areas of the middle and top cross-sections given that the roadway is 5.0 m wide and the sides slope at gradient 1 to 1.

Solution

Cross-section (middle)

$hl = 1.7$	$Dl = 4.2$	Area = 19.50 m²
$c = 2.5$	$Dr = 6.1$	
$hr = 3.6$		

Cross-section (top)

$hl = 2.9$	$Dl = 5.4$	Area = 32.45 m²
$c = 3.5$	$Dr = 7.5$	
$hr = 5.0$		

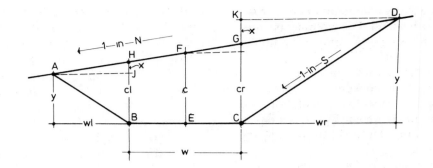

Figure 15.16

(c) *Cross fall section* In Fig. 15.16 a roadway cutting ABCD is to be formed in ground which has a regular gradient across the longitudinal line of the roadway. From the fieldwork, the reduced ground level F on the centre line is known, together with the formation level E, side slope values 1 in *S*, and transverse ground slope 1 in *N*.

Since the figure is not a geometrically regular figure, it must be split into figures whose areas are easily calculated. Perpendicular lines are drawn through B and C to split the figure into two triangles AHB and GDC, and one trapezium BCGH. Horizontal lines are also drawn through A and D to meet the perpendiculars at J and K respectively.

The central height *c* is the difference between reduced level F and formation level E. Thus *w* and *c* are known, and the area of the trapezium BCGH = $(c \times w)$m².

The areas of the triangles AHB and GDC are of course ½(base × altitude)

therefore   area $\triangle$AHB = $\frac{1}{2}(cl \times wl)$
and   area $\triangle$GDC = $\frac{1}{2}(cr \times wr)$

In Fig.15.16 the following information is known:
(i)   formation width = 5 m
(ii)   central height $c = 4$ m
(iii)   side slopes are 1 to 2
(iv)   transverse ground slope is 1 to 5.

Since the ground slopes at 1 to 5, $cr = 1/5$ of $w/2$ greater than *c*, while *cl* is the same value less than *c*. Therefore

$cr = c + (1/5 \times 2.5)$ and $cl = c - (1/5 \times 2.5)$
    $= 4.0 + 0.5$ $= 4.0 - 0.5$
    $= 4.5$ m $= 3.5$ m

In $\triangle$ABJ $\frac{y}{wl} = \frac{1}{2}$, therefore $y = wl/2$ and in $\triangle$AHJ

$\frac{x}{wl} = \frac{1}{5}$ therefore $x = wl/5$

Now $cl = (y + x) = \frac{wl}{2} + \frac{wl}{5}$

$= \frac{5wl + 2wl}{5 \times 2}$

$= \frac{wl(5 + 2)}{5 \times 2}$

Substitution of the general *S* for 2 and *N* for 5 shows

that $cl = wl \frac{N + S}{NS}$

and $wl = cl \frac{NS}{N + S}$, $\therefore wl = 3.5 \times \frac{(5 \times 2)}{(5 + 2)} = 5.00$ m.

Similarly in $\triangle$KDC, $\frac{y}{wr} = \frac{1}{2}$ therefore $y = wr/2$

and in $\triangle$GDK $\frac{x}{wr} = \frac{1}{5}$ therefore $x = \frac{wr}{5}$.

Now $cr = (y - x) = \frac{wr}{2} - \frac{wr}{5}$

$= \frac{5wr - 2wr}{5 \times 2}$

$= \frac{wr(5 - 2)}{5 \times 2}$

Again substitution of *S* for 2 and *N* for 5 shows that

$$cr = \frac{wr \times (N - S)}{N \times S}$$

and

$wr = cr \times \frac{(NS)}{(N - S)}$, therefore $wr = 4.5 \times \frac{(5 \times 2)}{5 - 2} = 15.0$ m

A general formula can be derived showing that

horizontal length = vertical length $\times \frac{(NS)}{(N \pm S)}$

The positive sign applies when the ground slope and side slope run in opposite directions, i.e., one up and one down; while the negative sign applies when both gradients are in the same direction, i.e., either both up or both down.

The areas of the three component figures in Fig. 15.11 are

(i) trapezium HGCB
   area $= (c \times w)$
       $= 4 \times 5$
       $= 20 \text{ m}^2$

(ii) triangle GDC
   $= cr/2 \times wr$
   $= \frac{1}{2}(4.5 \times 15.0)$
   $= 33.75 \text{ m}^2$

(iii) triangle ABH
   area $= cl/2 \times wl$
       $= \frac{1}{2}(3.5 \times 5)$
       $= 8.75 \text{ m}^2$

Total cross-sectional area: 62.5 m².

Example 16 In Fig. 15.17, a roadway is to be built on ground having a transverse ground slope of 1 in 8. The road is 8.0 m wide, has a central height of 3.5 m, and 1 to 4 side slopes. Calculate the cross-sectional area of the embankment.

Solution
$c = 3.5 \text{ m}$, $w = 8.0 \text{ m}$ (given) therefore $w/2 = 4.0 \text{ m}$
$cr = c + (\frac{1}{8} \times 4.0)$ and $cl = c - (\frac{1}{8} \times 4.0)$
   $= 4.0 \text{ m}$ $= 3.0 \text{ m}$

$wr = \dfrac{4.0 \times NS}{N - S}$ $wl = \dfrac{3.0 \times NS}{N + S}$

$= \dfrac{4.0 \times 32}{4}$ $= \dfrac{3.0 \times 32}{12}$

$= 32.0$ $= 8.0 \text{ m}$

Area HGCB $= c \times w$
       3.5×8.0
       $= 28 \text{ m}^2$

Area ABH $= \dfrac{cr}{2} \times wr$
       $= 2 \times 32.0$
       $= 64.0 \text{ m}$

Area CGD $= \dfrac{cl}{2} \times wl$
       $= 1.5 \times 8.0$
       $= 12.0 \text{ m}^2$

Total area $= 104.0 \text{ m}^2$

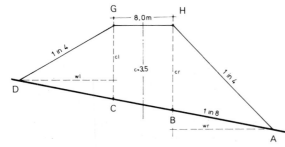

Figure 15.17

Example 17 Figure 15.18 shows a side hill section where the ground slope gradient is known. Given the following information, calculate the area of cutting and area of filling:
(a) Formation width $w = 15$ m.
(b) Central height $c = 0.5$ m.

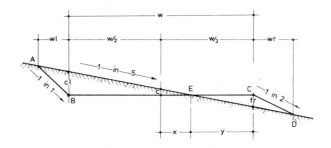

Figure 15.18

(c) Side slopes

   Cutting = 1 in 1
   Embankment = 1 in 2

(d) Ground slope 1 in N = 1 in 5.

   Cutting $cl = c + (\frac{1}{5}$ of $w/2)$ m as before
       $= 0.5 + (\frac{1}{5} \times 7.5)$ m
       $= 2.0 \text{ m}$

Point of no cutting or filling occurs at x metres from the centre line of the formation:

   $x = 5 \times c$
     $= 2.5$ m therefore $y = 5.0$ m

Filling 'fr' $= \frac{1}{5}$ of y
       $= 1.0 \text{ m}$

By principle of converging gradients:

$wl = \dfrac{cl \times (5 \times 1)}{(5 - 1)}$ and $wr = \dfrac{fr \times (5 \times 2)}{(5 - 2)}$

$= \dfrac{2.0 \times 5}{4}$ $= \dfrac{1.0 \times 10}{3}$

$= 2.5 \text{ m}$ $= 3.33 \text{ m}$

   Slope of cutting = 1 in 1
       therefore $hl = 2.5$ m
   Slope of embankment = 1 in 2
       therefore $hr = 1.67$ m

Area of cutting = triangle ABE

$$= \tfrac{1}{2}\left(\frac{w}{2} + x\right) \times hl$$

$$= \tfrac{1}{2} \times 10.0 \times 2.5$$

$$= \underline{12.5 \text{ m}^2}$$

Area of filling = triangle ECD
$$= \tfrac{1}{2}y \times hr$$
$$= \tfrac{1}{2} \times 5.0 \times 1.67$$
$$= \underline{4.17 \text{ m}^2}$$

Calculation of volume (by Simpson's rule)

Once the various cross-sections have been calculated the volume of material contained in the embankment is calculated by Simpson's Volume rule. The rule is the same as the rule for area except that cross-sectional areas are substituted for ordinates in the formula (Fig. 15.19).

$$\text{Volume} = \frac{d}{3}\left[A_1 + A_5 + 2 \times A_3 + 4 \times (A_2 + A_4)\right] \text{ m}^2$$

Example 18 In Fig. 15.19 levels were taken along the centre line xy of a proposed level roadway at 100 m intervals and the central heights of the embankment found to be as follows:

Chainage (m)	0	100	200	300	400
Central height c (m)	2.2	2.3	4.4	1.3	0.9

The roadway is to be 6 m wide and the embankment is to have side slopes of 1 in 2. The volume is calculated thus:

Width of embankment base at any chainage $= (6 + 2c + 2c)$ metres $= (6 + 4c)$ metres

therefore cross-sectional area at any chainage (trapezium) $= \dfrac{(6 + 4c) + 6}{2} \times c$

$$= (6 + 2c)c \text{ m}^2$$

Area at
$$0 \text{ m} = (6 + 2 \times 2.2) \times 2.2 = 22.88 \text{ m}^2$$
$$100 \text{ m} = (6 + 2 \times 2.3) \times 2.3 = 24.38 \text{ m}^2$$
$$200 \text{ m} = (6 + 2 \times 4.4) \times 4.4 = 65.12 \text{ m}^2$$
$$300 \text{ m} = (6 + 2 \times 1.3) \times 1.3 = 11.18 \text{ m}^2$$
$$400 \text{ m} = (6 + 2 \times 0.9) \times 0.9 = 7.02 \text{ m}^2$$

$$\text{Volume} = \frac{100}{3}\left[22.88 + 7.02 + (2 \times 65.12)\right.$$

$$\left. + 4(24.38 + 11.18)\right]$$

$$= \underline{10\,079.3 \text{ m}^3}$$

The rule applies only when there is an ODD number of cross-sections. Should there be an EVEN number, Simpson's rule is used to calculate the nearest odd number of sections and the Prismoidal rule applied to the remainder of the earthworks.

Calculation of volume (by Prismoidal rule)

A prismoid is defined as any solid having two plane parallel faces, regular or irregular in shape, which can be joined by surfaces either plane or curved on which a straight line may be drawn from one of the parallel ends to the other. Examples of prismoids are shown in Fig. 15.20.

Consider Fig. 15.20(a). In order to determine the volume by Simpson's rule, it is necessary to split the

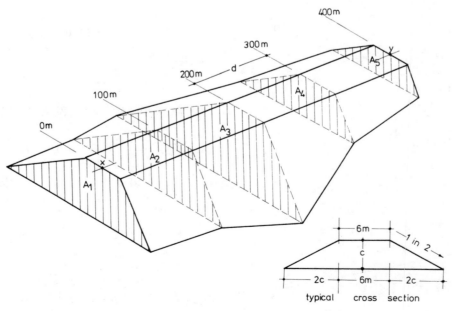

Figure 15.19

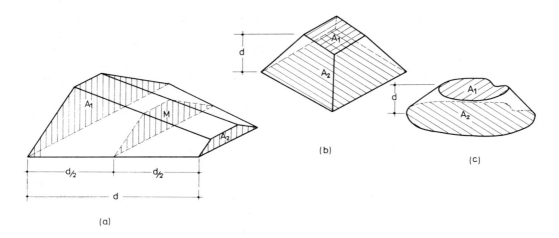

Figure 15.20

figure such that there is an odd number of equi-distant cross-sections. Three is the minimum number which fulfils this condition.

Calling the mid-section M, the volume by Simpson's rule, is

$$\text{Vol.} = (\tfrac{1}{3} \times d/2) \, [A_1 + A_2 + 2(\text{zero}) + 4M]$$
$$= (d/6) \, [A_1 + A_2 + 4M]$$

This is Simpson's Prismoidal rule which can be used to find the volume of any prismoid, provided the area M of the central section is determined. NOTE that the area of M is NOT the mean of areas A_1 and A_2.

Example 19 Figure 15.21 shows a proposed cutting where the following information is known:

(a) Length of cutting   30 m
(b) Formation width   8 m
(c) Depth at commencement   8 m
(d) Depth at end   5 m
(e) Side slopes   1 in 1

Using the prismoidal formula, calculate the volume of material to be removed.

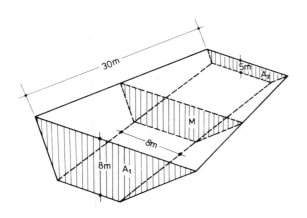

Figure 15.21

Solution
(a) Section A_1

$$\begin{aligned}\text{Formation width} &= 8 \text{ m} \\ \text{Top width} &= (8 + 2c) \text{ m} \\ \text{Central depth } c &= 8 \text{ m} \\ \text{Therefore top width} &= (8 + 16) \text{ m} \\ &= \underline{24 \text{ m}}\end{aligned}$$

(b) Section A_2

$$\begin{aligned}\text{Formation width} &= 8 \text{ m} \\ \text{Top width} &= (8 + 2c) \text{ m} \\ \text{Central depth } c &= 5 \text{ m} \\ \text{Therefore top width} &= (8 + 10) \text{ m} \\ &= \underline{18 \text{ m}}\end{aligned}$$

(c) Section M

$$\begin{aligned}\text{Formation width} &= 8 \text{ m} \\ \text{Top width} &= (8 + 2c) \text{ m} \\ \text{Central depth } c &= \text{Average of depths of } A_1 \\ &\quad \text{and } A_2 \\ &= \tfrac{1}{2}(8 + 5) \text{ m} \\ &= 6.5 \text{ m} \\ \text{Therefore top width} &= 8 + 13 \text{ m} \\ &= 21 \text{ m} = \text{Average of widths of} \\ &\qquad A_1 \text{ and } A_2\end{aligned}$$

(d) Cross-sectional areas (trapezia)

$$\begin{aligned} A_1 &= \tfrac{1}{2}(8 + 24) \times 8 &= 128 \text{ m}^2 \\ A_2 &= \tfrac{1}{2}(8 + 18) \times 5 &= 65 \text{ m}^2 \\ M &= \tfrac{1}{2}(8 + 21) \times 6.5 &= 94.25 \text{ m}^2 \end{aligned}$$

(e) Volume $= 30/6[128 + 65 + (4 \times 94.25)] \text{ m}_3$
$$= \underline{2850 \text{ m}^3}$$

Example 20 A proposed service roadway, 5.5 m wide, is to be built along a centre line XY. The embankment is to have side slopes of 1 to 2. Given the following data, calculate the volume of material required to construct the embankment.

Chainage (m)	40	60	80	100	120	140
Centre depth (c)	1.00	1.50	1.70	0.90	4.10	1.10

Solution

Chainage (m)	Depth (m)	Width (m)	Area (m²)	
	c	w	(w + 2c)	(w + 2c)c
5 sections (40 m–120 m)				
40	1.00	5.50	7.50	7.50
60	1.50	5.50	8.50	12.75
80	1.70	5.50	8.90	15.13
100	0.90	5.50	7.30	6.57
120	4.10	5.50	13.70	56.17
2 sections (120 m–140 m)				
120	4.10	5.50	13.70	56.17
130 (mid)	2.60	5.50	10.70	27.82
140	1.10	5.50	7.70	8.47

Volume: Sections (40 m–120 m) by Simpson's rule.

$$\text{Vol.} = \frac{20}{3} \, [7.50 + 56.17 + 2(15.13) + 4(12.75 + 6.57)]$$

$$= \frac{20}{3} \, [63.67 + 30.26 + 77.28]$$

$$= \underline{1141.40 \text{ m}^3}$$

Sections (120 m–140 m) by Prismoidal rules.

$$\text{Vol.} = \frac{20}{6} \, [56.17 + 4(27.82) + 8.47]$$

$$= \underline{586.4 \text{ m}^3}$$

$$\text{Total volume} = \underline{1727.8 \text{ m}^3}$$

3. Volumes of large-scale earthworks

Whenever the volumes of large-scale earthworks have to be determined, e.g., the formation of sports fields, reservoirs, large factory buildings, the field work consists of covering the area by a network of squares and obtaining the reduced levels. The volume is then determined either from the grid levels themselves or from the contours plotted therefrom. The work is described in detail in Chapter 8.

(a) Volumes from spot levels

Figure 15.22 shows a small section of a grid. The total area is to be excavated to a formation level of 90.000 m to form a car park. The sides of the excavation are to be vertical.

The solid formed by each grid square is a vertical truncated prism, that is, a prism where the end faces are not parallel.

Volume of each prism = mean height × area of base

Mean height of each truncated prism above 90.00 m level is

prism 1 = (1.0 + 3.0 + 2.0 + 2.0) ÷ 4 = 2.0 m
2 = (3.0 + 4.0 + 3.0 + 2.0) ÷ 4 = 3.0 m
3 = (2.0 + 3.0 + 2.0 + 1.0) ÷ 4 = 2.0 m
4 = (2.0 + 2.0 + 1.0 + 3.0) ÷ 4 = 2.0 m

Area of base of each truncated prism = 10 × 10
= 100 m²

Therefore

Volume of 1 = 100 × 2.0 = 200 m³
2 = 100 × 3.0 = 300 m³
3 = 100 × 2.0 = 200 m³
4 = 100 × 2.0 = 200 m³

Total volume of excavation = 900 m³

Alternatively the volume can be found thus:

Volume = mean height of excavation × total area

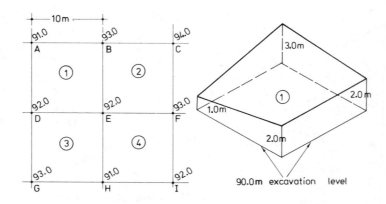

Figure 15.22

The mean height of the excavation is the mean of the mean heights of the truncated prisms. It is *not* the mean of the spot levels.

Mean height excavation = (2.0 + 3.0 + 2.0
      + 2.0) ÷ 4 = 2.25 m
Total area of site = 20 × 20 = 400 m²
Therefore Total volume = 2.25 × 400 = 900 m³

When examined closely, it is seen that spot level A is used only once in obtaining the mean height of the excavation; spot level B twice, while spot level E is used four times in all.

This mean height, hence the volume, can be readily found by tabular solution as in Table 15.5.

Grid station	Height above formation level	No. of times used	Product
A	1.0	1	1.0
B	3.0	2	6.0
C	4.0	1	4.0
D	2.0	2	4.0
E	2.0	4	8.0
F	3.0	2	6.0
G	3.0	1	3.0
H	1.0	2	2.0
I	2.0	1	2.0
		Sum 16	Sum 36.0

Mean height of excavation = 36.0/16 m
      = 2.25 m as before

Table 15.5

The various spot height are tabulated in column 2 and the number of times they are used, in column 3. Column 4 is the product of columns 2 and 3. The mean height of the excavation is found by dividing the sum of column 4 by the sum of column 3.

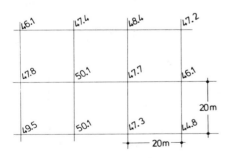

Figure 15.23

Example 21 Figure 15.23 shows spot levels at 20-metre intervals over a site which is to be excavated to 47.00 m to accommodate three tennis courts.

Calculate the volume of material to be removed assuming that the excavation has vertical sides.

Height above formation level	No. of times used	Product
−0.9	1	−0.9
0.4	2	0.8
1.4	2	2.8
0.2	1	0.2
0.8	2	1.6
3.1	4	12.4
0.7	4	2.8
−0.9	2	−1.8
2.5	1	2.5
3.1	2	6.2
0.3	2	0.6
−2.2	1	−2.2
	24	25.0

Mean height of excavation $=\left(\dfrac{25}{24}\right)$ m = 1.042 m

Total Area of site = 60 × 40
      = 2400 m²

Therefore Volume of excavation = 1.042 × 2400
      = 2500 m³

(b) Volumes from contours

Figure 15.24 shows a mound which has been contoured.

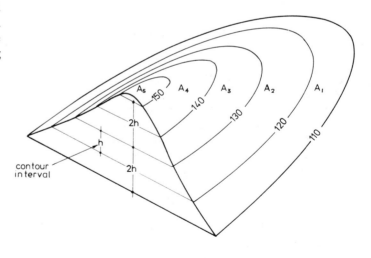

Figure 15.24

If the mound is to be removed the volume of material can be calculated by considering the solid to be split along the contours into a series of prismoids. The volume can then be calculated by successive applications of the prismoidal rule or, where circumstances are favourable, by direct application of Simpson's rule.

When using the prismoidal rule, three contours are taken at a time and the central one is used as the mid-area. The accuracy of the volume depends basically on the contour vertical interval. Generally, the closer the contour interval the more accurate is the volume.

Taking the prismoid formed by contours 110 m and 130 m, the areas enclosed by the contours are determined from the plan by planimeter. The mid-area enclosed by the 120 m contour is likewise determined and the volume of the prismoid is therefore

$$\text{Volume} = \frac{2h}{6}[A_1 + 4A_2 + A_3]$$

Similarly the volume between contours 130 m and 150 m is

$$\text{Volume} = \frac{2h}{6}[A_3 + 4A_4 + A_5]$$

Adding these results gives the volume between the 110 m and 150 m contours:

$$\text{Volume} = \frac{2h}{6}[A_1 + 4A_2 + A_3]$$
$$+ \frac{2h}{6}[A_3 + 4A_4 + A_5]$$
$$= \frac{h}{3}[A_1 + A_5 + 2A_3$$
$$+ 4(A_2 + A_4)]$$

which is the volume by Simpson's rule.

The part of the solid lying above the 150 m contour is not included in the above calculations. It must be approximated to the nearest geometrical solid and calculated separately. In general, the nearest regular solid is a cone or pyramid where the volume = ⅓ base area × height.

Example 22 Figure 15.25 shows ground contours at 1-metre vertical intervals. ABCD is a proposed factory building where the floor level is to be 32.00 m. The volume of material to be excavated is required. The side slopes of any earthworks are 1 in 2.
(a) The earthwork contours are drawn at 1-metre vertical intervals, that is, 2 metres horizontally apart.
(b) The surface intersections are found and the outline of the cutting drawn (broken line).

(c) The area enclosed by each contour is obtained by planimeter. The 32 m contour is bounded by A1CD; the 33 m contour by all points numbered 2, the 34 m contour by points numbered 3, the 35 m contour by points numbered 4, while the 36 m contour (point 5) has no area.

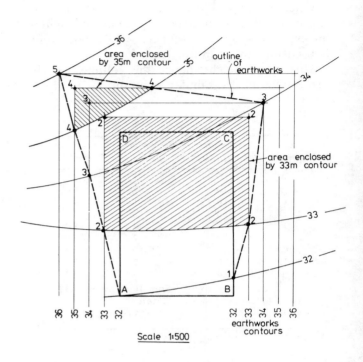

Figure 15.25

(d) The respective areas are:

Contour	32	33	34	35	36
Area (m²)	315	294.5	125.00	30.0	0.0

(e) Volume by Simpson's rule:

$$V = \tfrac{1}{3}[315.0 + 0.0 + (2 \times 125.00) + 4(294.5 + 30.0)] \text{ m}^3$$
$$= \underline{621.0 \text{ m}^3}$$

Example 23 Calculate the volume of water which can be stored in a reservoir, 23.3 m deep, given the following information:

Contour value	130.00 (Reservoir full)	120.00	110.00
Area enclosed by contour (m²)	610 000	150 000	1100

Note: 1 m³ = 1000 litres.

Solution

(a) Volume between 130 m and 110 m contours (prismoidal rule)

$$= \frac{20}{6}[610\,000 + 4(150\,000) + 1100]$$

$$= \underline{4\,037\,000 \text{ m}^3}$$

(b) Depth of reservoir = 23.3 m
Therefore depth below 110.0 m contour = 3.3 m
Therefore volume between 110 m contour and bottom (cone)

$$= \tfrac{1}{3} \times 1100 \times 3.3$$

$$= \underline{1210.0 \text{ m}^3}$$

(c) Total capacity = 4 038 210 m³

$$\underline{4.038 \times 10^9 \text{ litres}}$$

Test questions

1 A small site is in the shape of a quadrilateral, the lengths of the sides being as follows:

AB 325 m AD 195 m DB 410 m DC 392 m CB 260 m

The site is lettered anticlockwise. Calculate the area enclosed by the survey lines, in square metres.

2 The site in question 1 is bounded by fences along the sides AD, DC, and CB and by a roadway along the side AB. The offsets to the fences and roadway are:

Line chainage (m)		Offset (m)	Line chainage (m)		Offset (m)
AB 0	(A)	2.0 right	AD 0	(A)	5.6 left
200		2.5 right	50		6.3 left
325	(B)	1.0 right	100		2.1 left
			150		4.0 left
			195	(D)	0.0
DC 0	(D)	0.0 left	CB 0	(C)	2.0 left
100		10.4 left	100		4.0 left
200		12.6 left	200		4.0 left
300		8.4 left	260	(B)	2.0 left
392	(C)	4.0 left			

Calculate the area of the site in square metres.

3 Stations M, N, O, P, and Q form a closed traverse. The following coordinates refer to the stations:

Station	Total latitude	Total departure
M	+2000	+2000
N	+3327	+1242
O	+4093	+2048
P	+3141	+3035
Q	+1192	+3572

Calculate the area in hectares enclosed by the stations.

4 Figure 15.26 shows an irregular parcel of ground bounded by a kerb on the south side and a fence on the north (scale 1.5000).
   Calculate the area by:
(a) Counting squares
(b) Simpson's rule
(c) Trapezoidal rule
(d) Planimeter.

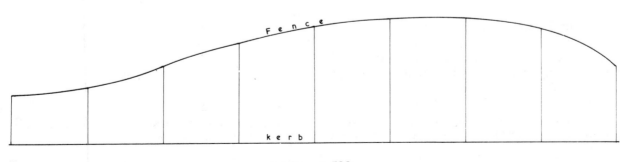

scale   1.500

Figure 15.26

5 Calculate the cross-sectional areas of Figs 15.27(a), (b), (c).

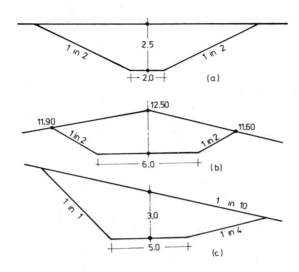

Figure 15.27

6 Calculate the volume of earth required to form the embankment shown in Fig. 15.28.

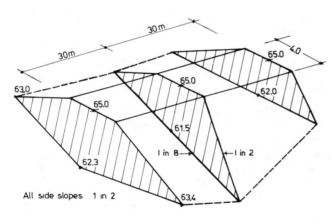

Figure 15.28

7 The central heights of an embankment at two points A and B 90 m apart are 4.0 m and 6.5 m respectively. The embankment is built on ground where the maximum slope is 1 in 10 at right angles to the line of the embankment.

Given that the formation width of the embankment is 6 m and the sides slope at 1 in 2, calculate the volume of material in cubic metres contained between A and B.

8 A sewer 0.75 m wide is to be excavated along a line AB. The sides are to be vertical. Given the following data, calculate the volume of material to be excavated to form the sewer track.

Chainage (m)	0	20	40	60	80	100	120	140
Depth (m)	1.20	1.70	0.95	2.21	2.27	2.21	0.95	1.82

9 Figure 15.29 shows a rectangular grid with levels at 10 m intervals. The whole area is to be covered with waste material to form a car park at formation level 86.5 m. Calculate the volume of material to be deposited.

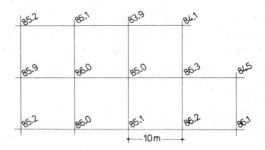

Figure 15.29

10 Figure 15.30 shows contours over an area where it is proposed to erect a small building with a formation level of 23.00 m AOD. Draw on the plan the outline of any earthworks required and thereafter calculate the volume of material required to be cut and filled to accommodate the building.

All earthworks have side slopes of 1 in 1.

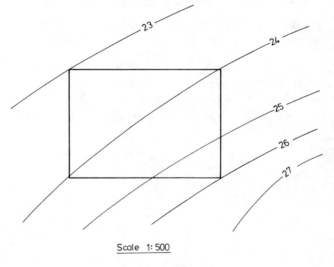

Scale 1:500

Figure 15.30

11 Cross-sections of a proposed roadway cutting have been plotted and the undernoted areas obtained by planimeter.

Chainage (m)	0	20	40	60	80	100	120
Area (m²)	52.20	58.35	70.40	72.40	60.00	40.50	27.30

Calculate the volume of material removed from the cutting.

16. Curve ranging

In construction surveying curves have to be set out on the ground for a variety of purposes. A curve may form the major part of a roadway, it may form a kerb line at a junction or may be the shape of an ornamental rose bed in a town centre. Obviously different techniques would be required in the setting out of the curves mentioned above but in all of them a few geometrical theorems are fundamental and it is wise to begin the study of curves by recalling those theorems.

Curve geometry

In Figure 16.1 A, B, and C are three points on the circumference of a circle.

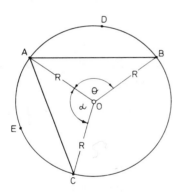

Figure 16.1

1. AB and AC are chords of the circle subtending angles θ and α respectively at the centre O.

ADB and AEC are arcs of the circle. Their lengths are

$$2\pi R\left(\frac{\theta}{360}\right)^{\circ} \text{ and } 2\pi R\left(\frac{\alpha}{360}\right)^{\circ} \text{ respectively.}$$

More conveniently their lengths are $R\theta$ and $R\alpha$ respectively where θ and α are expressed in radians.

2. In Fig. 16.2, lines ABC and ADE are tangents to the circle at B and D respectively. AB = AD and

angles ABO and ADO are right angles.

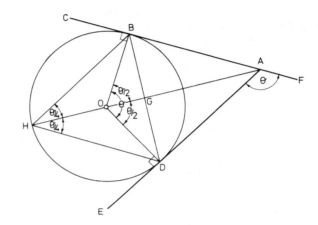

Figure 16.2

3. Since the opposite angles of a cyclic quadrilateral are supplementary, the figure ABOD must be cyclic as angles ABO and ODA together make 180°.
4. The exterior angle of a cyclic quadrilateral equals the interior opposite angle therefore angle FAD = angle BOD = θ.
5. Join OA, the perpendicular bisector of chord BD. Angle OGB is therefore a right angle and angle BOG = θ/2.

Angle ABG + angle GBO = 90°
and angle BOG + angle GBO = 90°
Therefore angle ABG = angle BOG = θ/2

that is, the angle ABG between the tangent AB and chord BD equals half the angle BOD at the centre.

6. Produce AO to the circumference at H and join HB.

Angle BOG is the exterior angle of triangle BOH
Therefore angle BOG = angle OHB + angle OBH

But angles OHB and OBH are equal since triangle BOH is isosceles

Therefore angle OHB = $\frac{1}{2}$ angle BOG

$\qquad$ = θ/4

Similarly angle OHD = θ/4

therefore angle BHD = θ/2

that is, the angle BHD at the circumference subtended by the chord BD equals half the angle BOD at the centre subtended by the same chord.

Also the angle ABD between tangent and chord equals the angle BHD at the circumference.

Curve elements

In Fig. 16.3 the centre lines AI and IB of two straight roadways called simply the straights meet at a point I called the intersection point. The two straights deviate by the angle θ which is called the deviation angle. Alternatively the angle may be called the deflection angle or 'intersection angle'.

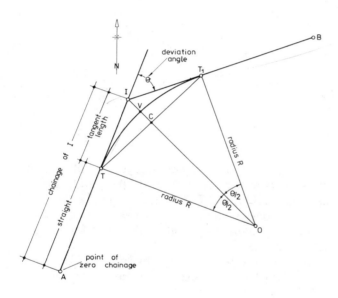

Figure 16.3

Clearly it is desirable to avoid having a junction at I so the straights are joined by a circular curve of radius R.

The straights are tangential to the curve at the tangent points T and T_1 and lengths IT and IT_1 known as the tangent lengths are equal.

Before setting the curve on the ground, the exact location of the tangent points must be known and in order to locate them a theodolite is set at point I and the deviation angle θ is measured together with the lengths of the lines AI and IB. Treating station A as the point of zero chainage, the chainage of the point I is the length AI.

The radius R is usually a multiple of 50 metres and is supplied by the architect or designer.

Knowing only the deviation angle and radius, the tangent lengths and curve length are derived thus.

Angle ITO = Angle OT_1I = 90°

therefore $ITOT_1$ is a cyclic quadrilateral

therefore Angle TOT_1 = θ

Join I to O

Angle TOI = angle IOT_1 = θ/2

(a) *Tangent lengths IT and IT_1*

$\qquad$ In triangle ITO

$$\frac{IT}{R} = \tan θ/2$$

$\qquad$ therefore IT $= R \tan θ/2$

Chainage of T = Chainage of I − IT

(b) *Length of curve TT_1*

Curve $TT_1 = R × θ$ radians

Chainage of T_1 = Chainage T + Curve length
(The chainage of the second tangent point is *always* derived via the curve)

This information enables the beginning and end of the curve to be located.

Example 1 In Fig. 16.3 the bearings and lengths of AI and IB are

| AI | N20°E | 450.30 m |
| IB | N70°E | 275.00 m |

The radius of the curve joining the straights is 300 m. Calculate the chainages of the tangent points.

Solution

(i) Deviation angle θ	= 70° − 20°
	= 50°
(ii) Chainage I	= 450.30 m
(iii) Tangent length IT	= $R \tan θ/2$
	= 300 tan 25°
	= 139.89 m
(iv) Chainage T	= 450.30 − 139.89
	= 310.41 m

(v) Curve length

$$= R \times \theta \text{ radians}$$
$$= 300 \times 0.872\,66$$
$$= 261.80 \text{ m}$$

(vi) Chainage T_1

$$= 310.41 + 261.80$$
$$= 572.21 \text{ m}$$

Other curve elements are frequently required and are calculated from the values of R and θ.

(c) *Long chord TT₁*

The long chord is the straight line joining T and T_1. The line IO is the perpendicular bisector of TT_1 at C.

In triangle TCO

$$\frac{TC}{R} = \sin \theta/2$$

therefore $TC = R \sin \theta/2$

therefore $TT_1 = 2R \sin \theta/2$.

(d) *Major offset CV*

Frequently called the mid-ordinate or versine, the length CV is the greatest offset from the long chord to the curve.

$$CV = R - OC$$

In triangle TCO

$$\frac{CO}{R} = \cos \theta/2$$

therefore $CO = R \cos \theta/2$

therefore $CV = R - R \cos \theta/2$

$$= R(1 - \cos \theta/2)$$

(e) *External distance VI*

The length of VI is the shortest distance from the intersection point to the curve

$$VI = IO - R$$

In triangle ITO

$$\frac{IO}{R} = \sec \theta/2$$

therefore $IO = R \sec \theta/2$

therefore $VI = R \sec \theta/2 - R$

$$= R(\sec \theta/2 - 1)$$

Example 2 Two straights AI and IB deviate to the left by 80° 36′. They are to be joined by a circular curve such that the shortest distance between the curve and intersection point is 25.3 m (Fig. 16.4).

Calculate (i) the radius of the curve
            (ii) the lengths of the long chord and major offset.

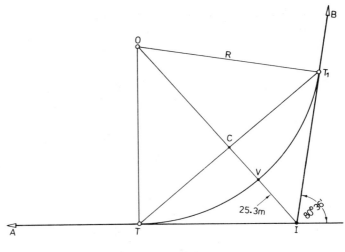

Figure 16.4

Solution

(i)
$$VI = R(\sec \theta/2 - 1)$$

i.e., $25.3 = R(\sec 40° 18′ - 1)$

$$= R(1.311\,186 - 1)$$

therefore $R = \dfrac{25.3}{0.311\,186} \text{ m}$

$$= 81.30 \text{ m}$$

(ii) Long chord TT_1
$$= 2R \sin \theta/2$$
$$= 162.60 \sin 40° 18′$$
$$= 162.60 \times 0.646\,790$$
$$= 105.17 \text{ m}$$

(iii) Major offset VC
$$= R(1 - \cos \theta/2)$$
$$= 81.30(1 - 0.762\,668)$$
$$= 81.30 \times 0.237\,332$$
$$= 19.30 \text{ m}$$

Designation of curves

In the UK, curves are designated by the length of the radius. Since there is generally some scope in the choice, the radius is usually a multiple of 50 metres.

The curve can also be designated by the degree of curvature which is defined as the number of degrees subtended at the centre by an arc 100 m long. The degree of curvature is given as a number of whole degrees. In Fig. 16.5 the angle $\theta = 5°$, that is, the degree of curvature is 5°. The relationship between radius and degree of curvature is as follows:

Arc length $AB = R \times \theta$ radians

$$= R \times \theta \times \frac{\pi}{180} \ (\theta \text{ in degrees})$$

$$\text{therefore } R = \frac{100 \times 180}{\theta \times \pi} \text{ m}$$

$$= \frac{5729 \cdot 8}{\theta} \text{ m}$$

$$= \underline{1145 \cdot 96 \text{ m}}$$

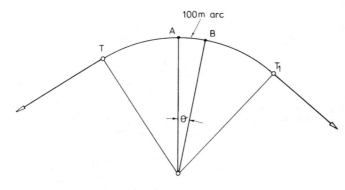

Figure 16.5

Problems in curve location

1. Inaccessible intersection point

It happens frequently on site that the intersection point cannot be occupied because of some obstacle. In Fig. 16.6 the intersection point I is in a built up area.

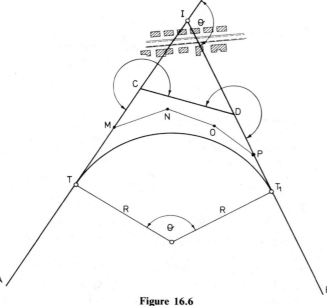

Figure 16.6

The relative position of the straights AI and IB must be obtained by traversing between them. The best conditions for surface taping determine the route of the traverse. In Fig. 16.6 the route AMNOP might provide the best conditions in which the traverse coordinates would be calculated and the position of station I deduced from a solution of triangle IMP.

The simplest traverse between the straights is one straight line between two stations C and D. Angles ACD and CDB are measured together with distances AC and CD. The chainage of C is therefore known.

Example 3 In Fig. 16.6 the following data were derived from traverse ACDB:

Angle ACD 252° 15′ 00″	AC = 559.28 m
Angle CDB 227° 25′ 00″	CD = 256.50 m

Calculate the chainage of the tangent point T if the straights are to be joined by a 300 m radius curve.

Solution
In triangle ICD:

$$\text{Angle C} = 72° \ 15′ \ 00″$$
$$\text{Angle D} = 47° \ 25′ \ 00″$$
$$\text{therefore Angle I} = 60° \ 20′ \ 00″$$

By Sine rule $\dfrac{IC}{\sin 47°25′} = \dfrac{CD}{\sin 60° 20′}$

$$\text{therefore } IC = \frac{256.50 \times 0.738\,259}{0.868\,920}$$

$$= \underline{217.35 \text{ m}}$$

$$\text{Chainage of I} = AC + 217.35$$

$$= \underline{776.63 \text{ m}}$$

$$\text{Deviation angle } \theta = 180° \ 00′ \ 00″ - 60° \ 20′ \ 00″$$

$$= \underline{119° \ 40′ \ 00″}$$

$$\text{Tangent length IT} = R \tan \theta/2$$

$$= 300 \tan 59° \ 50′$$

$$= 300 \times 1.720\,474$$

$$= \underline{516.142 \text{ m}}$$

$$\text{therefore chainage of T} = \text{chainage I} - 516.142$$

$$= \underline{260.49 \text{ m}}$$

2. Curve tangential to three straights

In Fig. 16.7 three straights are to be joined by a circular curve, the radius of which is unknown and has to be calculated.

One condition must be fulfilled, namely, that

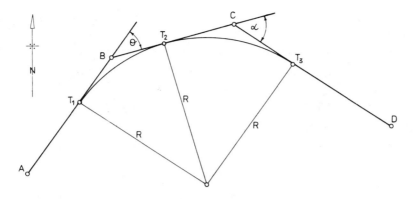

Figure 16.7

each straight be a tangent to the curve of radius *R*.

The problem is reasonably simple if one considers straights AB and BC only.

BT_1 and BT_2 are equal tangent lengths deviating by angle θ

therefore $BT_1 = BT_2 = R \tan \theta/2$.

Considering straights BC and CD only:

CT_2 and CT_3 are equal tangent lengths deviating by angle α

therefore $CT_2 = CT_3 = R \tan \alpha/2$.

The length BC is known and is also equal to $(BT_2 + CT_2)$

therefore $BC = (BT_2 + CT_2) = R \tan \theta/2 + R \tan \alpha/2$

hence $R = BC \div (\tan \theta/2 + \tan \alpha/2)$

Example 4 The following data refer to Fig. 16.7.

Straight	Bearing	Distance (m)
AB	34°	735.70
BC	74°	210.50
CD	124°	640.40

Calculate: (a) the radius of the curve joining the straights;
(b) the length of curve.

Solution

(a)   Angle $\theta = 74° - 34° = 40°$

         Angle $\alpha = 124° - 74° = 50°$

           $R = BC \div (\tan \theta/2 + \tan \alpha/2)$
           as before
           $= 210.50 \div (\tan 20° + \tan 25°)$

           $= 210.50 \div 0.830\ 277\ 9$

           $= \underline{253.53 \text{ m}}$

(b)   Angle $T_1OT_3 = (40° + 50°) = 90°$

      therefore curve $= \pi/2 \times R$ metres
               $= 398.245 \text{ m}$

3. Curve passing through three known points

P, Q, and R are three points whose coordinates are known. They are to be joined by a curve of unknown radius, the length of which is required. The circle which passes through the points is the circumscribing circle of triangle PQR. Therefore

Angle QPR (at the circumference)

     $= \frac{1}{2}$ angle QOR (at the centre)
     $=$ angle SOR (SO bisects QR)

Angle QPR and SR can be calculated from the coordinates

therefore $OR = SR \text{ cosec } S\hat{O}R$

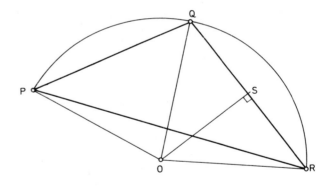

Figure 16.8

Perhaps a simpler solution is provided by the Sine rule which states that

$$\frac{p}{\sin P} = \frac{q}{\sin Q} = \frac{r}{\sin R}$$

     $= 2 \times$ radius of circumscribing circle

Side $QR = p$ and $Q\hat{P}R$ can be found from the coordinates.

Example 5 In Fig. 16.8 the following are the known coordinates of points P, Q, and R:

Point	Latitude	Departure
P	171.3 N	247.6 E
Q	205.4 N	332.0 E
R	122.1 N	390.4 E

Calculate the radius of the curve which passes through all three points

Solution

(a)
$$\text{Tan bearing PQ} = \frac{\Delta \text{ dep.}}{\Delta \text{ lat.}}$$

$$= \frac{332.0 - 247.6}{205.4 - 171.3}$$

$$= \frac{84.4}{34.1}$$

$$\text{Bearing PQ} = \text{N } 68° \text{ E}$$

$$= 68° \text{ WCB}$$

(b)
$$\text{Tan bearing PR} = \frac{\Delta \text{ dep.}}{\Delta \text{ lat.}}$$

$$= \frac{390.4 - 247.6}{122.1 - 171.3}$$

$$= \frac{142.8}{-49.2}$$

$$\text{Bearing PR} = \text{S } 71° \text{ E}$$

$$= 109° \text{ WCB}$$

(c)
$$\text{Distance QR} = \sqrt{(\Delta \text{ dep.}^2 + \Delta \text{ lat.}^2)}$$

$$= \sqrt{(83.3^2 + 58.4^2)}$$

$$= 101.73 \text{ m}$$

(d)
$$\text{Angle QPR} = 109° - 68°$$

$$= 41°$$

(e) In triangle PQR $\dfrac{p}{\sin P} = 2 \times \text{Radius}$

$$\text{therefore Radius} = \frac{101.73}{2 \times \sin 41°}$$

$$= 77.53 \text{ m}$$

Setting out

Where curves are long and of large radius (over 100 m) a theodolite must be used to obtain the desired accuracy of setting out.

Small radius curves can be set out quickly and accurately by using tapes only.

Small radius curves

Method 1. Finding the centre

In Fig. 16.9 kerbs have to be laid at the roadway junction. Consider the right-hand curve. The deviation angle α is measured from the plan and the tangent lengths IT and IT_1 ($= R \tan \alpha/2$) calculated. The procedure for setting the curve is then as follows:

(a) From I, measure back along the straights the distances IT and IT_1.

(b) Hammer in pegs at these points and mark the exact positions of T and T_1 by nails.

(c) Hook a steel tape over each nail and mark the centre O at the point where the tapes intersect when reading R. Hammer in a peg and mark the centre exactly with a nail.

(d) Any point on the curve is established by

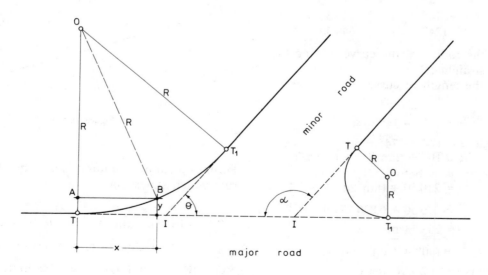

Figure 16.9

hooking the tape over the peg O and swinging the radius. This method is widely used where the radius of curvature is less than 30 m.

Method 2. Offsets from the tangent
When the deviation angle is small (less than 50°) the length of the curve short and the centre inaccessible, the curve can be set out by measuring offsets from the tangent. In the left-hand curve of Fig. 16.9, y is an offset from the tangent at a distance x metres from tangent point T. In the figure, the line AB is drawn parallel to the tangent until it cuts the radius. The length AT = y and the length AO = $(R - y)$.

In triangle OAB OA = $\sqrt{(OB^2 - AB^2)}$ (by Pythagoras)

i.e., $(R - y) = \sqrt{(R^2 - x^2)}$

therefore $y = R - \sqrt{(R^2 - x^2)}$

Thus the offset y can be calculated for any distance x along the tangent and can be set by eye or by optical square.

Example 6 Given that the deviation angle $\theta = 50°$ and the radius $R = 60$ m, calculate the offsets from the tangent at 5–metre intervals.

(a) Tangent lengths IT = IT_1 = 60 tan 25°
$$= 27.98 \text{ m}$$

(b) Offsets at 5 m = $60 - \sqrt{(60^2 - 5^2)}$ = 0.210 m
10 m = $60 - \sqrt{(60^2 - 10^2)}$ = 0.840 m
15 m = $60 - \sqrt{(60^2 - 15^2)}$ = 1.905 m
20 m = $60 - \sqrt{(60^2 - 20^2)}$ = 3.430 m
25 m = $60 - \sqrt{(60^2 - 25^2)}$ = 5.456 m
27.98 m = $60 - \sqrt{(60^2 - 27.98^2)}$ = 6.923 m

The procedure for setting out the curve is then as follows:
(i) From I, measure back along the straights the distances IT and IT_1 and drive in pegs to establish the exact positions of T and T_1.
(ii) Establish pegs at 5 m intervals along the straights between I and the tangent points.
(iii) Using an optical square set out the appropriate offsets y at right angles to the tangents and drive in a peg at each point.

Method 3. Offsets from the long chord
This method is suitable for curves of small radius. The curve is established by measuring offsets y at right angles to the long chord TT_1 at selected distances from the tangent points (Fig. 16.10).
VC is the major offset y at the mid-point C of the long chord and OC is constant, k.

Major offset $y = (R - k)$
In triangle OTC $k = \sqrt{(R^2 - x^2)}$
therefore $y = R - \sqrt{(R^2 - x^2)}$

Any other offset $y_n = (AB - k)$
In triangle ABO, AB = $\sqrt{(R^2 - x_n^2)}$
therefore $y_n = \sqrt{(R^2 - x_n^2)} - k$

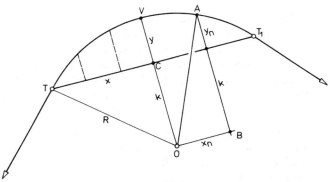

Figure 16.10

Any offset y can be calculated for any distance along the long chord, and can be set out by eye or prism square.

Example 7 A roadway kerb in Fig. 16.10 has a radius of curvature of 40 m. The length of the long chord is 60 m. Calculate the offsets from the chord at 10 m intervals.

Major offset = $R - \sqrt{(R^2 - x^2)}$
$$= 40 - \sqrt{(1600 - 900)}$$
$$= 40 - 26.46$$
$$= 13.54 \text{ m}$$
$$k = 26.46 \text{ m}$$

Offset $y_{10}(x = 20$ m from C)
$$y_{10} = \sqrt{(R_2 - x^2)} - k$$
$$= \sqrt{(1600 - 400)} - 26.46$$
$$= 34.64 - 26.46$$
$$= 8.18 \text{ m}$$

Offset $y_{20}(x = 10$ m from C)
$$y_{20} = \sqrt{(R_2 - x^2)} - k$$
$$= \sqrt{(1600 - 100)} - 26.46$$
$$= 38.73 - 26.46$$
$$= 12.27 \text{ m}$$

The offsets at 40 m and 50 m from T are the same lengths as the offsets at 20 m and 10 m respectively.
The procedure for setting out the curve is as follows:
(a) Locate T and T_1 and measure the distance between them. It should equal 60 metres.
(b) At 10 m intervals along the long chord drive in pegs.
(c) Set out the offsets at right angles to the long chord using a prism square and drive in pegs to mark the curve.

Method 4. Offsets from chords produced

Length of chord In this and subsequent methods of setting out curves, chords have to be chosen such that the difference in length between the chord and arc is as small as possible.

The chord length should not be greater than one-tenth of the length of the radius. The error caused by assuming that the chord equals the arc is 1 in 2400 which is acceptable for much construction work. If greater accuracy is required the ratio of chord to radius must be reduced. The ratio of 1:20 gives errors of the order of 1 in 10 000.

Procedure for setting out the curve In Fig. 16.11(a) T is the tangent point set out as before by measuring back the distance IT ($= R \tan \theta/2$) from the intersection point I.

The procedure for setting out the curve is as follows:

(a) Select a length, c, less than one-tenth the length of the radius and lay off the distance from the point T along the straight towards I. Mark the point B.

(b) Calculate the offset y_1 and swing the tape from T through the arc y_1 to establish point C on the curve. Drive in a peg and mark C accurately with a nail.

(c) Extend the chord TC for a further distance of c metres and mark point D.

(d) Calculate the second offset y_2 and swing the tape from C through the arc y_2 to establish point F on the curve.

(e) Repeat operations (c) and (d) to establish further pegs H, etc., on the curve. The offset distance in each case is y_2.

Lengths of offsets In these calculations, the arc and chord lengths are assumed equal.

In Fig. 16.11(a), C is joined to O to form angle TOC and O is joined to X making OX the perpendicular bisector of TC.

Therefore angle TOX = angle XOC = α

Angle BTC is the angle between the tangent BT and the chord CT

$$\text{therefore BTC} = \tfrac{1}{2} \text{ angle TOC}$$
$$= \text{angle TOX} = α$$

therefore sector BTC is similar to sector TOX.

$$\text{Now } \frac{BC}{TC} = \frac{TX}{TO}$$

$$\text{that is, } \frac{y_1}{c} = \frac{c/2}{R}$$

$$\text{therefore } y_1 = \frac{c^2}{2R} \text{ (1st offset)}$$

If the tangent ZE is drawn through C:

$$\text{angle ZCT} = α$$
Also angle ZCT = angle DCE
   (vertically opposite angles)

therefore angle DCE = α

Angle ECF is the angle between tangent EC and chord FC

therefore angle ECF = α
   DCF = angle DCE + angle ECF
    $= 2α$
    = angle COF (angle at centre)

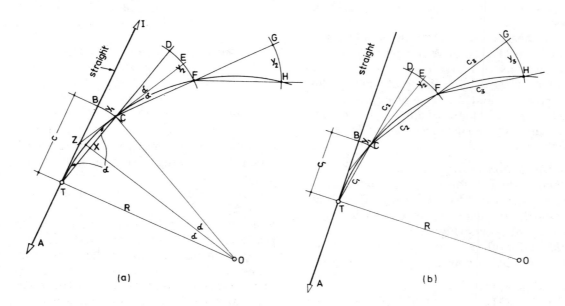

(a)   (b)

Figure 16.11

therefore sector DCF is similar to sector COF.

$$\text{Now } \frac{DF}{CF} = \frac{CF}{CO}$$

$$\text{that is } \frac{y_2}{c} = \frac{c}{R}$$

$$\text{therefore } y_2 = \frac{c^2}{R} \text{ (2nd \& subsequent}$$
$$\text{offsets)}$$

Similarly GH and other offsets are equal to c^2/R.

This method is fairly accurate and is not restricted to setting out curves of small radius.

Large radius curves

Method 1. Offsets from chords produced
The pegs are usually set on the curve at intervals of one tape length measured from the point of zero chainage A (Fig. 16.11(b)).

The first chord is nearly always less than one tape length. In the figure the initial chord is c_1 metres long and all others are the standard one-tape length of c_2 metres.

$$\text{As before, offset } y_1 = \frac{c_1^2}{2R} \text{ (initial sub-chord)} \quad (1)$$

Sector BTC is similar to sector DCE

$$\text{therefore } \frac{DE}{BC} = \frac{c_2}{c_1}$$

$$\text{therefore offset DE} = y_1 \times \frac{c_2}{c_1}$$

$$= \frac{c_1^2 \times c_2}{2R \times c_1}$$

$$= \frac{c_1 c_2}{2R}$$

$$\text{As before} \quad EF = \frac{c_2^2}{2R}$$

$$\text{Offset } y_2 = DF = (DE + EF)$$

$$= \frac{c_1 c_2}{2R} + \frac{c_2^2}{2R}$$

$$= \frac{c_2(c_1 + c_2)}{2R} \text{ (1st full chord)} \quad (2)$$

$$\text{As before} \quad y_3 = \frac{c_3^2}{R} \quad (3)$$
$$\text{(2nd \& subsequent full}$$
$$\text{chords)}$$

A general expression can be deduced from formula (2) which can be applied to find any offset, namely:

$$y_n = \frac{c_n(c_n + c_{(n-1)})}{2R} \quad (4)$$

Example 8 A 300-metre radius curve is to be set out by offsets from chords. The chainages of the first and second tangent points are 327.5 m and 425.3 m respectively. Calculate the lengths of the offsets to set out pegs at even chainages of 20 m.

Solution
Length of curve = (425.3 − 327.5) = 97.8 m
First peg on curve occurs at 340 m
therefore initial sub-chord = (340.0 − 327.5) = 12.5 m

Curve is composed of (i) initial sub-chord 12.5 m
                      (ii) four 20 m chords 80.0 m
                      (iii) final sub-chord   5.3 m
                                             _____
                                              97.8 m

From formulae (1)–(4)
First offset $= 12.5^2/600$
                          = 0.260 m
Second offset $= 20(20 + 12.5)/600$
                          = 1.083 m
Third, fourth, and fifth offsets $= 20^2/300$
                          = 1.333 m
Final offset $= 5.3(5.3 + 20)/600$
                          = 0.223 m

The setting-out information is tabulated thus:

Chord no.	Chord length (m)	Chord chainage (m)	Offset (m)
1	12.5	340.0	0.260
2	20.0	360.0	1.083
3	20.0	380.0	1.333
4	20.0	400.0	1.333
5	20.0	420.0	1.333
6	5.3	425.3	0.223
	97.8		

Table 16.1

Method 2. Tangential angles
The method involves the use of tape and theodolite and is the common method of setting out large radius curves when accuracy is required.

In Fig. 16.12, the tangent point T at the beginning of the curve has been established as in previous methods.

BC and CD are equal standard chords, c_2 and c_3, chosen such that their length is less than one-twentieth the length of the radius.

TB is the initial sub-chord, c_1, which is shorter than the standard chords owing to the chainage of T being uneven and c_4 is the final sub-chord which is also shorter than the standard chords.

Angles α_1, α_2, α_3, and α_4 are the tangential angles or, as they are sometimes called, the deflection angles. Their values must be calculated in order to set out the curve.

Assuming for a moment that the tangential angles are known, the curve is set out as follows:

(a) Set the theodolite at T and sight intersection point I on zero degrees.

(b) Release the upper clamp and set the theodolite to read α_1 degrees.

(c) Holding the end of a tape at T, line in the tape with the theodolite and drive in a peg B at a distance of c_1 metres from T.

(d) Set the theodolite to read $(\alpha_1 + \alpha_2)$ degrees.

(e) Hold the rear end of the tape at B and with the tape reading c_2 metres, i.e., a standard chord length, swing the forward end until it is intersected by the line of sight of the theodolite. This is the point C on the curve.

(f) Set the theodolite to read $(\alpha_1 + \alpha_2 + \alpha_3)$ degrees. Repeat the operation (e) to establish point D on the curve.

In most cases this operation will be repeated several more times to establish a number of pegs on the curve at one standard chord interval.

In this example peg D is the last standard chord.

(g) Set the theodolite to read $(\alpha_1 + \alpha_2 + \alpha_3 + \alpha_4) = \theta/2$ degrees.

(h) Holding the rear end of the tape at D swing the forward end until the reading of c_4 metres is intersected by the line of sight of the theodolite. This establishes tangent point T_1.

Calculation of tangential angles Angle ITB is the angle between tangent TI and chord TB.

Angle TOB is the angle at the centre subtended by chord TB.

Therefore angle ITB = ½ angle TOB = α

In Fig. 16.12(b) OX is the perpendicular bisector of chord TB. Therefore angle TOX = angle XOB = α.

In triangle TOX

$$\sin \text{TOX} = \frac{\text{TX}}{\text{TO}}$$

$$= \frac{c_1}{2}\Big/R$$

$$= \underline{\frac{c_1}{2R}}$$

The value of any tangential angle (α_1, α_2, α_3, and α_4)

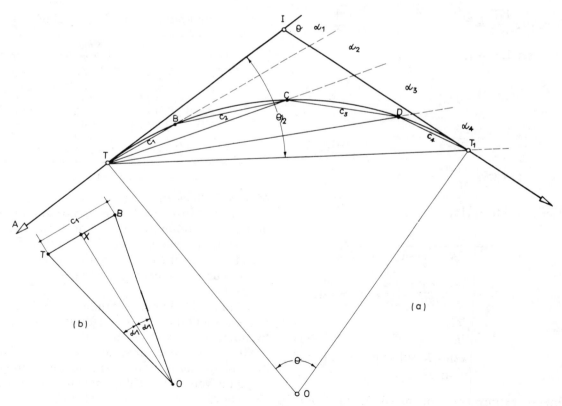

Figure 16.12

can similarly be found and the formula can be written in general terms:

$$\sin \alpha = \frac{c}{2R} \qquad (1)$$

When c is less than one-fifteenth of R, an accurate value of α can be determined thus:

$\sin \alpha = \alpha$ radians (since α is always small)

therefore α radians $= \dfrac{c}{2R}$

hence α degrees $= \dfrac{c}{2R} \times \dfrac{180}{\pi}$

and α minutes $= \dfrac{c}{2R} \times \dfrac{180}{\pi} \times 60$

$$\text{i.e., } \alpha = \left(\frac{c}{R} \times 1718.9 \right) \text{ minutes} \qquad (2)$$

In Fig. 16.12(a):

Standard-chord angle $\alpha_2 = \left(\dfrac{c_2}{R} \times 1718.9 \right)$ minutes

$$\text{or } \sin \alpha_2 = \frac{c_2}{2R}$$

Initial sub-chord angle $\alpha_1 = \left(\dfrac{c_1}{R} \times 1718.9 \right)$ minutes

$$\text{or } \sin \alpha_1 = \frac{c_1}{2R}$$

From these calculations it can be seen that angles α_1 and α_2 are proportional to their chord lengths and the most convenient way to calculate tangential angles is firstly to calculate the standard-chord angle, then by proportion to calculate the initial and final sub-chord angles.

$$\text{Final sub-chord angle} = \alpha_2 \times \frac{c_4}{c_2}$$

Number of chords Since each chord is assumed to be equal to the length of its arc, the number of chords is found by dividing the length of the curve by the length of a standard full chord.

Example 9 Two straights AI and IB have bearings of N 80° E and S 70° E respectively. They are to be joined by a circular curve of 300 metres radius. The chainage of intersection point I is 872.485 m.
Calculate the data for setting out the curve by 20 m standard chords.

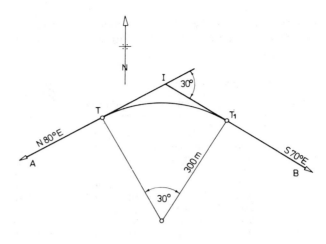

Figure 16.13

(a) Tangent length $= 300 \tan 15°$
$\phantom{\text{(a) Tangent length }} = \underline{80.385 \text{ m}}$

(b) Chainage of T $= (872.485 - 80.385)$ m
$\phantom{\text{(b) Chainage of T }} = \underline{792.100 \text{ m}}$

(c) Curve length $ = 300 \times \theta$ radians
$\phantom{\text{(c) Curve length }} = \underline{157.080 \text{ m}}$

(d) Chainage of $T_1 = \underline{949.180 \text{ m}}$

(e) Number of chords
Initial sub-chord
$= (800.00 - 792.10) = 7.90$ m

7 full chords of 20 m
$= (7 \times 20.0) = \underline{140.00}$
$\phantom{= (7 \times 20.0) } 147.90 \text{ m}$

therefore final sub-chord
$= (157.08 - 147.90) = \underline{9.18}$ m
$ \underline{157.08 \text{ m}} = \text{curve length}$

(f) Tangential angle for standard 20 m chord

$$= \left(\frac{20}{300} \times 1718.9 \right) \text{ minutes}$$

$ = 114.593$ minutes
$ = \underline{01° 54' 36''}$

(g) Sub-chord angles

$$\text{Initial} = \frac{114.593 \times 7.90}{20.0} \text{ minutes}$$

$ = 45.264$ minutes
$ = \underline{00° 45' 16''}$

$$\text{Final} = \left(\frac{114.593 \times 9.18}{20.0} \right) \text{ minutes}$$

$ = 52.598$ minutes
$ = \underline{00° 52' 36''}$

For this and every other example, the setting-out information is presented in tabular fashion as in Table 16.2.

Chord No.	Length (m)	Chainage (m)	Chord angle	Tangential angle
(T)		792.10		
1	7.90	800.00	00° 45′ 16″	00° 45′ 16″
2	20.00	820.00	01° 54′ 36″	02° 39′ 52″
3	20.00	840.00	01° 54′ 36″	04° 34′ 28″
4	20.00	860.00	01° 54′ 36″	06° 29′ 04″
5	20.00	880.00	01° 54′ 36″	08° 23′ 40″
6	20.00	900.00	01° 54′ 36″	10° 18′ 16″
7	20.00	920.00	01° 54′ 36″	12° 12′ 52″
8	20.00	940.00	01° 54′ 36″	14° 07′ 28″
9(T_1)	9.18	949.18	00° 52′ 36″	15° 00′ 00″
	157.08	157.08	15° 00′ 04″	

Table 16.2

Notes: There is a discrepancy of 4″ in the final tangential angle due to the rounding off of the angles to whole seconds.

Left-hand curves When curves are to turn to the left the tangential angles must be subtracted from 360°. For example, if the straights in Example 9 had deviated to the left by 30° the final tangential reading would have been

$$(360° \ 00′ \ 00″ - 15° \ 00′ \ 00″) = \underline{345° \ 00′ \ 00″}$$

Method 3. Using two theodolites
In the methods of setting out discussed so far linear measurements were used.

When the ground conditions are unsuitable for measuring, the curve may be set out by the use of two theodolites and linear measurements are dispensed with altogether.

In Fig. 16.14 the tangent points T and T_1 have been located on the ground. The following procedure is used to set out the points on the curve.

(a) Set No. 1 theodolite at T, sight I on 00° 00′ 00″.
(b) Calculate tangential angle α, for initial subchord TC and set theodolite to this reading.
(c) Set No. 2 theodolite at T_1 and sight T on 00° 00′.
(d) Set the theodolite to read α.
(e) The intersection of the lines of sight of the two theodolites is a point on the curve, since IT̂C is the angle between tangent TI and chord TC which equals angle TT₁C, the angle at the circumference subtended by the chord TC. A peg is driven in at this point.
(f) Further tangential angles are set on both theodolites, the intersection of the sight lines in all cases being a point on the curve. This method has the advantage over the others that each point is set out completely independently.

Obstructions

Very often it is impossible to set out all the points on the curve from the tangent point because of some obstruction. In Fig. 16.15 the third point E on the curve has been set out by the tangential angle $(\alpha_1 + \alpha_2 + \alpha_2)$ = say 5°.

Owing to the trees, point F cannot be set from T. In such a case the theodolite is removed to E and a sight taken back to the tangent point T with the horizontal circle of the theodolite reading $180° - (\alpha_1 + \alpha_2 + \alpha_2) = 175°$ (say).

If a tangent is drawn to the circle at E, angle XTE equals angle TEX, that is angle TEX = 5°.

The theodolite is then set to read (175° + angle TEX) = (175° + 5°) = 180° in which case, the line of sight is along tangent EX. The telescope is swung through a further 180° to point along the continuation of the tangent EY and the horizontal circle reads zero.

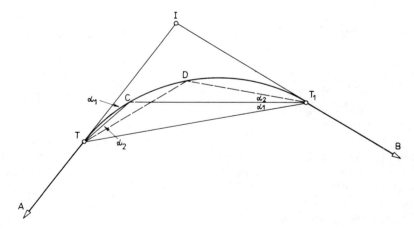

Figure 16.14

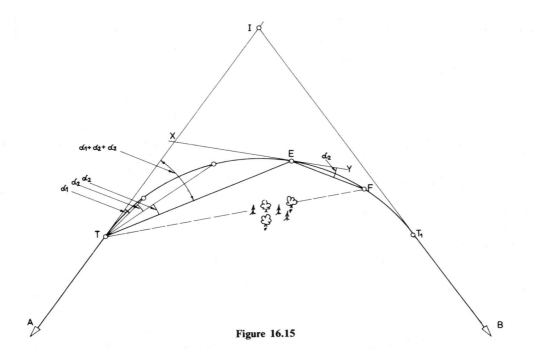

Figure 16.15

Point E is then treated as being a tangent point and F is set out by setting the circle to read α_2 degrees and measuring the standard chord length EF.

Vertical curves

Whenever roads or railways change gradient, a vertical curve is required to take traffic smoothly from one gradient to the other.

When the two gradients form a hill, the curve is called a summit curve and when the gradients form a valley a sag or valley curve is produced (Fig. 16.16).

The gradients are expressed as percentages. A gradient of 1 in 50 is a 2 per cent gradient, that is, the gradient rises or falls by 2 units in 100 units. Similarly a gradient of 1 in 200 is a 0.5 per cent gradient.

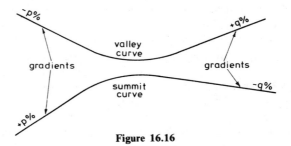

Figure 16.16

Since the change of gradient from slope to curve is required to be smooth and gradual, parabolic curves are chosen. This form of curve is flat near the tangent point and calculations are reasonably simple. The form of the curve is $y = ax^2 + bx$ where y is the height of the curve above or below the first tangent point at a distance x therefrom (Fig. 16.17).

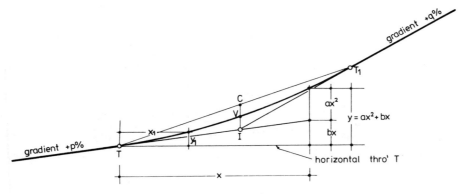

Figure 16.17

Strictly speaking, the vertical offsets y should be at right angles to the gradient but since all distances are reduced to horizontals and verticals in surveying it is acceptable to treat the offsets as verticals without error.

Properties of the simple parabola

1. The distance between the points T and T_1 as measured along (a) the curve TT_1, (b) the tangents TIT_1, (c) the chord TT_1 are so close in length that they are considered equal (Fig. 16.17).
2. The intersection point I is treated as being midway between the points T and T_1; thus the lengths IT and IT_1 are equal. The curve is in fact often called an equal tangent parabola.
3. The height IV is called the correction in gradient and equals the height VC. In other words, the parabola bisects the length CI.

Setting-out data

The following information is required for any setting-out calculations:
1. The length of curve. The length of the curve is dependent upon: (a) the gradient of the straights; (b) the sight distance.
(a) Generally the steeper the approach gradients, the greater will be the centrifugal effect caused by the change of gradient from the slope to the curve. The curve length must be increased to reduce this effect when gradients are steep.
(b) The sight distance is the length required for a vehicle to stop from the moment a driver sees an obstruction over the brow of a summit curve. The sight distance includes thinking, braking, stopping, and safety margin distances.
   The length of curve is taken from tables published by the Ministry of Transport.
2. The gradients of the slopes and the reduced level of one chainage point, preferably the intersection point.

Calculation of data

A simple way to understand the calculations involved in vertical curves is to follow the steps in a typical example.

Example 10 In Fig. 16.18(a) a gradient of +1 per cent (i.e., 1 in 100) meets a gradient of +4 per cent (i.e., 1 in 25) at intersection point I, the chainage and reduced level of which are 500 m and 261.30 m respectively. A 100 m long vertical curve is to be inserted between the straights. Calculate the corrected grade elevations (i.e., levels on the curve) at 25 m intervals.

Step 1 Calculate the reduced levels of the initial tangent point T, the final tangent point T_1 and the intersection point I.
In Figure 16.18

(i) IT = IT_1 = 100/2 = 50 m
(ii) Red. lev. I = 261.30 m (given)
(iii) Red. lev. T = Red. lev. I − 1 per cent of 50
$$= 261.30 - (0.01 \times 50)$$
$$= \underline{260.80 \text{ m}}$$

(iv) Red. lev. T_1 = Red. lev. I + 4 per cent of 50
$$= 261.30 + (0.04 \times 50)$$
$$= \underline{263.30 \text{ m}}$$

Step 2 Calculate the tangent levels, i.e., the levels which would obtain on the left-hand gradient if it were extended above or in this case below the right-hand gradient, towards the final tangent point. Mathematically any tangent level is

Tan. lev. = (red. lev. T + bx)
where b = left-hand gradient p per cent
x = distance from T

Tan. lev. at 25 m = red. lev. T + bx
$$= 260.80 + 1 \text{ per cent of } 25 \text{ m}$$
$$= 261.05$$
Tan. lev. 50 m (I) = 260.80 + 1 per cent of 50 m
$$= 261.30$$
Tan. lev. 75 m = 260.80 + 1 per cent of 75 m
$$= 261.55$$
Tan. lev. 100 m (E) = 260.80 m + 1 per cent of 100 m
$$= 261.80 \text{ m}$$

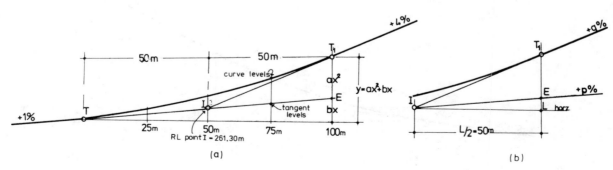

Figure 16.18

Step 3 Calculate the grade corrections at the required chainage points. The grade correction is the value of ax^2 which when added to or subtracted from the various tangent levels in Step 2 will give the level on the curve.

The term x is, of course, the distance of the chainage point from the initial tangent point. The value of a is unknown and has to be found in order to calculate these grade corrections.

In Fig. 16.18(b) the reduced levels of T_1 and E are 263.30 and 261.80 respectively, therefore (red. lev. T_1 − red. lev. E) = $(263.30 − 261.80) = 1.50$ m. This value of 1.50 m is really the grade correction at point E, i.e., the value which is applied to tangent level E to produce the curve level T, therefore

$$\text{grade correction } 1.50 \text{ m} = ax^2 \text{ (where } x = 100 \text{ m)}$$
$$= a \times 100^2$$
$$\text{therefore } a = 1.50 \div 100^2$$
$$= 1.50 \times 10^{-4}$$

Value of ax^2 at 25 m intervals from T:

Chainage	ax^2
25 m	$(1.50 \times 10^{-4}) \times 25^2 = 0.094$ m
50 m	$(1.50 \times 10^{-4}) \times 50^2 = 0.375$ m
75 m	$(1.50 \times 10^{-4}) \times 75^2 = 0.844$ m
100 m	$(1.50 \times 10^{-4}) \times 100^2 = 1.500$ m

An alternative method of calculating a exists. In Fig. 16.18(b):

Red. lev. T_1 = red. lev. I + q per cent of $L/2$

$$= \text{RL I} + \frac{qL}{200}$$

Red. lev. E = red. lev. I + p per cent of $L/2$

$$= \text{RL I} + \frac{pL}{200}$$

Now ax^2 = red. lev. T_1 − red. lev. E

$$= \frac{qL}{200} - \frac{pL}{200}$$

$$= \frac{(q - p)L}{200}$$

Since $x = L$ at point T_1

$$aL^2 = \frac{(q - p)L}{200}$$

$$\text{therefore } a = \frac{(q - p)}{200} \times \frac{L}{L^2}$$

$$= \frac{q - p}{200L}$$

The formula applies in all situations where a is required. The proper sign convention for positive and negative gradients p or q must of course be used.

In the example $q = +4$ per cent, $p = +1$ per cent, and $L = 100$ m, therefore

$$a = \frac{4 - 1}{200 \times 100} = \frac{3}{20\,000} = 1.5 \times 10^{-4}$$

Step 4 Calculate the curve level at the various chainage points. The curve level at any point is the algebraic addition of the tangent level $(T + bx)$ and grade correction (ax^2)

i.e. curve level = tangent level + grade correction

therefore curve level 25 m = 261.05 + 0.094
$$= 261.144 \text{ m}$$

curve level 50 m = 261.30 + 0.375 = 261.675 m

75 m = 261.55 + 0.844 = 262.394 m

100 m = 261.80 + 1.500 = 263.300 m

In all examples the calculations are performed in tabular fashion as below.

Ch. (m)	Tangent level (m) T + (bx)	Grade correction (m) (ax²)	Curve level (m) (T + ax² + bx)
0(T)	260.80	0	260.800
25	261.05	0.094	261.144
50	261.30	0.375	261.675
75	261.55	0.844	262.394
100(E)	261.80	1.500	263.300

Example 11 A rising gradient of 1 in 40 is to be connected to a falling gradient of 1 in 75 by means of a vertical parabolic curve 400 m in length. The reduced level of the intersection point of the gradients is 26.850 m above Ordnance Datum.

Calculate: (a) the reduced levels of the tangent points, (b) the reduced levels at 50 m intervals along the curve.

(SCOTEC—OND Building)

Solution
See Fig. 16.19
(i)   IT = IT_1 = 400/2   = 200 m
(ii) Red. lev. I (given)   = 26.850 m
(iii) Red. lev. T   = 26.850 − 200/40
    = 21.850 m
(iv) Red. lev. T_1 = 26.850 − 200/75
    = 24.183 m
(v)   Red. lev. E   = 26.850 + 200/40
    = 31.850 m
(vi) $ET_1 = ax^2$ at 400 m = 24.183 − 31.850
    = −7.667
    therefore $a = -7.667/400^2$
    = -4.792×10^{-5}

Ch (m)	Tangent level (m) T + (bx)	Grade correction (m) ax^2	Curve level (m) (T + ax^2 + bx)
0(T)	21.850	0	21.850
50	23.100	−0.120	22.980
100	24.350	−0.479	23.871
150	25.600	−1.078	24.522
200	26.850	−1.917	24.933
250	28.100	−2.995	25.105
300	29.350	−4.313	25.037
350	30.600	−5.870	24.730
400(T₁)	31.850	−7.667	24.183

Highest (or lowest) point on any curve

The highest or lowest point on any curve is the turning point, that is, the position where the gradient of the tangent is zero.

The gradient of the tangent is found by differentiating y with respect to x in the equation $y = ax^2 + bx$.

$$\frac{dy}{dx} = 2ax + b$$

When $dy/dx = 0$
$$x = -b/2a$$

In Example 11:
(i) $b = +2.5$ per cent $= +2.5 \times 10^{-2}$
$a = -4.792 \times 10^{-5}$

therefore $x = \dfrac{-2.5 \times 10^{-2}}{2 \times -4.792 \times 10^{-5}}$

$= 10^3 \times 0.260\ 85$

$= \underline{260.851\ \text{m from T}}$

(ii) Level of highest point
$= 21.850 + ax^2 + bx$
$= 21.850 - 3.261 + 6.521$
$= \underline{25.110\ \text{m}}$

Test questions

1 (a) A 40 m radius roadway kerb is to be set out at the junction of two roadways by the method of offsets from the long chord.

Given that the length of the long chord is 30 m, calculate the lengths of the offsets from the chord at 5 m intervals.

(b) Two straight roadways, AB and BC, intersecting at B, have bearings of N 5° 30′ E and N 6° 30′ W respectively. They are to be joined by a circular curve of 500 m radius.

The curve is to be set out by the method of tangential angles. Pegs are to be set out at intervals of 20 m of 'through' chainage from the point A.

Given that the chainage of point B is 842.75 m from A, calculate:
(i) the chainages of the tangent points of the curve;
(ii) the chainages of the setting-out pegs along the curve;
(iii) the tangential angles required to set out the curve from the initial tangent point.

(ONC, Topographic Studies)

2 Two straights AX and XB are to be connected by a 400 m radius circular curve. The bearings and lengths of the straights are as follows:

Straight	Whole circle bearing	Length (m)
AX	73° 10′	197.5
XB	81° 40′	—

Calculate:
(a) the intersection angle between the straights;
(b) the tangent lengths;
(c) the curve length;
(d) the chainages of the tangent points;

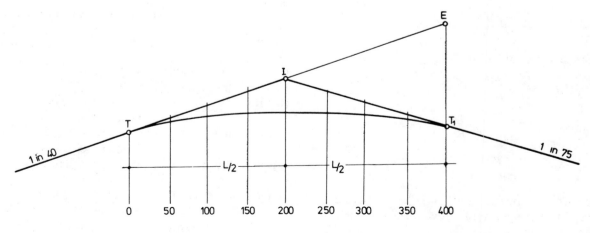

Figure 16.19

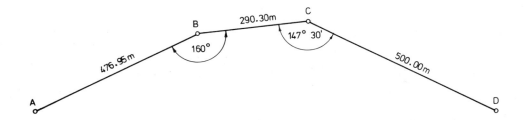

Figure 16.20

(e) the setting-out information to enable the curve to be set out at 20 m intervals of through chainage. (Present the information in the form of a setting out table.)

(ONC, Topographic Studies)

3 The following data refer to three straight sections of roadway:

Straight	Bearing	Length (m)
AB	N 75° E	610.00 m
BC	S 65° E	450.86 m
CD	N 45° E	343.10 m

The straights are to be connected by two curves of equal radius such that there is a straight portion of 100 m between them along the straight BC.

Calculate: (a) the radius of the curves;
(b) the chainages of the four tangent points.

4 Figure 16.20 shows two straights AB and CD which are to be joined by a curve of 330 m radius. The intersection point I is inaccessible and a traverse ABCD produced the results shown in the figure.

Calculate: (a) the tangent lengths;
(b) the chainages of the tangent points;
(c) the setting-out information to set out the curve at even chainages of 20 metres.

5 Two straight roadways AI and IB are to be joined by a circular curve of radius 40 metres. The *TOTAL* coordinates of A, I, and B are shown in Table 16.3.

Station	Total latitude	Total departure
A	+ 75.38	+111.20
I	+154.60	+146.81
B	+128.70	+165.35

Table 16.3

Calculate:
(a) the distance from A and B to the tangent points of the curve;
(b) the *TOTAL* coordinates of each tangent point.

(OND, Building)

6 A vertical curve is to be used to connect a rising gradient of 1 in 60 with a falling gradient of 1 in 100 which intersect at a point having a reduced level 65.25 metres AOD.

Given that the curve is to be 150 metres long, calculate:

(a) the reduced levels of the tangent points;
(b) the levels at 30 metre intervals along the curve and the depths of cutting required;
(c) the distance from the tangent point on the 1 in 60 gradient to the highest point on the curve, and the reduced level of this point.

(OND, Building)

7 Two straights AB and BC having gradients of −1:40 and +1:50 respectively, meet at point B at a level of 40.00 m AOD and chainage 1500.0 m in the direction AB. The gradients are to be joined by a vertical parabolic arc 240 m in length.

Calculate the level and chainage of the lowest point on the curve.

(OND Building)

17.　Setting out

Setting out could be defined as the reverse process of surveying. The positions and levels of buildings, drains, sewers, etc., already marked on a plan, are transferred to the ground by a variety of methods and by specially manufactured instruments.

Dwelling houses are still largely traditionally built and small inaccuracies in the setting out can usually be tolerated. Large factory buildings, multi-storey buildings, schools, etc., are nowadays largely prefabricated and little if any inaccuracy can be tolerated in the setting out. Consequently the methods of setting out vary considerably.

Setting out—horizontal control

1. Small buildings

The exact position of the building is governed by the building line as defined by the local authority. In Fig. 17.1 the building line is parallel to the roadway and the frontages of all the buildings lie along the line.

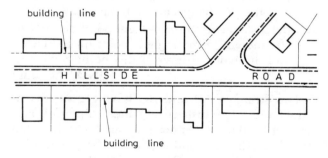

Figure 17.1

In the figure, buildings of several different shapes are present but when setting out, each building is reduced to a basic rectangle, enabling checks to be easily applied.

Figure 17.2 shows a typical dwelling house where the building line has been fixed from the centre line of the roadway.

The building may be set out using either
(a) steel tape;

(b) some surveying instrument, usually a site-square, or optical square, as follows:

(i) Determine the dimensions of a basic rectangle ABCD to enclose the house.

(ii) Measure the length AB (13.0 m) and mark the points by nails driven into wooden stakes.

(iii) Using a basic 3:4:5 right angle, measure the lengths AD and BC (8.0 m) and establish pegs at C and D.

(iv) Check the lengths of the diagonals AC and BD (15.264 m). Both measurements should be equal thus proving that the building is square.

Although the method of setting out a right angle using a 3:4:5 triangle is theoretically sound, in practice it tends to lead to inaccuracies in positioning.

By calculating the length of the diagonal of the rectangle and using two tapes, the setting out can be accomplished much more accurately and speedily as follows:

(i)   As before, measure the length AB and mark the positions A and B by nails driven into the wooden stakes.

(ii) Calculate the diagonal size of the rectangle using the theorem of Pythagoras:

$$AC = \sqrt{(13.0^2 + 8.0^2)} = 15.264 \text{ m}$$

(iii) Hold the zero of tape No. 1 against point A; hold the zero of tape No. 2 against point B and stretch them out in the direction of the point C.

(iv) At the intersection of 15.264 m of the first tape and 8.000 m of the second tape, mark point C on a peg.

(v)   Repeat for point D by measuring AD = 8.00 m with tape No. 1 and BD = 15.264 m with tape No. 2.

(vi) Check that DC = 13.000 m.

Profile boards
During the excavation of the foundations, the pegs A, B, C, and D will be destroyed and it is necessary to establish subsidiary marks on profile boards (Fig. 17.2).

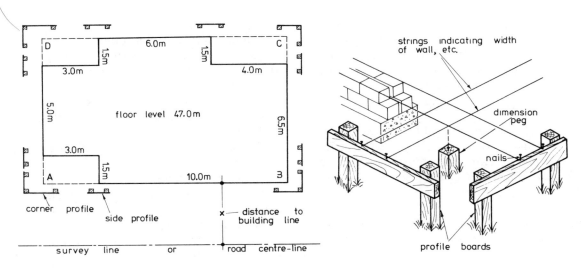

Figure 17.2

Profile boards are stout pieces of timber 150 mm by 25 mm cut to varying lengths. The boards are nailed to 50 mm square posts hammered into the ground, well clear of the foundations. Once the profiles have been established, builder's lines are strung between them and accurately plumbed above the pegs A, B, C, and D, and nails are driven into the boards to hold the strings and mark the positions of the walls, foundations, etc.

Setting out on sloping ground
When setting out buildings on sloping ground, it must be remembered that the dimensions taken from the plan are horizontal lengths and consequently the tape must be held horizontally and the method of step taping used.

The diagonals must also be measured horizontally and in practice considerable difficulty is experienced in obtaining checks under such conditions. Besides, the method is laborious and time-consuming.

It is possible to dispense with measuring the diagonals if a sitesquare is used to set out the right angles at A and B (Fig. 17.3). The instrument is

capable of setting out right angles with an accuracy of 1 in 2000. It consists of two small telescopes fixed rigidly at right angles on a small tripod. When the sitesquare is set up at peg A the observer simply sights peg B through one telescope and lines in peg D through the other.

Similarly peg C is set out from B and a check is provided on the work by erecting the instrument at D and checking that angle CDA is right angled.

2. Large buildings

In setting out large buildings which are mainly pre-fabricated, accuracy is absolutely essential. The factory-built components cannot be altered on site and even though some allowance is made for fitting on site, faulty setting out causes loss of time and money in corrective work.

In setting out such buildings, cognisance must be taken of the effects of calibration, temperature, and tension on tape measurements, and the effects of instrumental errors in angular observations.

Tape measurements
Fabric-based tapes should never be used for setting out. Good quality steel tapes are always employed and due allowance made for the following potential sources of error.

(a) *Calibration* Calibration errors can be ignored when good quality tapes are used. After long use, however, the tape should be tested against a standard. If the tape has been broken and repaired it should not be used for accurate setting out.

(b) *Temperature* The length of a steel tape varies with temperature. The tape is the standard length at 20° C only. If left lying in direct sunlight the tape may reach an abnormally high temperature. In winter the temperature may well be at freezing point. Chapter 4 dealt with the corrections to be

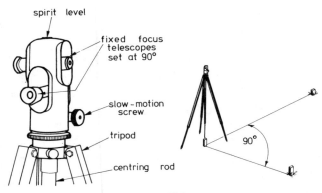

Figure 17.3

applied for differences in temperature but a fairly accurate estimation is obtained by allowing 1 mm per 10 metres per 10° C difference in temperature. For example, when using a 30 m steel tape on a winter day when the temperature is 0° an allowance of $-(1 \times 3 \times 2) = -6$ mm should be made for each tape length measured.

(c) *Tension* Steel tapes should be used on the flat with a tension of 4.5 kg. Without a spring balance few people can judge tensions and tests have shown that errors of 10 mm in 30 m can be caused by exerting excessive pressure on the tape.

When the tape is allowed to sag there is even more error. The natural tendency is to apply excessive tension to correct the sag, and in many cases the error due to stretching of the tape is greater than that due to sag.

A constant-tension handle has now been developed which applies a compromise tension. The tension applied is such that the effect of sag is compensated by the effect of stretching the tape.

(d) *Slope* It should be clear from (c) above that sagging of the tape should be avoided if possible, which suggests that measurements should be made along the ground. This, however, introduces slope errors and measurements of slope must be made and corrections applied.

Taping procedure
For highly accurate measurements the following procedure should be carried out:
(a) Use a good quality tape which has never been broken.
(b) Lay the tape on wooden blocks if measuring on concrete, etc., to allow air to circulate around the tape.
(c) Whenever possible do not allow the tape to sag.
(d) Measure the temperature.
(e) Use a spring balance or constant-tension handle.
(f) Measure the ground slope.
(g) Compute the various corrections and apply the correction to the distance set out.

Angular measurements
Angular measurements must always be made on both faces of the instrument because of the effect of instrumental maladjustments which were fully explained in Chapter 12.

Procedure for setting out
Figure 17.4 shows the plan of a large factory building and office block. The columns of the office block are of reinforced concrete, the cladding being prefabricated panels. The factory columns are of steel supporting steel latticed roof trusses.

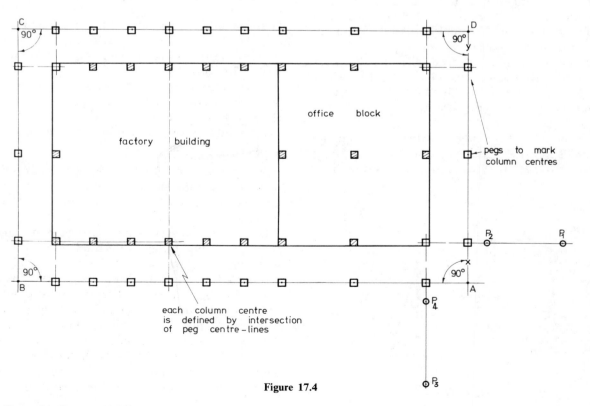

Figure 17.4

The column centres must be placed at exactly the correct distance apart and must be perfectly in line.

The following procedure is necessary to ensure that the requirements are met:

(a) Establish a line AB from the site traverse stations at a predetermined distance x metres from the centre line of the left-hand columns. Measure the distance accurately.

(b) Set out peg D by face left and face right observations from A positioning D at some predetermined distance from the centre line of the right-hand columns.

(c) Set out C by double face observations from D making CD = AB.

(d) Finally check angle C and distance CB to ensure that ABCD is a perfect rectangle.

(e) Set out the column centres along each line on stout profiles or preferably on pegs embedded in concrete. The centres must be set out by steel tape with due allowance being made for temperature, etc.

(f) The column centres are defined by the intersections of wires strung between the appropriate reference marks or by setting the theodolite at a peg on one side of the building, sighting the appropriate peg on the other side and lining in the columns directly from the instrument.

No hard and fast rules can be formulated for setting out since conditions vary greatly from site to site and some ingenuity is called for on the part of the engineer.

Checking verticality

In Fig. 17.4 the office block rises to a height of five storeys and the columns must be checked at every storey for verticality.

Several methods are available to the engineer using (a) theodolite, (b) plumb-bob, (c) special instruments.

1. *Theodolite methods*

(a) When setting out the framework for horizontal control further marks are established at the points P_1 to P_4, that is, on two lines at right angles to each other. Four pegs are required for each corner.

In Figs 17.4 and 17.5 the pegs are shown for the bottom right-hand corner only.

The theodolite is set up on face left over the outer peg P_1 and sighted onto the inner peg P_2. The telescope is then raised to any level on the building and a mark made on the outside of the column. The procedure is repeated on face right. If the theodolite is in good adjustment the two marks will coincide. If not, the mean position is correct.

The instrument is removed to peg P_3 and the

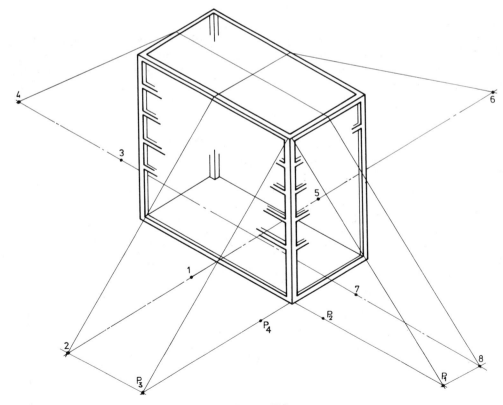

Figure 17.5

whole of the above procedure repeated to establish a second mark on the column. Thus, it is possible to determine the amount of deviation of the column from the vertical. Each corner is checked in this way.

(b) The centre lines of the building are established and extended on all four sides of the building. Two permanent marks are placed on each line. The marks are shown by the numbers 1 to 8 in Fig. 17.5.

The theodolite is placed on each of the outer marks, the telescope is raised to the required floor level and a mark established on all four sides by double face observations.

The intersection of lines strung between the appropriate marks locates the centre of the building. From the lines measurements may be made to the corners, etc.

(c) *Diagonal eyepiece method.* Most theodolites have an eyepiece which is interchangeable with a diagonal eyepiece shown in Fig. 17.6.

The diagonal eyepiece allows the telescope to be placed vertically and direct plumbing is possible to any height with high accuracy.

The engineer must arrange for a hole 500 mm square to be left in each floor so the theodolite can sight vertically upwards.

Sights are taken to a specially designed target shown in Fig. 17.6.

The following procedure is carried out when establishing a vertical line by diagonal eyepiece:

(i) The theodolite is set accurately over a predetermined mark on the ground floor and is centred and levelled in the usual manner. The horizontal circle is set to zero.

(ii) The altitude spirit level is centralized and the vertical circle set to read zenith (the reading varies with the type of instrument). The telescope is therefore pointing vertically.

(iii) An assistant on the upper floor moves to the Perspex target and frame over the hole until the horizontal crosshair lies along the centre line of the target. The frame is nailed to the floor using masonry nails and the ends of the target centre line lightly marked on the frame with a pencil.

(iv) The instrument is turned until the horizontal circle reads 180°; the altitude spirit level is re-centralized and the vertical circle set to zenith.

(v) If the instrument is in good adjustment the horizontal cross hair will lie along the centre line of the target. If not, the assistant is directed to move the target until it does.

(vi) The centre line of the target is again pencilled lightly on the frame and the mean position clearly marked. This mean line is the true vertical plane.

(vii) The operations are repeated with horizontal circle readings of 90° and 270°. Two more sets of marks are lightly pencilled on the frame and the mean position clearly marked, to establish a second vertical plane at right angles to the first. The intersection of the planes so marked is the true plumb point above the instrument.

2. *Plumb-bob method*

Heavy plumb-bobs are suspended from adjustable reels on piano wire over marks on the floor level. The positions of the marks are generally on the

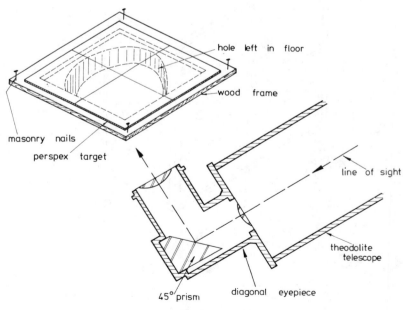

hole left in floor

wood frame

masonry nails

perspex target

line of sight

theodolite telescope

45° prism

diagonal eyepiece

Figure 17.6

centre line at a known distance *x* metres from the centre of the building.

The plumb-bobs have usually to be immersed in barrels of water to damp the oscillations of the wire set up by wind currents. The barrels therefore sit over the marks on the floor and the marks have to be referenced to some form of staging built around them.

The reels are adjusted on the upper floor until the wires are correctly positioned relative to the marks on the staging. The wires form a baseline from which measurements may be made to the various columns on the upper floor.

3. *Automatic plumbing*

General instruments are now marketed which resemble theodolites but which set out a vertical line of sight automatically.

The Hilger & Watts' Autoplumb is a good example. The optical system is shown in Fig. 17.7.

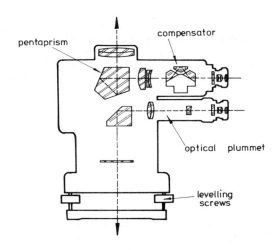

Figure 17.7

The Autoplumb is fitted with an optical plummet for downward sighting and a high power telescope for upward sighting.

Arrangements must be made for leaving a hole in each floor for upward sighting. Sights are taken to a target similar to that used with the diagonal eyepiece method.

The procedure for setting vertical with the Autoplumb is identical to that of the theodolite diagonal eyepiece method.

Setting out—vertical control

Whenever any proposed level is to be set out, sight rails (profiles) must be erected either at the proposed level in the case of a floor level or at some convenient height above the proposed level in cases of foundation levels, formation levels, and invert levels.

Suitable forms of sight rails or profiles are shown in Fig. 17.8. The rails should be set at right angles to the centre lines of drains, sewers, etc.

A traveller or boring rod is really a mobile profile which is used in conjunction with sight rails. The length of the traveller is equal to the difference in height between the rail level and the proposed excavation level. Figure 17.8 shows the traveller in use in a trench excavation and on a site clearance excavation.

In Fig. 17.8 the floor level of a building, the invert levels of a drain and the formation levels of a car parking area are to be set out. The following specifications are to be complied with:

1. Building: Floor level — 47.000 m

2. Car Park Area:
   Formation level in centre — 45.900 m
   Cross fall X to Y — 1 in 50
   Width XY — 5.50 m

3. Drain:
   Invert level at B (chainage 85.00 m) — 44.80 m
   Gradient B to C — 1 in 40 rising
   Chainage C — 100.00 m

Preliminary work

1. Suitable locations are chosen for the profile pegs. The uprights are firmly driven into the ground and their positions relative to centre lines are noted.

2. A levelling is made of all the ground levels at these points from the nearest bench mark. The relevant field data are given in Table 17.1.

BS	IS	FS	HPC	Red. lev.	Remarks
2.040	—	—	47.800	45.760	TBM
—	1.84	—	47.800	45.960	Ground level A
—	1.69	—	47.800	46.110	Ground level B
—	1.73	—	47.800	46.070	Ground level C
—	1.77	—	47.800	46.030	Ground level D
—	—	1.59	47.800	46.210	Ground level E

Table 17.1

3. All relevant proposed levels are calculated from the specification data.

Drain
Peg B (chn 85.00)   invert level   44.80 m
Peg C (chn 100.00)
Rise B to C = $\frac{1}{40} \times (100.00 - 85.00)$ = +0.375 m

therefore invert level   45.175 m

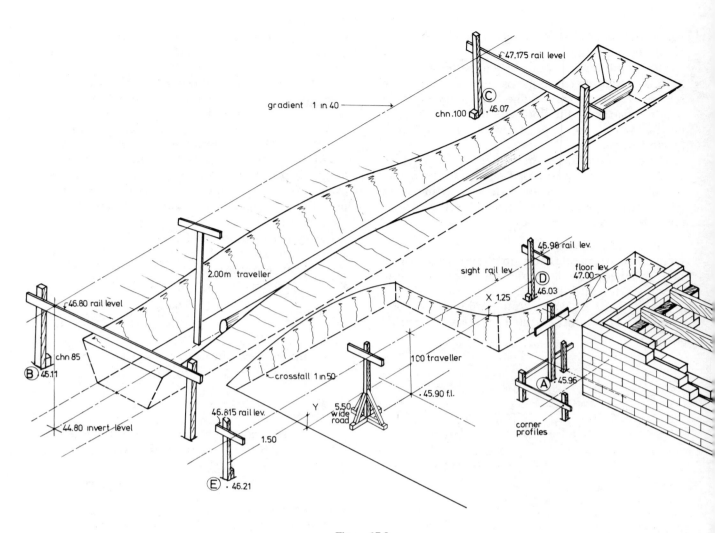

Figure 17.8

Car Parking Area

Formation level of centre line	45.90 m
Rise to D = $\frac{1}{50} \times (2.75 + 1.25)$	+0.080

Formation level D = 45.980 m

Fall to D to E = $\frac{1}{50} \times (1.25 + 5.50 + 1.50)$ −0.165

Formation level E = 45.815 m

4. A comparison is made between ground levels and proposed levels in order to determine suitable sight rail levels and traveller lengths. The sight rails should be a comfortable height above ground and the traveller lengths should be multiples of 0.5 m whenever possible.

At peg A the rail should be set at floor level.

No traveller is required. The rail will be 1.04 m approximately above ground.

At pegs B and C the rails will be 0.69 m and 1.105 m above ground if a 2.0 m traveller is chosen.

At pegs D and E the rails will be 0.95 m and 0.605 m above ground if a 1.0 m traveller is chosen.

Table 17.2 shows this setting out information.

Peg	Ground level	Proposed level		Traveller			Rail level	Height above ground
A	45.960	47.00	+	0.00	=	47.00		1.04
B	46.110	44.80	+	2.00	=	46.80		0.69
C	47.070	45.175	+	2.00	=	47.175		1.105
D	46.030	45.980	+	1.00	=	46.980		0.95
E	46.210	45.815	+	1.00	=	46.815		0.605

Table 17.2

Field work in setting out sight rails

1. The setting out table is taken into the field (Fig. 17.9) and used in conjunction with a levelling table ruled for HPC reduction (Table 17.3).
2. A level is set up in a suitable location from which all pegs and the nearest bench mark can be observed. A backsight reading (2.150 m) is taken to

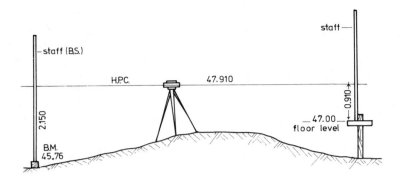

Figure 17.9

the bench mark and entered on the levelling table on line 1.

3. The HPC is calculated: HPC = 45.760 + 2.150 = 47.910. The value is entered in the HPC column on line 1.

4. The difference between the HPC and the level of sight rail A is (47.910 − 47.000) = 0.910 m. The reading is entered in the field book in the IS column on line 2. In order to set the sight rail level at 47.000 m high, a staff is held against the profile peg A and moved slowly up the peg until the observer reads 0.910 m. The bottom of the staff is then at 47.000 m and is marked in pencil against the peg.

5. A sight rail is nailed to the peg such that the top of the rail is level with the pencil mark.

6. Usually sight rails are erected at each corner of a building and the floor level transferred to the walls of the building by the bricklayers using a spirit level.

Pegs B, C, D, and E are set out in the same way after calculation of the appropriate staff readings. Each staff reading is found by subtracting the rail level from the HPC and is entered in the field book as an intermediate sight or foresight if it is the final reading. Table 17.3 shows all four readings entered in their correct table positions on lines 3, 4, 5, and 6 respectively.

BS	IS	FS	HPC	Red. level	Remarks
2.150	—	—	47.910	45.760	TBM
—	0.910	—	—	47.000	Sight rail A
—	1.110	—	—	46.800	Sight rail B
—	0.735	—	—	47.175	Sight rail C
—	0.930	—	—	46.980	Sight rail D
—	—	1.095	—	46.815	Sight rail E

Table 17.3

Large-scale excavations

Consider Fig. 17.10 in which a large area of high ground is to be reduced to formation level 50.00 m to form a sports arena. The following is the usual procedure adopted for setting out the sight rails.

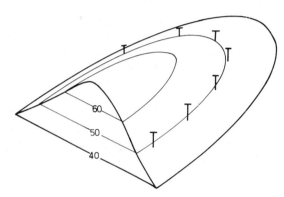

Figure 17.10

1. The sight rail level is calculated. This is normally 1.00 m above formation level = 51.00 m.

2. A levelling is made from a nearby TBM and the 50 m contour is traced on the ground. Sight rail uprights are driven in around the contour at intervals. If large earth-moving plant is being used it is wise to move the uprights outside the area of excavation.

3. The cross pieces are nailed to the uprights at 51.00 m level using the collimation system of levelling as in Table 17.1.

4. The length of traveller in this case is (51.00 − 50.00) = 1.00 m. The length of the traveller is written on the back of the sight rail in paint or waterproof chalk.

Example 1 Calculate the staff readings required to set out sight rails at the 0 m and 60 m chainages of a drain where the following data have been obtained.

Invert level at 0 m chainage 24.210 m
Ground level at 0 m chainage 25.690 m
Gradient 1 in 40 falling from 0 m to 60 m
BS reading (0.870 m) to TBM (RL 26.550 m)

Solution

Height of collim. = 26.550 + 0.870 = 27.420 m
Suitable sight rail level at A   = 26.690 m
Invert level at 0 m chainage   = 24.210 m
therefore length of traveller   = 2.480 m

This length of traveller is unsuitable and a better length is 2.500 m, so:

Sight rail level at A = (24.210 + 2.500) = 26.710 m
Fall from 0 m to 60 m = (60/40) m   = 1.500 m

Sight rail level at B   = 25.210 m
Staff reading A = (27.420 − 26.710)   = 0.710 m
Staff reading B = (27.420 − 25.210)   = 2.210 m

Vertical control using laser instruments

In setting out the proposed levels of construction works, a levelling instrument is set up in order to project a plane of collimation over the site. Measurements are then made downwards from the plane of collimation.

Since the early 1980s, laser instruments have been developed for general use on building projects. The instruments can either sweep out a plane of infra-red laser light over the sight or set out a beam of visible laser light on any predetermined gradient.

Figure 17.11 shows the LB 2 self-levelling laser level developed by Laser Alignment Inc. of Michigan, USA. There are several other similar types of level currently in use, namely Gradomat, and the Stolz Baulaser.

Rechargeable nickel–cadmium batteries provide the power to generate an invisible infra-red laser beam. This beam is rotated in a complete circle around the instrument

The instrument is either set on a tripod or free standing and is levelled by two levelling screws and spirit level. This brings the instrument within its automatic self-levelling range. The instrument automatically shuts off if the self-levelling unit is jarred out of its position and has to be relevelled.

The unit is switched on and the beam is swept out across the site over a range of 100 metres. Being invisible, the beam has to be detected at any point of setting out. This is accomplished by the Rodeye 2 sensor unit (Fig. 17.11). It weighs less than 300 grammes, is charged by a 9 volt rechargeable battery and is easily carried in the pocket. The Rodeye is either hand held against a levelling staff or against some vertical object such as a column or wall. In use it is moved slowly vertically until an indication is received in the sensor window that the sensor is on the line of the beam. A mark is made against the column or levelling staff and the setting out level is

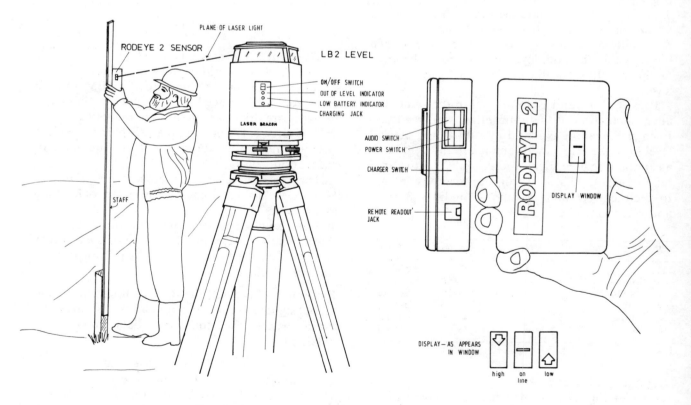

Figure 17.11

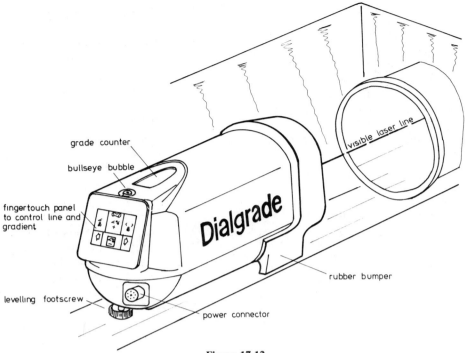

Figure 17.12

measured up or down from this mark. The Rodeye incorporates an audio signalling device which can be used to indicate the vertical position of the sensor. The instrument has an accuracy of ±1 mm.

The instrument may also be laid on its side and when properly levelled it will sweep out a vertical plane of laser light. In this position, it is ideal for setting columns and formwork in their true vertical positions..

Figure 17.12 shows Spectra Physics Dialgrade laser alignment unit for use in setting out drains and sewers, etc. The instrument is connected to a 12 volt d.c. electrical supply which produces a 2.0 MW helium neon laser beam. The instrument can be used in a free standing position or may be tripod mounted or attached to a bar clamped between the walls of a sewer pipe.

In operation, the unit is roughly levelled using the bullseye level. When switched on, the wide levelling range allows the instrument's levelling motors to adjust it automatically from there. Using the finger touch panels, the gradient of the sewer is dialled into the unit and the laser beam is projected along the trench or through a pipe at the correct gradient. Gradients of ±7 per cent may be set out with the Dialgrade. The instrument has proved to be robust, waterproof, and reliable in rugged site conditions.

Test questions

1 A drain is to be set out using an Engineer's tilting level from the following information (see Fig. 17.13):

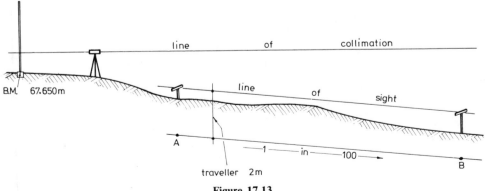

Figure 17.13

Length of drain AB—150 m
Gradient AB—falling from A to B at 1 in 100
Invert level A 64.350 m
Length of traveller 2.000 m

A backsight of 1.200 m has been taken to a nearby bench mark, the reduced level of which is 67.650 m.

Calculate the staff readings necessary to locate sight rails over A and B.

2 The following readings were taken along the line of a drain.

BS	IS	FS	Ht of coll.	RL	Dist. (m)	Remarks
0.860				97.240		TBM
2.240		0.550			0	MH 5
	2.590				48.000	MH 4
	1.690				64.000	MH 3
1.322		0.975			85.600	MH 2
		0.937			118.400	MH 1

(a) Copy the field book entries and complete the booking.

(b) The invert level of MH 1 is to be 96.811 m and the pipe runs are to slope down at the following gradients:

MH 1 to MH 2	1 in 125
MH 2 to MH 3	1 in 115
MH 3 to MH 4	1 in 100
MH 4 to MH 5	1 in 80

Calculate the invert levels of each manhole.

(c) How should the depth and gradient of the drainage trench be controlled?

(City & Guilds of London Institute
Const. Tech. Cert.)

18. Trigonometrical levelling

Principle

The principle of trigonometrical levelling is illustrated in Fig. 18.1, where the height of a building is required.

A theodolite is set up some distance from the building and a vertical angle θ is measured to the top.

The horizontal distance AB from the theodolite to the building is measured using a steel tape, and from these field quantities the height BC is calculated trigonometrically.

This height is really the height of the building above the horizontal plane through the theodolite telescope. The height BD from the horizontal plane to the ground must therefore be added to obtain the true height of the building.

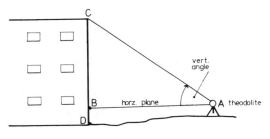

Figure 18.1

Fieldwork

1. Set the theodolite in a position such that the distance from it to the building is about one to two times the height of the building. If the theodolite is set any closer, the effect of instrumental errors on the vertical angle will be magnified.
2. Measure the vertical angle to the top of the building on both faces, and make due allowance for index error.
3. Set the telescope to the horizontal position (face left) and make a mark on the building where the horizontal plane strikes the wall. Repeat this operation on face right and, if the marks do not coincide, find the mean position.

4. Measure from the mean position to the base of the building as required using a tape or rod. This is dimension BD.
5. Measure the horizontal distance AB from the theodolite to the building using a steel tape.

In Fig. 18.1, the various field quantities are as follows:

> Mean vertical $\theta + 28°\ 30'\ 00''$
> Horizontal distance AB   39.48 m
> Height BD   1.82 m

Basic calculation

Using Fig. 18.1 and the associated field results:

$$BC = AB \tan \theta$$
$$= 39.48 \tan 28°\ 30'\ 00''$$
$$= \underline{21.44 \text{ m}}$$

$$BD = 1.82 \text{ m}$$

therefore height of building CD $= (21.44 + 1.82)$ m
$$= \underline{23.26 \text{ m}}$$

Problems in trigonometrical levelling

Problem 1

Frequently the horizontal plane through the telescope will strike the building in such a position that the height BD will be too great for physical tape or rod measurement (Fig. 18.2). In such a case the fieldwork has to be amended as follows.

Fieldwork
1. Measure the height of the instrument i and mark this height if possible on the building.
2. Measure the vertical angle θ as before.
3. Measure vertical angle α to the newly marked position on the wall.
4. Measure the slope length EF. This length is equal to AD.

Example 1   In Fig. 18.2 the various field quantities are as follows:

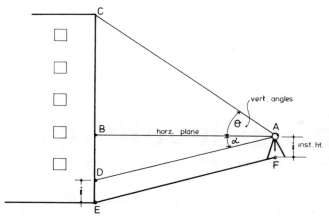

Figure 18.2

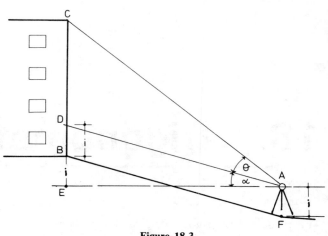

Figure 18.3

Vertical angle θ = +26° 15′ 00″
Vertical angle α = −4° 30′ 00″
Slope length FE = 51.22 m
Instrument height i = 1.36 m

Calculate the height of the building.

Solution

Height of building = CE
= CB + BD + DE
= CB + BD + 1.36 m

In triangle ABD, AB = AD cos α
= 51.22 cos 4° 30′ 00″
= 51.06 m

and BD = AD sin α
= 51.22 sin 4° 30′ 00″
= 4.02 m

In triangle ABC, BC = AB tan θ
= 51.06 tan 26° 15′ 00″
= 25.18 m

therefore height of building = 25.18 + 4.02 + 1.36
= 30.56 m

Problem 2

When the theodolite telescope is at a lower level than the base of the building, the horizontal plane through the telescope will strike the ground before reaching the building (Fig. 18.3)

Fieldwork

The fieldwork is basically the same as in the previous example. The theodolite is set at F and its height i is measured. Vertical angle θ is measured to the top of the building, and vertical angle α to a mark on the building at height i above the base. The slope length BF is measured using a steel tape.

Example 2 In Fig. 18.3 the field measurements are as follows:

Vertical angle θ = 32° 10′ 00″
Vertical angle α = 10° 05′ 00″
Slope length BF = 48.32 m
Instrument height i = 1.29 m

Calculate the height of the building.

Solution

Height of building = CB
= CE − BE
= CE − (DE − DB)
= CE − DE + i

Slope length BF = length AD
In $\triangle$ ADE AE = AD cos α
= 48.32 cos 10° 05′
= 47.57 m

DE = AD sin α
= 48.32 sin 10° 05′
= 8.46 m

In $\triangle$ ACE CE = AE tan 32° 10′
= 47.57 tan 32° 10′
= 29.92 m

Height of building = CE − DE + i
= 29.92 − 8.46 + 1.29
= 22.75 m

Alternatively:
In $\triangle$ CAD, angle DAC = 32° 10′ − 10° 05′
= 22° 05′

angle ACD = 90° − 32° 10′
= 57° 50′

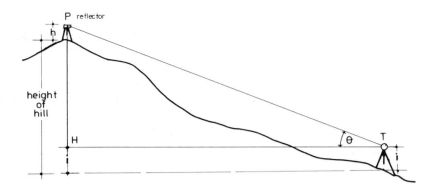

Figure 18.4

Using the Sine rule:

$$CD = \frac{\sin DAC \times AD}{\sin ACD}$$

$$\frac{\sin 22° \ 05' \times 48.32}{\sin 57° \ 50'}$$

$$= 21.46 \text{ m}$$

height CB $= 21.46 + 1.29$

$$= 22.75 \text{ m}$$

Problem 3

When the object does not have a vertical side, for example, a hill, it becomes impossible to measure a distance from the theodolite to a point vertically below the top (Fig. 18.4).

Fieldwork
The simplest solution to such a problem is to obtain the vertical angle from the theodolite position to a prism erected on top of the hill and to measure the slope length between the two electronically. The height i of the theodolite and height h of the prism above their respective ground points must also be measured.

Calculation
The height of the hill is computed trigonometrically from a solution of triangle TPH. If the reduced level of the theodolite station is known, the level of the top of the hill can also be calculated.

Example 3 In Fig. 18.4 the fieldwork quantities are as follows:

$$\begin{aligned} \text{Instrument height } i &= 1.51 \text{ m} \\ \text{Prism height } h &= 0.71 \text{ m} \\ \text{Vertical angle } \theta &= +10° \ 31' \ 10'' \\ \text{Slope length TP} &= 318.29 \text{ m} \\ \text{Reduced ground level at T} &= 93.23 \text{ m} \end{aligned}$$

Calculate the height of the hill.

Solution

$$\begin{aligned} \text{Height of hill} &= i + HP - h \\ \text{In triangle TPH, HP} &= TP \sin \theta \\ &= 318.29 \sin 10° \ 31' \ 10'' \\ &= 58.11 \text{ m} \\ \text{Height of hill} &= 1.51 + 58.11 - 0.71 \\ &= 58.91 \text{ m} \\ \text{Reduced level of hilltop} &= RL \text{ at } T + \text{height of hill} \\ &= 93.23 + 58.91 \\ &= 152.14 \text{ m} \end{aligned}$$

Problem 4

When electronic equipment is not available, there is no practical means of measuring the slope or plan length to the top of the hill, except for the very laborious method of ground taping. Distances must therefore be obtained by methods of triangulation or traversing.

Figure 18.5(a) shows an elevated point H which is visible from two coordinated stations A and B. The reduced level of point H is required.

From the coordinates, the plan length AB is 310.26 m and the reduced levels of the stations are A = 86.32 m and B = 91.60 m.

Fieldwork
A theodolite is set over station A and a target over station H, and their respective heights above the ground are measured.

The vertical angle between station A and target H is obtained as before.

The horizontal angle from H to theodolite station B, that is, angle H_1AB, is then obtained by reiteration.

The operations are repeated at station B and the results booked as below.

Instrument station	Instrument height	Target station	Target height	Vertical angle	Horizontal circle reading	Horizontal angle
A	1.350	H	2.00	+08° 30′ 00″	00° 00′ 00″	
A	—	B	—	—	65° 30′ 00″	65° 30′ 00″
B	—	A	—	—	00° 00′ 00″	
B	1.500	H	2.00	+07° 10′ 00″	59° 40′ 00″	59° 40′ 00″

Calculation

Figure 18.5(b) is a plan view of the problem. Horizontal lengths AH and BH are required from a solution of the triangle ABH.

(a)   In Δ ABH, angle HAB = 65° 30′ 00″
              angle ABH = 59° 40′ 00″
                          ―――――――
                          125° 10′ 00″
       therefore angle BHA = 54° 50′ 00″

$$AH_1 = \frac{AB \sin B}{\sin H}$$

$$= \frac{310.26 \sin 59° 40′ 00″}{\sin 54° 50′ 00″}$$

$$= 327.57 \text{ m}$$

and $$BH_2 = \frac{AB \sin A}{\sin H}$$

$$= \frac{310.26 \sin 65° 30′ 00″}{\sin 54° 50′ 00″}$$

$$= 345.36 \text{ m}$$

(b) The reduced level of point H above the stations

A and B is computed. It is unlikely that the results will be identical owing to small unavoidable observation errors. The mean is accepted as the most probable value of reduced level H.

In Δ AHH₁ (Fig. 18.5(c)):

$$HH_1 = AH_1 \tan 08° 30′ 00″$$
$$= 327.57 \tan 08° 30′ 00″$$
$$= 48.96 \text{ m}$$

$$RL \text{ ‘H’} = RLA + i + 48.96 - h$$
$$= 86.32 + 1.35 + 48.96 - 2.00$$
$$= 134.63 \text{ m}$$

In Δ BHH₂ (Fig. 18.5(c)):

$$HH_2 = BH_2 \tan 07° 10′ 00″$$
$$= 345.36 \tan 07° 10′ 00″$$
$$= 43.43 \text{ m}$$
$$RL \text{ ‘H’} = RLB + i + 43.43 - h$$
$$= 91.60 + 1.50 + 43.43 - 2.00$$
$$= 134.53 \text{ m}$$

$$\text{Mean RL ‘H’} = \tfrac{1}{2}(134.63 + 134.53)$$
$$= 134.58 \text{ m}$$

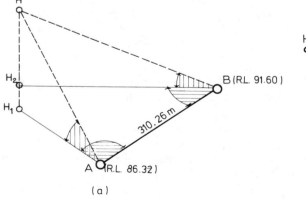

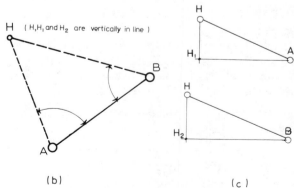

(a)   (b)   (c)

Figure 18.5

Problem 5

In the previous examples, the lines of sight from theodolite to object are assumed to be straight lines, whereas in fact all lines of sight are curved owing to the curvature of the earth. Furthermore, all lines of sight are affected by refraction as was explained on pages 80–81.

The combined correction due to curvature and refraction was shown to be $(0.0673L^2)$ m, where L is in kilometres, therefore

$$\text{correction at 500 m} = 0.0673 \times (0.5)^2$$
$$= \underline{0.017}$$

For distances below 500 m, the curvature and refraction errors are ignored in most circumstances.

When distances exceed 500 m, the problems which arise are generally beyond the ability of the technician surveyor to solve, and trigonometrical levelling involving long sights does not appear in any construction surveying syllabus.

Test questions

1 The following vertical angles were measured in the triangulation scheme of Fig. 18.6:

Instrument station	Target station	Vertical angle
A	E	+04° 20′
E	D	−02° 30′
D	C	+05° 15′
C	B	−04° 25′
B	A	−06° 26′
A	C	+07° 28′
E	C	+03° 31′

Given that the reduced level of station A is 156.20 m, and that the height of the target was adjusted to be the same height as the theodolite for each sight, calculate the reduced level of the stations B, C, D, and E.

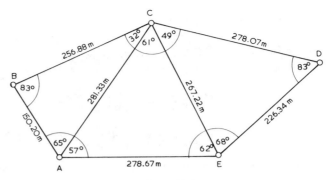

Figure 18.6

2 In a traverse survey of three lines the vertical angles were measured reciprocally as follows:

Line	Vertical angle	Length (m)
AB	+03° 35′ 20″	87.35
BA	−03° 36′ 00″	
BC	+01° 10′ 00″	153.25
CB	−01° 10′ 40″	
CD	+05° 35′ 20″	93.57
DC	−05° 36′ 00″	

Calculate the reduced levels of stations B, C, and D, given that the reduced level of A was 135.70 m AOD, and that the target heights were made equal to the instrument heights on all sightings.

3 Figure 18.7 shows two theodolite stations A and B set up some distance from a tall building. All relevant survey observations for finding the height of the building are shown in the figure. Calculate the height of the building CD.

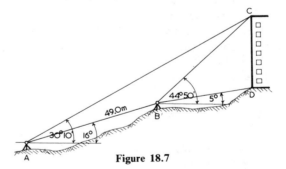

Figure 18.7

19. Electromagnetic distance measurement (EDM) and radiation detail surveying

In any form of surveying, the accurate measurement of length is at best difficult. Each method of determining the length of a line has its own particular drawbacks, and surveyors have long wished for an instrument which could measure distance as easily as a theodolite measures angles.

Modern developments in electronics have now made possible the measurement of distance using an electromagnetic signal. The measurement is accomplished in seconds with a very high degree of accuracy.

The instruments were first introduced during the 1950s and at present (1988) there are some sixty different instruments on the market.

Basic concept of measurement

The basic concept is simple. An EDM instrument capable of transmitting an electromagnetic signal is set up over a survey station at one end of a survey line (Fig. 19.1). The signal is directed to a reflector or second transmitter at the other end of the line, where it is reflected or instantly retransmitted back to the transmitter. The transit time of the double journey is measured by the transmitter and, since the speed of light is accurately known, the distance is calculated from the formula:

Distance between stations (m) = velocity of signal (m/s) × transit time (s)

i.e., $$D = V \times t \qquad (1)$$

The electromagnetic signal which is transmitted is in the form of radio waves—infra-red light, visible light, or laser beam—all of which have different properties, although they travel at the same speed.

In order to understand the complexities of EDM even in this simplified version, it is necessary to have a basic understanding of the properties of electromagnetic radiation and the methods employed in measuring the time interval.

Properties of the signal

A direct analogy can be made between the movement of electromagnetic waves and that of waves caused by dropping a stone into a pool of water.

(a) Water waves

(i) *Wavelength and phase difference*
In Fig. 19.2, a stone has been dropped into a pool at a point X, and the pattern of waves is seen emanating from that point to the bank at point Y. There are

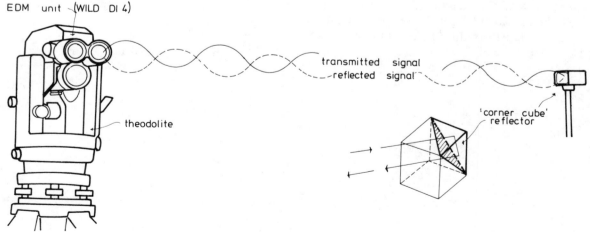

EDM unit (WILD DI 4)

transmitted signal
reflected signal

theodolite

'corner cube' reflector

Figure 19.1

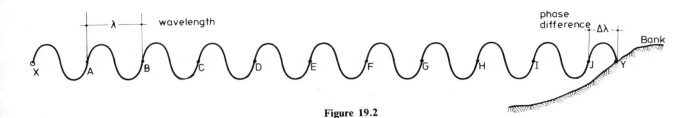

Figure 19.2

ten whole waves and a fraction of a wave between X and Y, the fraction in this case being a half wavelength.

The distance between any two similar points is obviously the length of the wave, so XA = AB, etc. = wavelength.

When wave formation is being considered, the waves are said to be in phase when a complete number of them occurs between starting and finishing points. In this case there is an incomplete portion of a wave finishing on Y, and this particular wave formation is said to exhibit a 'phase difference' of 0.5 wavelength.

(ii) *Frequency*

If the wave takes 2.1 seconds to travel from X to Y, its frequency (*f*) of occurrence is

$$f = 10.5 \text{ times per } 2.1 \text{ seconds}$$
$$= 5.0 \text{ times per second}$$

Technically the wave makes 5 cycles per second. The SI unit of frequency meaning one cycle per second is 1 hertz, so the wave in question has a wavelength of λ metres occurring with a frequency of 5 Hz.

The various multiples of one hertz are derived in the usual manner by prefixing kilo, mega, and giga to the word hertz, as in Table 19.1.

1 hertz = 1 Hz
10^3 hertz = 1 kilohertz = 1 kHz
10^6 hertz = 1 megahertz = 1 MHz
10^9 hertz = 1 gigahertz = 1 GHz

Table 19.1

If the wavelength λ measures 0.5 metre, the distance XY is found thus:

$$\text{distance XY} = \text{wavelength } \lambda \times \text{frequency } (f)$$
$$= 0.5 \times 10.5$$
$$= 5.25 \text{ m}$$

The velocity (*v*) of travel of this wave is therefore:

$$\text{velocity} = \text{distance } D \div \text{time } t$$
$$= 5.25 \div 2.1$$
$$= 2.5 \text{ metres per second}$$
$$= 2.5 \text{ m/s or m s}^{-1}$$

Taking a different view of the problem the distance D_{XY} can be calculated without a knowledge of the wavelength provided the frequency (*f*) and transit time (*t*) are known. The 'fieldwork' consists simply of counting the number of wavelengths between X and Y, and the problem (though impracticable) can be presented as follows.

Example 1 A stone dropped into a pool at a point X generates a wave which travels through the water with a frequency of 5 Hz and at a velocity of 2.5 m/s.

A count of the waves shows that 10.5 waves occur by the time the wave reaches the bank Y. Calculate the distance XY.

Solution

$$\text{Number } (n) \text{ of waves} = 10.5$$
$$\text{Frequency } (f) = 5 \text{ Hz}$$
$$\therefore \text{ Time } (t) = 10.5/5$$
$$= 2.1 \text{ s}$$

$$\text{Now, velocity } (v) = 2.5 \text{ m/s (given), therefore}$$
$$\text{Distance } D_{XY} = \text{velocity } (v) \times \text{time } (t)$$
$$= 2.5 \times 2.1$$
$$= 5.25 \text{ m}$$

From above
$$D = vt$$
$$= v \times \frac{n}{f} \quad (2)$$

This method and the derived formula are important in EDM and indeed form the basis of the system.

(b) Electromagnetic waves

Light, infra-red rays, and radio waves are all forms of electromagnetic radiation and, like heat and sound, are forms of energy.

(i) *Wavelength and frequency*

Electromagnetic waves behave in much the same manner as waves in water. Figure 9.3 shows part of the electromagnetic spectrum. The wavelengths of the various bands vary from 10 000 m long waves to 0.001 mm visible light waves, the corresponding frequencies being 30 kHz and 30×10^{10} kHz respectively.

Electromagnetic distance measurement (EDM) 257

Only a narrow band of these waves can be used to measure distances to the standard of accuracy required in surveying.

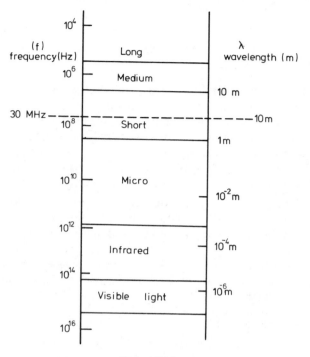

Figure 19.3

In the sections following, it will be pointed out that a phase difference of 1/1000th part of a wavelength can be resolved by EDM instruments. For most construction surveys, an accuracy of ±1 cm is acceptable, therefore the derived wavelength is 1000 × 1 cm = 10 m which from Fig. 19.3 corresponds to a frequency of 30 MHz.

Table 19.2 lists a few of the EDM instruments in common use and shows the frequency and approximate wavelength of the measuring signal.

As previously stated, the range of suitable measuring frequencies shown in Table 19.2

Instrument	Frequency	Approximate wavelength
Wild M.D.60		
fine measurement	149.8483 MHz	2 m
coarse measurement	14.977337 MHz	20 m
Kern DM 500		
fine measurement	14.9854 MHz	20 m
coarse measurement	149.854 kHz	2000 m
Wild D.I.3		
fine measurement	7.4927 MHz	40 m
coarse measurement	74.927 kHz	4000 m
Geodimeter 6	30 MHz	10 m
Tellurometer CA 1000	(19–25) MHz	(16–12) m

Table 19.2

represents only a small section of the complete electromagnetic spectrum. Unfortunately this range of frequencies is not suitable for direct transmission through the atmosphere by EDM instruments, because the waves tend to fade and scatter, and suffer from interference.

Very high frequency waves are not so prone to these effects, and it is possible to impress a low-frequency measuring wave onto a wave of higher frequency and transmit them together. The high-frequency wave acts as a carrier for the low-frequency wave and is said to be modulated by this process. Among others, infra-red and visible light waves are suitable carriers.

In simpler language, the visible light wave is analogous to the thin strip of steel from which a tape is manufactured. The steel is 'modulated' by the metric graduations stamped on it and carries them when the tape is stretched during linear measurement.

(ii) *Velocity*

All electromagnetic waves travel through outer space with the same velocity (c) of 299 792.5 km/s. When travelling through earth's atmosphere, the velocity (v) is retarded. Variations in temperature, pressure and humidity affect the velocity, with the result that the value of v is not quite constant.

This is analogous to measuring lines with a steel tape which continually changes its length, so some standard must be set for the EDM instrument in the same way as standards of 20° C temperature and 44.5 newtons tension are set for steel tapes.

The normal standardizing values are 760 mm Hg pressure and 12° C temperature, and under those conditions it can be shown that electromagnetic signals are travelling at 99.97 per cent of their velocity (c) in vacuum.

The velocity (v) through earth's atmosphere is therefore (299 792.5 × 99.97 per cent) = 299 708.0 km/s approximately.

If at the time of measurement the values of temperature, pressure, and humidity differ from the standard values, corrections must be made to the measurement of the line.

Principle of distance measurement

The basic concept of measurement already outlined now becomes a definite principle in which a modulated electromagnetic wave of known frequency (f) is transmitted to a distant reflector and returned to the transmitter. The transmitting instrument is capable of counting the number (n) of wavelengths with an accuracy of 1/1000th part of a wavelength, using one of the methods outlined later in this chapter.

The value of (n/f) is computed, either manually, or automatically by the instrument, and multiplied by the 'standardized' speed of the signal through the atmosphere.

The result is the slope length of the line measured.

EDM systems

The systems developed from the transmission of electromagnetic waves can be conveniently divided into two classes, namely:
1. microwave system (long range).
2. electro-optical system (medium and short range).

1. Microwave system

As the name suggests, this group of EDM instruments uses microwaves to measure distances from 20 m to a maximum of 150 km, with an accuracy of about 3–4 mm per km. An instrument typical of this class is the Wild DI 60 (Fig. 19.4) which operates on frequencies of about 15 MHz.

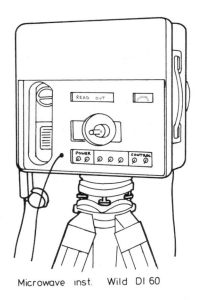

Microwave inst. Wild DI 60

Figure 19.4

Fieldwork

In Fig. 19.5 the length of line AB is known to be 835.30 m. In order to check the length using microwave methods, two instruments are required. The instruments may be identical as with the Wild M.D. 60, or may be different and not interchangeable as with the Tellurometer CA 1000.

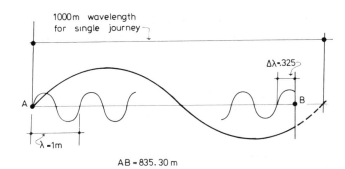

Figure 19.5

(i) One instrument called the 'master' is set at A and pointed in the direction of B. The second instrument called the 'remote' is set at B and reciprocally pointed to A. It is not critical that the two instruments are exactly in alignment since the microwave beam has a spread of about 10°. The instruments are fitted with powerful radiotelephones, so communication between the parties is always possible over long distances. Using the telephones, the parties make contact and tune their respective instruments to each other.

(ii) The measuring switch is activated, and a microwave of wavelength w metres is transmitted to the remote instrument where it is instantly retransmitted back to the master instrument. The signal has therefore traversed the double distance AB, and in doing so will have completed a certain number of whole wavelengths and a fractional part of a wavelength. From the earlier water-wave analogy, the fractional part of the wave is called the phase difference, and it is the phase difference that is accurately measured by the instrument (Table 19.3).

(iii) The wavelength is then changed to w_2, and the phase difference is again measured. This process is repeated a certain number of times on different wavelengths, the number depending on the make of instrument and length of the line. These wavelength changes are made automatically with the wild M.D. 60, the whole measuring procedure taking approximately thirty seconds.

A typical set of wavelengths and the phase difference obtained on the measurement of line AB are shown in Table 19.3.

Wavelength λ	Equivalent λ for single journey	Phase difference ($\Delta\lambda$)
w_1 2 m	1 m	0.325
w_2 20 m	10 m	0.532
w_3 200 m	100 m	0.353
w_4 2000 m	1000 m	0.835

Table 19.3

Measurement of atmospheric conditions

At the finish of the measuring procedure, measurements of the prevailing temperature, pressure, and humidity must be made, since they may differ from the standard. The measurements are made using wet-and-dry bulb thermometers and a barometer.

These measurements are made at both ends of the line and are usually assumed to represent the meteorological conditions along the whole length of the line. If the line is long, however, there is no guarantee that the measured conditions actually prevail along the complete length, and the uncertainties are thought to be responsible for errors of about 2 mm per km.

In this example, the measurements of the actual temperature and pressure are assumed to be $3°$ C and 800 mm Hg respectively.

Calculation of the distance

(i) *Number (n) of wavelengths* From Table 19.3, the number of wavelengths of length w_1 ($\lambda = 2$ m) can be successively approximated. For example, using wavelength w_4 ($\lambda = 2000$ m) where the phase difference $\Delta\lambda = 0.835$, the number of *complete* wavelengths is found as follows:

$$\Delta\lambda = 0.835$$
$$\text{but } w_4 = 1000w_1$$
$$\text{therefore } \Delta\lambda \times 1000 = 0.835 \times 1000$$
$$= 835$$

The *total* number of wavelengths (complete and partial) is derived thus

$$1000 \times \Delta\lambda \ (0.835) = 835.0$$
$$100 \times \Delta\lambda \ (0.353) = 35.3$$
$$10 \times \Delta\lambda \ (0.532) = 5.32$$
$$1 \times \Delta\lambda \ (0.325) = 0.325$$
$$\text{therefore No. of wavelengths} = 835.325$$

(ii) *Transit time (t)* From the formula (2) on water-wave analogy, the transit time for a wave is:

$$t = n/f$$

$$= \frac{(835.325)}{(149.848\ 300\ 0 \times 10^6)} \text{ second}$$

$$= 5.574\ 471 \times 10^{-6} \text{ second}$$

(iii) *Distance AB* The double distance AB is computed by multiplying the value of the transit time (t) by the velocity (v) of the waves through the atmosphere. The standard value of v at $12°$ C temperature and 760 mm Hg has been shown to be 299 708.0 km/s. Therefore

$$2 \times AB = (5.574\ 471 \times 10^{-6} \times 299\ 708.0 \times 1000)\text{m}$$
$$= 1670.650 \text{ m}$$
$$\text{therefore } AB = 835.356 \text{ m}$$

These calculations are made automatically in the most recent microwave instruments and are displayed digitally as an 8-figure number.

Atmospheric correction for $3°$ C temperature and 800 mmHg is -0.009 m, therefore

$$
\begin{aligned}
AB = &\quad 835.356 \\
- &\quad\ \ 0.009 \\
\hline
&\quad 835.347 \text{ m} \\
= &\quad 835.347 \text{ m}
\end{aligned}
$$

2. Electro-optical system

The instruments employed in this system of measurement can be conveniently divided into two classes, depending upon which part of the spectrum they use for signal transmission. The instruments which use visible light form the medium-range class, while those using infra-red light form the short-range class. The classes have many common elements.

The measuring signals are carried on a narrow, highly focused beam of light which has to be directed optically to the distant target by means of an in-built telescope. Alternatively, the short-range EDM units may be mounted on the telescope of a theodolite by means of a bracket specially designed to point the EDM unit exactly along the line of the theodolite telescope, wherever it is pointed (Figs 19.1, 19.8 and 12.13).

The medium-range Geodimeter 710 model is a purpose-built instrument which combines an electronic digital theodolite with an EDM unit as a calculator. All measurements are presented in digital form.

Fieldwork

The fieldwork is substantially the same for all kinds of electro-optical distance measurement.

(i) The transmitter is placed at one end of the line being measured and is accurately aligned by means of the telescope onto a corner-cube reflector at the other end (Fig. 19.1). The reflector is reciprocally aligned back to the transmitter by means of the open sights on top of the mounting. The alignment is not critical. The important property of the reflector is that it reflects light back along any line on an exactly parallel course. In general one corner-cube reflector is effective for ranges up to about 600 m. For longer lengths a bank of three, six or nine prisms is required.

(ii) The signal is transmitted on a known frequency to the reflector, from which it is returned to the instrument and the phase difference is measured.

(iii) The frequency is changed either manually or automatically by the instrument, and the measuring procedure repeated, enabling the number of wavelengths to be counted and the slope distance to be calculated.

(iv) Measurements of the prevailing atmospheric conditions are made, and appropriate corrections are applied to the slope length by means of the nomograph supplied with some instruments.

(v) The vertical angle between instrument and reflector is measured in the usual manner to enable the plan length of the line to be calculated.

These are the common steps in measuring the line. The generation of the signal and the method used to count the number of wavelengths differs with the two groups of instrument.

(a) *Medium-range instruments*

The 'Geodimeter 6' typifies the medium-range group of instruments and is shown diagrammatically in Fig. 19.6.

(i) *Signal generation* The light source is a normal 5V lamp powered by battery which is used for measuring lengths up to 5 km long in daylight or 15 km long during darkness. For longer distances of up to 25 km, a mercury-vapour lamp powered by a generator is used.

(ii) *Wavelength count* The phase difference for any measured line is obtained on three different but very closely related frequencies of around 30 MHz. Thus the wavelength is 10 m for the double journey along the line.

Using half the wavelengths gives an equivalent value of 5 m for a 'one way' journey. The three wavelengths used for the measurement are:

$$w_1 = 5.000\ 000 \text{ m}$$
$$w_2 = 4.987\ 532 \text{ m}$$
$$w_3 = 4.761\ 904 \text{ m}$$

These lengths are chosen such that

$$400w_1 = 401w_2 = 2000 \text{ m}$$
and
$$20w_1 = 21w_2 = 100 \text{ m}$$

Using the same line as that used in the microwave system, namely AB = 835.300 m, the phase differences resulting from measurements on the wavelengths w_1, w_2, and w_3 are $\Delta w_1 = 0.300$, $\Delta w_2 = 2.382$, and $\Delta w_3 = 1.967$ respectively.

$$\text{Distance AB} = nw_1 + \Delta w_1 \qquad (3)$$
$$= nw_2 + \Delta w_2 \qquad (4)$$
$$= nw_3 + \Delta w_3 \qquad (5)$$

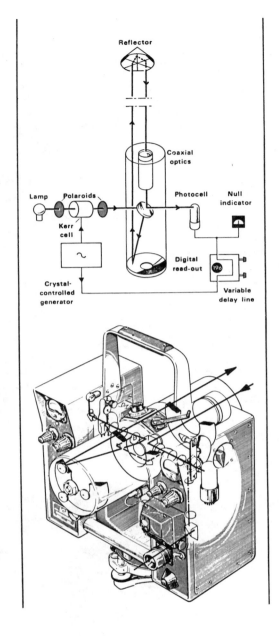

Figure 19.6

From (3) and (4):
$$n(w_1 - w_2) = \Delta w_2 + \Delta w_1$$
and since
$$400w_1 = 401w_2$$
$$w_2 = (400/401)w_1$$

therefore $n\left(w_1 - \dfrac{400}{401}w_1\right) = 2.382 - 0.300$

therefore $\dfrac{nw_1}{401} = 2.082$

$$nw_1 = 834.9$$
$$= \underline{835}$$

(d) Triple prism with tilting holder

(c) DM-S2 Independent setting

(b) DM-S3 Telescope setting

(a) DM-S2 Yoke setting

Figure 19.7

This value will be repeated every 2000 m.

From (3) and (5):

$$n(w_1 - w_3) = \Delta w_3 - \Delta w_1$$

and since

$$20w_1 = 21w_2$$

$$w_2 = (20/21)w_1$$

therefore $n\left(w_1 - \dfrac{20}{21}w_1\right) = 1.967 - 0.300$

therefore $\dfrac{nw_1}{21} = 1.667$

$$nw_1 = \underline{\underline{35}}$$

This value will be repeated every 100 m.

The total number of wavelengths is therefore $835 + \Delta w_1 = (835.30)$.

(b) Short-range instruments

(i) *Signal generation* All modern short-range instruments emit an infra-red carrier wave generated by a gallium arsenide (GaAs) diode. The wavelength is less than 1 micrometre. The power is supplied by nickel–cadmium dry-cell batteries or by a 12 volt car battery. The beam is invisible and harmless, and will produce the correct distance even when it is broken by traffic.

(ii) *Wavelength count* All instruments of this class are completely automatic. The number of wavelengths is counted by some electromechanical device and the slope distance is displayed digitally.

Modern infra-red EDM instruments are compact and lightweight and can be readily telescope mounted or yoke mounted on a theodolite or can be independently mounted on a detachable theodolite tribrach (Fig. 19.7(a) (b) (c)).

The reflector is mounted on a telescopic rod which can be extended and set to instrument height or may be tripod mounted. One, three or nine prism assemblies are used for different distances (Fig. 19.7(d)).

Operation of an infra-red instrument

The Kern DM 502 is a typical instrument of this class. It is small, compact and weighs less than 2 kg. The unit is easily attached to a Kern theodolite by means of a spring clip (Fig. 19.8). Power is supplied from a nickel–cadmium battery strapped to the tripod, and is connected to the EDM unit by a flexible sheathed cable.

The measuring operations are as follows:
(a) Switch on the power.
(b) Set the theodolite line of sight onto the distant reflector.
(c) Set the function switch to 'measure'.
(d) Press the starting button marked 'measure'.

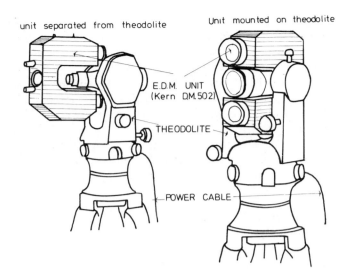

Figure 19.8

The instrument then measures the phase difference on two different frequencies and, using an electromechanical device, automatically computes the slope distance and displays it digitally. The complete operation takes about 8 seconds.

A correction has to be made for the prevailing atmospheric conditions if they differ from the standard conditions.

The effect of changing humidity has little effect at infra-red frequencies, but the corrections due to differences in temperature and pressure may be significant.

Settable switches enable this information to be entered into the instrument.

Comparison of microwave and electro-optical systems

(i) Application

The microwave system is used to best advantage in large-scale geodetic surveys, for example, the measurement of baselines in triangulation schemes or the fixation of points by trilateration methods.

Measurement is possible in light rain or mist when the stations might not be intervisible, and in this respect the telephone link is a great advantage.

The main disadvantage of the system lies in the fact that two expensive instruments are necessary and two trained operators are required to use them.

Electro-optical instruments are used in all surveying fields except that of long-range geodetic surveying. Their versatility, accuracy, and saving in time and manpower have made them invaluable for all short-range survey measurements.

They are used for purposes of traversing and trigonometrical levelling in and around construction sites; for baseline measurement in minor triangulation schemes; for measuring shaft depths and tunnels in mining, and for measuring previously inaccessible distances across rivers, etc., and have virtually replaced tacheometry as a means of detail surveying.

Most instruments have a tracking device which allows them to be used for setting out all manner of construction and engineering works.

(ii) Range and accuracy

Reference has been made throughout this chapter to the range of the various classes of instrument. These are summarized in Table 19.4.

The accuracy of EDM equipment comprises two elements, namely:
(a) the instrumental limitation,
(b) the influence of atmospheric irregularities.

Most instruments have an instrumental error of about ±5 mm, although a very recent development of the Kern Mekometer has reduced this error to ±0.2 mm.

Atmospheric vagaries of temperature, pressure, and humidity produce errors varying from 1 mm to 10 mm per km. The effect of humidity on microwave measurements is significant.

The errors are summarized in Table 19.4.

Radiation surveys

The introduction of EDM instruments has radically altered the accepted methods of detail surveying.

It should be apparent to readers of this book that the object of any survey is to establish the tridimensional coordinates (easting x, northing y, and reduced level z) of a series of points which when plotted together form a plan.

The routes to the coordinates are many and varied (see Figure 19.9).

Instrument	Type	Accuracy	Range
Tellurometer CA 1000	microwave long-range	<15 mm ± 5 mm/km	min. 50 m max. 30 km
Wild M.D. 60	microwave long-range	±1 cm ± 3 mm/km	min. 20 m max. 150 km
Geodimeter 114	electro-optical medium-range	±5 mm ± 1 mm/km	max. 25 km
Kern DM 503	electro-optical short-range	±3 mm ± 2 mm/km	max. 5 km
Wild D.I. 4	electro-optical short-range	±5 mm ± 5 mm/km	max. 2 km
Hewlett-Packard 3800 B	electro-optical short-range	±5 mm ± 7 mm/km	max. 3 km

Table 19.4

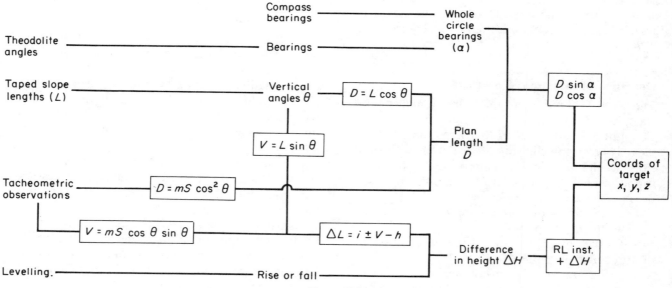

Figure 19.9

These calculations are considerably simplified when an instrument is available which can directly measure the three quantities x, y, z (Fig. 19.9), namely bearing, plan length and difference in height. Theodolite/EDM combinations can provide these quantities either singly or in combination.

Radiation—fieldwork

Figure 19.10 shows a garage forecourt which is to be surveyed. The bearing of the roadway fronting the garage is already known from a map. In a radiation survey the theodolite is set up in a central position from which all points to be surveyed are visible. The known bearing of the roadway is set on the circle and a reference point (RO) is sighted.

The instrument height is measured and the staffman sets the reflector on the rod to this height. He directs the assembly towards the instrument station at every point to be fixed. The points are sighted in succession from the theodolite. Thus they radiate from the instrument station.

At every pointing the horizontal angle, vertical angle and slope length are measured. Depending on the theodolite/EDM combination the plan length, difference in height, reduced level and coordinates of the target station are calculated either manually, semi-automatically or completely automatically.

The points may be surveyed in random groups (points 1–9 where all the points are grouped together) or may be surveyed in strings (points 17–20 and points 21–24) even though it would be more convenient to survey these points in the order 17, 24, 18, 23, etc. The string method is preferred whenever

the entries are made into an electronic field book.

The radiation measurements may be obtained and booked using the following theodolite/EDM combinations.

(a) *Theodolite and slope measuring EDM* (Fig. 19.11) Virtually every manufacturer of surveying instruments produces a simple EDM which can be mounted on the theodolite. Every instrument must be used in the manner prescribed by the manufacturer.

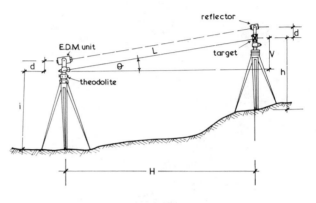

Figure 19.11

The Sokkisha RED 1A and Kern DM 104 are capable of measuring only the slope distance L. The vertical angle, θ, and the horizontal bearing, α, are measured using the theodolite (Fig. 19.12).

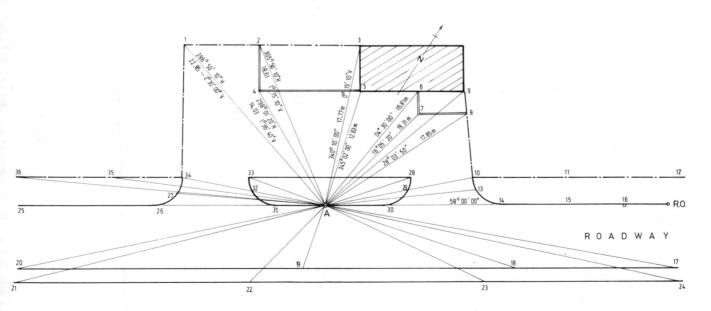

Figure 19.10

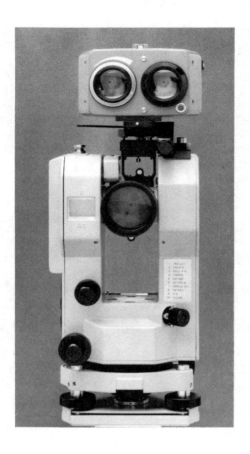

(a) Kern DM 104

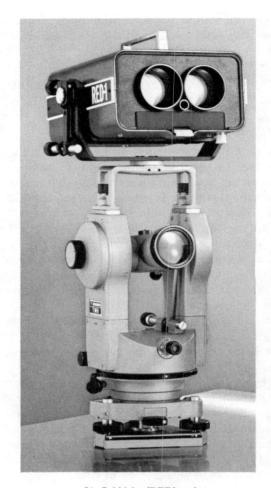

(b) Sokkisha 'RED' series

Figure 19.12

Calculation of coordinates

The *x*, *y*, and *z* coordinates must be calculated manually. In Fig. 19.10 the coordinates of any target point in relation to the instrument station are:

(a) Horiz. distance H = $L \cos \theta$
(b) Difference in height V = $L \sin \theta$
(c) Target easting (x) = theodolite easting + $H \sin \alpha$
(d) Target northing (y) = theodolite northing + $H \cos \alpha$
(e) Target red. level (z) = theodolite red. level + $i + V - h$

Example 2 Calculate the coordinates of stations 1–4

of Fig. 19.10 given that the coordinates of station A are 100.00E, 100.00N, and the reduced level is 50.00 m.

Solution See Table 19.5.

These same calculations are performed for every target point. The calculations are numerous and the normal practice is to computerize them. *Program 11 of Chapter 22 shows stations 1–4 of Fig. 19.10 calculated by computer and spreadsheet.*

Line	WCB	Slope length	Vert. angle	Plan length	Partial coords Dep.	Partial coords Lat.	Total coords E	Total coords N	Height diff.	Reduced level	Stn
							100.00	100.00		50.00	A
A–1	286° 50′ 10″	22.85	2° 30′ 00″	22.83	−21.80	6.61	78.20	106.61	+1.00	51.00	1
A–2	305° 50′ 10″	18.61	1° 15′ 10″	18.61	−15.09	10.90	84.91	110.90	+0.41	50.41	2
A–3	340° 10′ 00″	17.77	−00° 15′ 10″	17.77	−6.03	16.72	93.97	116.72	−0.08	49.92	3
A–4	298° 01′ 20″	14.03	−01° 36′ 40″	14.02	−12.38	6.59	87.62	106.59	−0.39	49.61	4

Table 19.5

(b) *Theodolite and automatic EDM* (Fig. 19.13)
Some EDM instruments can calculate and display the plan distance and difference in elevation using an in-built computer. The Geodimeter 216 measures slope distances up to 2.2 km with an eight prism assembly and vertical angles using a vertical angle sensor. An atmospheric correction factor is applied automatically to the slope length by simply entering the prevailing temperature and pressure on a correction dial.

Using the position selector switch, the slope length L, plan length H, or vertical height V, may be displayed automatically.

Figure 19.13

Calculation of coordinates

(a) Target easting (x) = theodolite easting $+ H \sin \alpha$
(b) Target northing (y) = theodolite northing $+ H \cos \alpha$
(c) Target red. level (z) = theodolite red. level $+ i + V - h$

Example 3 Calculate the coordinates of stations 5–8 of Fig. 19.10 relative to station A (100.00 m E, 100.00 m N, 50.00 m AOD).

Solution See Table 19.6.
Computer program 12 of Chapter 22 shows the same calculation.

(c) *Electronic tacheometer*
An electronic tacheometer or tachymeter is really an electronic theodolite and EDM which are controlled by the same microprocessor. The EDM may be in-built or detachable. The instruments can measure the slope length, vertical angle, and bearing of a line and display these parameters together with plan length, vertical height, and even the coordinates of the target point (Fig. 19.14).

Horizontal and vertical angles are measured electronically to a few seconds. Some form of gravity sensor is used in the measurement of vertical angles, leading to highly accurate measurement.

Distances up to about 2.5 km are measured electronically in 2–4 seconds and measurements may be made in single, repeat, or track mode.

All measurements of angles and distances are displayed on eight digit LCDs.

Data recording
Most electronic tacheometers can be linked to some form of data recorder or electronic field book. The angles and distance are stored automatically in the field book, resulting in a saving of time and reduction of errors.

Typical examples of this generation of microprocessors are the AGA Geodat, the Sokkisha SDR2, the Husky Hunter, and the Kern DIF41/Hp41 (Fig. 12.13).

The instruments are programmed to record the instrument station and instrument height; the horizontal angle, vertical angle and slope distance to a target; the target station and target height. The

Line	WCB	Length	Dep.	Lat.	E	N	Height diff.	Reduced level	Stn
					100.00	100.00		50.00	A
A–5	345° 02′ 00″	12.83	−3.31	12.39	96.69	112.39	−0.95	49.05	5
A–6	4° 30′ 00″	16.81	1.32	16.76	101.32	116.76	−0.75	49.25	6
A–7	16° 05′ 20″	19.31	5.35	18.55	105.35	118.55	−0.10	49.90	7
A–8	28° 03′ 50″	17.85	8.40	15.75	108.40	115.75	+0.34	50.34	8

Table 19.6

surveyor is prompted by the data recorder and enters information at the touch of a button. The instruments can record the information of hundreds of points.

The data may be processed in the field utilizing the survey programs of the data collector or may be down loaded into a computer system via an interface.

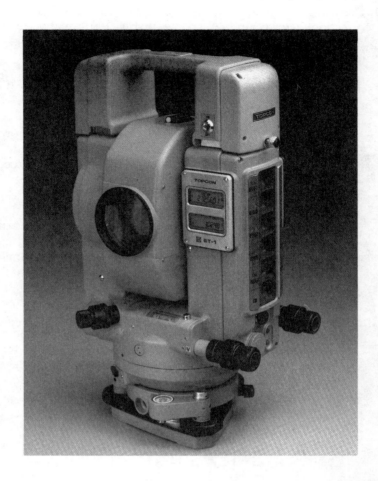

Figure 19.14

20. Surveys of existing buildings

The building technician and, in particular, the building surveyor will at some time in his career be concerned with the extension, repair, alteration, or demolition of existing buildings. In all of these cases, planning departments and building control departments of local authorities require accurate plans of the existing and proposed buildings, and it is the surveyor's task to take sufficient measurements to enable these to be made. The survey of even a small building will necessitate a large number of measurements being taken.

Besides being capable of conducting the survey, the surveyor must also be aware of (a) the Building Regulations and (b) current building construction practice.

(a) *Building Regulations*

Before any building works can proceed, the local authorities must be satisfied that the Building Regulations have been observed.

The surveyor does not require a detailed knowledge of these regulations but should certainly be aware of their implications.

(b) *Building construction*

The surveyor must have a sound knowledge of building construction and must understand thoroughly the construction of foundations, solid and cavity walls, roof, floors, and windows in order to be able to draw a building convincingly.

Classification of drawings

In all construction schemes, several classes of drawings are required, the classification depending upon the particular information which is to be disseminated to users.

In British Standards Institution BS 1192: 'Recommendations for Building Drawing Practice', the following classifications are recommended.

1. Design stage

Sketch drawings to show the designer's general intentions.

2. Production stage

(a) *Location drawings*

(i) Block plans to identify the site and locate outlines of buildings in relation to town plan wherever possible. It is recommended that this plan should be made from the appropriate OS 1:1250, although most authorities accept a scale of 1:2500.

(ii) Site plans to locate the position of buildings in relation to setting out points, means of access, general layout of site. The plans should also contain information on services, drainage, etc. The recommended scale of these drawings is 1:200, but again local authorities accept 1:250 or 1:500.

(iii) General location drawings to show the position occupied by the various spaces in building, the general construction, the overall dimensions of new extensions, alterations, etc. The recommended scale of these plans is 1:50 or 1:100.

(b) *Component drawings*

This classification includes ranges of components, details of components, and assembly drawings which are not really the concern of this chapter.

Principles of measurement

In general, the principles involved in measuring a building are those used in the measurement of areas of land. In particular, the principles of chain surveying are applied most often, since buildings can usually be measured completely by taping.

Occasionally a building is of such a complex nature that a theodolite is required.

When floor levels are to be related to outside ground and drainage levels, a level and staff are required.

Conducting the survey

Example 1 Figure 20.1 shows the location of a holiday cottage situated in pleasant rural surroundings on the bank of a loch (or lake). The cottage is old and requires renovation and extension. No drawings exist and a complete survey of the premises is to be made.

The following sections describe and illustrate some of the survey work required.

At the conclusion of the chapter the reader should attempt to plot the survey.

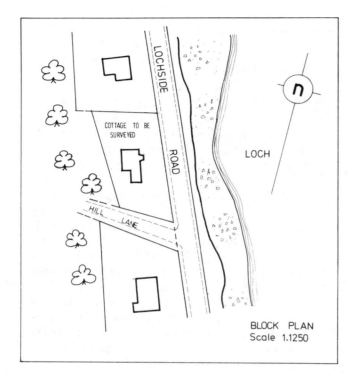

Figure 20.1

1. Preparation

Before surveying any properties, it is good practice to study the OS sheet and any old drawings which may exist. From these plans, a knowledge of the North direction can usually be gained, and a list of adjoining addresses compiled. The proprietors or tenants of these adjoining properties may have to be contacted before permission to build any extension will be given by local authorities.

From the study, it may also be possible to gain some knowledge of difficulties which may arise during the subsequent survey.

2. Reconnaissance

As in all surveys, time spent in reconnaissance is well spent. During the 'recce', attention is paid to the shape of the building, the number of floors, type of roof, position of doors and windows.

Squared paper is a necessity, and all sketches made during the reconnaissance should be roughly

to scale. The scale depends upon the complexity of the building, but in general 1:100 scale proves adequate.

In measuring buildings, the relationship between rooms is all-important, and a plan view of each floor is preferable to a room-by-room sketch.

Elevations will also be required, and here again the whole side of a building should be sketched in preference to a floor-by-floor elevation. The rule of working from the whole to the part is thereby adhered to.

Equipment

In most cases the method of taping will be used, in which case the following equipment will be required:
1. 20 m steel tape with locating hook.
2. 5 m steel tape.
3. 2 m folding rule
4. Plumb-bob and string, chalk, light hammer and short nails.
5. Measurement book or loose-leaf pad containing a supply of squared paper.
6. Soft pencils, hard pencils and eraser.

The ideal number of surveyors is two, one of whom should be fairly experienced.

3. Procedure

(a) *Site survey*

The site survey uses the principles of chain surveying, namely trilateration and offsetting. On most small sites the details can be surveyed by trilateration alone.

In many cases it is possible to use the building sides as base lines and extend them to the boundaries, supplementing the dimensions with additional diagonal checks (Fig. 20.2).

Whenever possible, running dimensions should be taken, as this procedure is normally physically easier and leads to fewer errors.

The drainage arrangements may be shown on this plan if the system is simple, otherwise a separate sketch is drawn.

(b) *Building location*

Plan During the reconnaissance survey, a plan view detailed sketch of the outside of the building is drawn on squared paper (Fig. 20.3).

The measuring procedure is arranged in a systematic orderly manner, so that every feature of the building is recorded.

The measurements are taken in a clockwise direction and, wherever possible, running sizes are taken along a complete side of the building.

The tape must be fitted with a locating hook, otherwise one surveyor will have to be employed

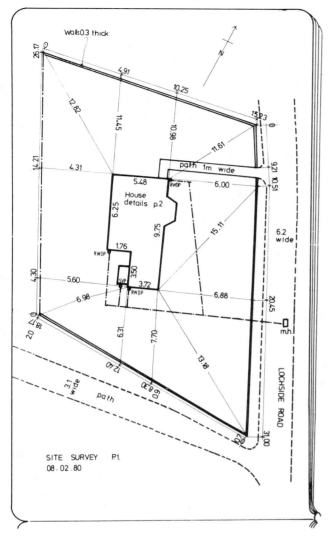

Walls 0.3 thick

path 1m wide

House details p.2

6.2 wide

LOCHSIDE ROAD

path 3.1 wide

SITE SURVEY P1.
08.02.80

Figure 20.2

and the width checked against the building location measurements.

The vertical dimensions from ground level to sill, sill to lintel, and lintel to soffit are measured next as running sizes when possible. Generally this is not possible without the aid of a ladder, and the dimensions are measured individually using the 2 m folding rule.

In the next section it will be seen that measurements must be made internally from the sill to floor level, and from lintel to ceiling level, thereby establishing relationships between floor, ceiling, and ground level.

Heights are then measured to the eaves, either from ground level or from a convenient window lintel. A ladder is usually necessary to reach the eaves but, if it is dangerous, a 4 or 5 m levelling staff is usually long enough to enable the measurement to be made.

From the paragraphs above it will be obvious that vertical measurement is difficult, sometimes dangerous. However, a simple, convenient, accurate method of determining heights exists and yet is seldom used.

A level is set up in a convenient position and, in a very few minutes, staff readings can be made to ground levels and window sills. The staff is then inverted and reading taken to lintels, soffits, and eaves. If the level has been judiciously placed, a reading can be taken through an open window to floor level, and an inverted reading taken to ceiling level.

The disadvantage of this method is, of course the fact that time must be spent in reducing the levels. However, the ease with which the levels are obtained (and their accuracy) is a great advantage and the method should be used whenever possible.

Table 20.1 and Fig. 20.4 show the levels obtained along the western gable and extension of the house.

(c) *Internal survey*

Plan Once again a neat accurate sketch of the interior of the building should be prepared on squared paper. In the case of a simple building, the sketch is combined with the building location sketch (Fig. 20.3).

The measuring procedure is arranged to start at the entrance hall and proceed in a clockwise direction around the hall. Each room is then taken in a clockwise direction and similarly measured.

Whenever possible, running dimensions are taken, but individual measurements of window openings, cupboards, etc., may have to be made in case paintings, pictures, or curtains are damaged by stretching a long tape along a wall. Furthermore, it

simply to hold the tape at the zero chainage point of each side. When hooked, the tape is run out along the side of the building; each feature is noted in turn and booked as in Fig. 20.3.

Bay windows present a problem, since the running chainage must be terminated against the window and individual dimensions taken round the window. In Fig. 20.3 these measurements are clearly shown. A check dimension is made by holding the tape in line with the front of the building, and measuring the running dimensions to the window, then to the back of the building. The projection of the window must also be obtained.

Elevation Elevations prove to be more difficult than plans, because of the inaccessibility of eaves, high windows, etc.

Windows are usually surveyed first of all. The overall width and height of each window are taken,

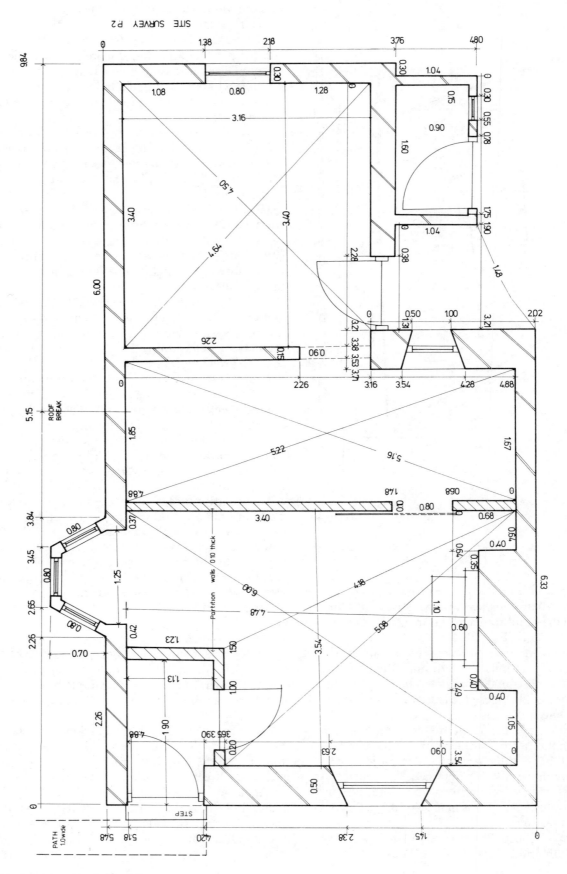

Figure 20.3

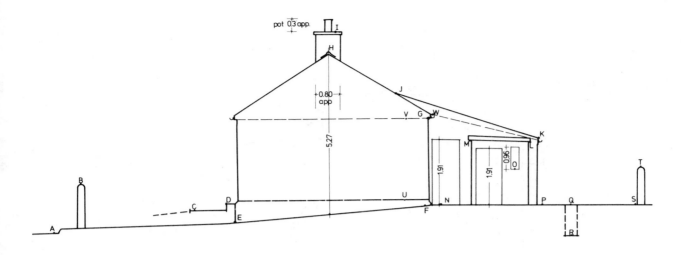

Figure 20.4

BS	IS	FS	Rise	Fall	Red lev.	Remarks
1.51					10.00	A Ground level
	0.62		0.89		10.89	B Top of wall
	1.29			0.67	10.22	C Path
	1.14		0.15		10.37	D Step
	1.41			0.27	10.10	E Ground level
	1.15		0.26		10.36	F Ground level
	−1.85		3.00		13.36	G Soffit
	−3.99		2.14		15.50	H Ridge
	−4.70		0.71		16.21	I Chimney stack
−2.65		−2.65		2.05	14.16	J Roof change
	−1.11			1.54	12.62	K Roof of extension
	−0.96			0.15	12.47	L Roof of WC
	−1.01		0.05		12.52	M Roof of WC
	1.15			2.16	10.36	N Floor level extension
	0.20		0.95		11.31	O Window WC
	1.21			1.01	10.30	P Ground level
	1.20		0.01		10.30	Q Manhole cover
	2.40			1.20	9.11	R Invert level
	1.31		1.09		10.20	S Ground level
	0.31		1.00		11.20	T Top of wall
	0.99			0.68	10.52	U Floor level living room
	−1.88		2.87		13.39	V Ceiling level living room
		−2.10	0.22		13.61	W Ceiling level extension
−1.14		−4.75	13.34	9.73	13.61	
−(−4.75)			−9.73		−10.00	
+3.61			+3.61		+3.61	

Table 20.1

may prove impossible to hook a tape handle to a point which is convenient for running dimensions.

Particular attention must be paid at door and window openings when obtaining the thickness of internal and external walls. A sound knowledge of building construction practice is very useful in this context.

The booking of the results is most important, and time must be taken to ensure that cross checks, etc., are properly dimensioned. Figure 20.3 shows the booking required in this example.

Internal heights Heights must be taken for the purposes of drawing cross-sections, and for determining ceiling heights in possible dormer extensions, etc.

A floor-to-ceiling height is taken in each room, in the middle of the room, and at each window. The dimensions should be measured from the ceiling

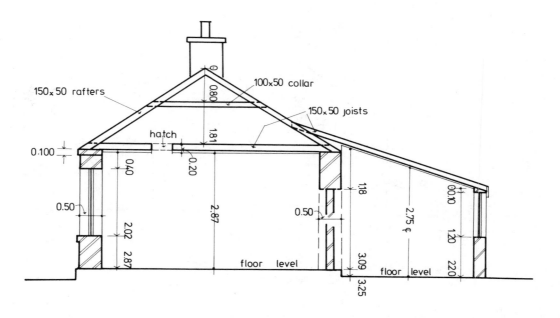

Figure 20.5

downwards to internal lintels and sills of windows (Fig. 20.5).

A spirit level is a useful item of equipment when measuring heights to sills, since sills are not horizontal. The spirit level is laid across the bottom rail of the window, and appropriate dimensions measured to the sill internally and externally.

All door heights are also measured.

When a building consists of more than one floor, the floor-to-floor heights are usually easily measured at the stair well. If difficulty is experienced, a levelling may have to be made up the stairs. Checks may be made externally by hanging a tape from a window on one floor down to a window on the next.

Measurements must also be taken from the floor of the uppermost room to the ceiling and into the roof space. These measurements are made through the ceiling hatch to the top of the ceiling joist, and from there to the apex of the roof.

(d) Sections, services, etc.
Depending upon the purpose of the survey, sections may be required through the roof space, ground floor window heads, and door thresholds.

The construction details are accurately sketched on squared paper and the relevant sizes obtained using a short 5 m tape. Figure 20.5 shows the relevant survey details for drawing a section through the roof space.

Services are traced individually, and separate sketches made to show the run of electrical cables and conduits, and all waste and soil pipes.

Plotting the survey

In drawing buildings, several views from different angles are required. These include a plan and elevations of all sides of the building. It is essential that these views be presented in a systematic manner.

BS 1192 'Building Drawing Practice' recommends that only orthographic first-angle projection be used for building drawings.

Plotting of the survey information begins with the plan view, and from it all other views are constructed.

In deciding where the plan view is to be placed on the sheet, the surveyor must decide which elevation is to be the principal view. Generally this is the front elevation.

If it is possible to place all of the views on one sheet, the principal view is placed left of centre (position 1) in the top half of the sheet (Fig. 20.6).

The plan view is then positioned immediately below the front view (position 2).

The side views from left and right are placed on the right (position 3) and left (position 4) respectively of the front view.

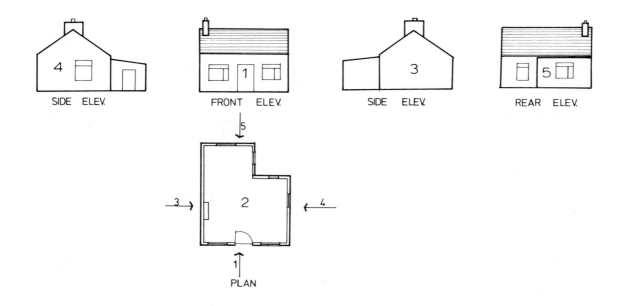

FRONT ELEV.

SIDE ELEV.

SIDE ELEV.

REAR ELEV.

PLAN

Figure 20.6

The view from the rear is usually placed on the extreme right (position 5) although, if more convenient, it may be placed on the extreme left.

Signs and symbols BS 1192 'Building Drawing Practice' recommends that the principal symbols used on building drawings be graphical. For a complete list of the symbols and abbreviations, readers are referred to this publication. However, in order to enable the reader to plot the exercise following, a few of these symbols are shown in Fig. 20.7.

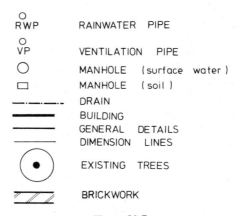

RWP	RAINWATER PIPE
VP	VENTILATION PIPE
	MANHOLE (surface water)
	MANHOLE (soil)
	DRAIN
	BUILDING
	GENERAL DETAILS
	DIMENSION LINES
	EXISTING TREES
	BRICKWORK

Figure 20.7

Exercise 1 Using the survey information of Figs. 20.1 to 20.5 and Table 20.1 draw:

(a) A site plan to a scale of 1:250.
(b) A general location drawing of the building to a scale of 1:50 showing the internal arrangements.
(c) An elevation of the western side of the building to a scale of 1:50.
(d) A north–south section through the building to a scale of 1:50.

Exercise 2 Figure 20.8 shows the front elevation of a three storey local authority flatted dwelling. The front elevation is to be renovated.

The various floor levels were located by measuring up the stairwell. The windows were observed to be vertically aligned and chimney dimensions were estimated.

Plot the front elevation to a scale of 1.50.

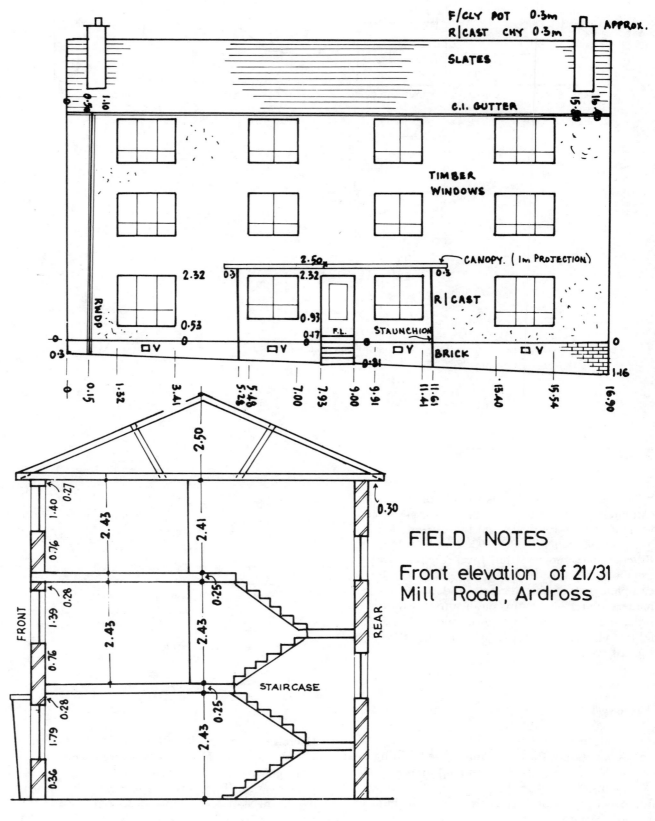

Figure 20.8

FIELD NOTES

Front elevation of 21/31
Mill Road, Ardross

21. Model answers to examination questions

Question 1

Using the conventional signs adopted by the Ordnance Survey on their 1:2500 scale plans, show, by sketches, how the following detail would be depicted on plans drawn to this scale.

(i)   quarry
(ii)   farm with outbuildings including a greenhouse and an orchard
(iii) a railway and level crossing
(iv)   a railway in a cutting crossed by a straight road.

(RICS Elementary Surveying, Building Surveying, General Practice Options)

Model answer
See Figs 21.1(a) and (b).

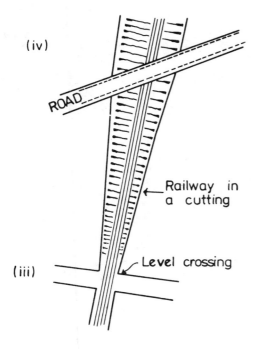

Figure 21.1(a)

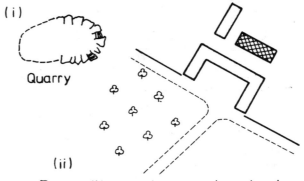

Farm with greenhouse and orchard

Figure 21.2(b)

Question 2

The field book, Figs 21.2(a)(b) (pages 278–279), gives details of a chain survey. On the drawing paper provided, plot the stations lines and detail using a scale of 1:500.

(Institute of Building Licentiate Examination)

Solution method
(i)   Make a rough sketch of the main framework lines from the given information (Fig. 21.3(a)).

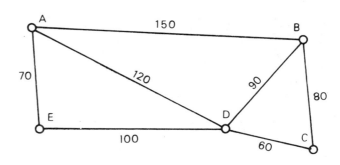

Figure 21.3(a)

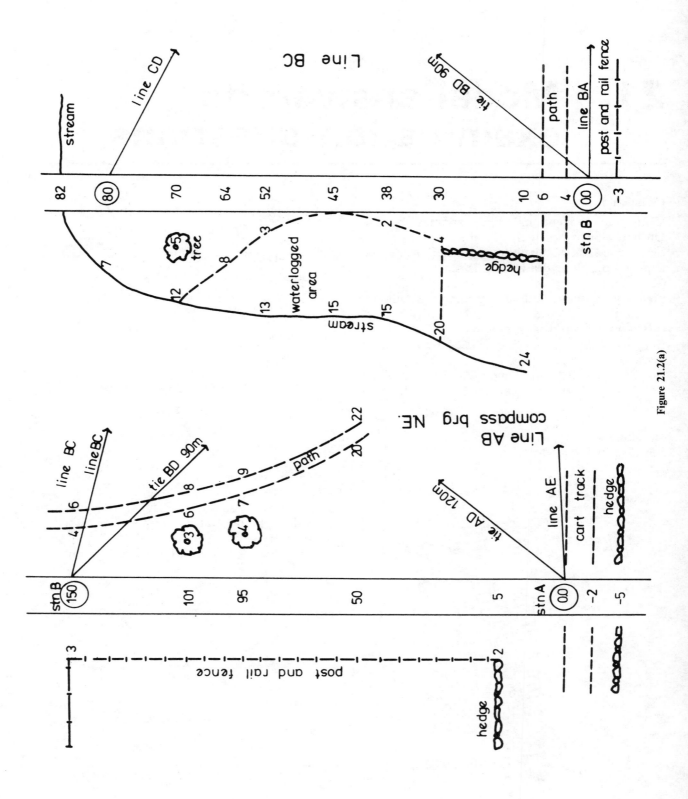

Figure 21.2(a)

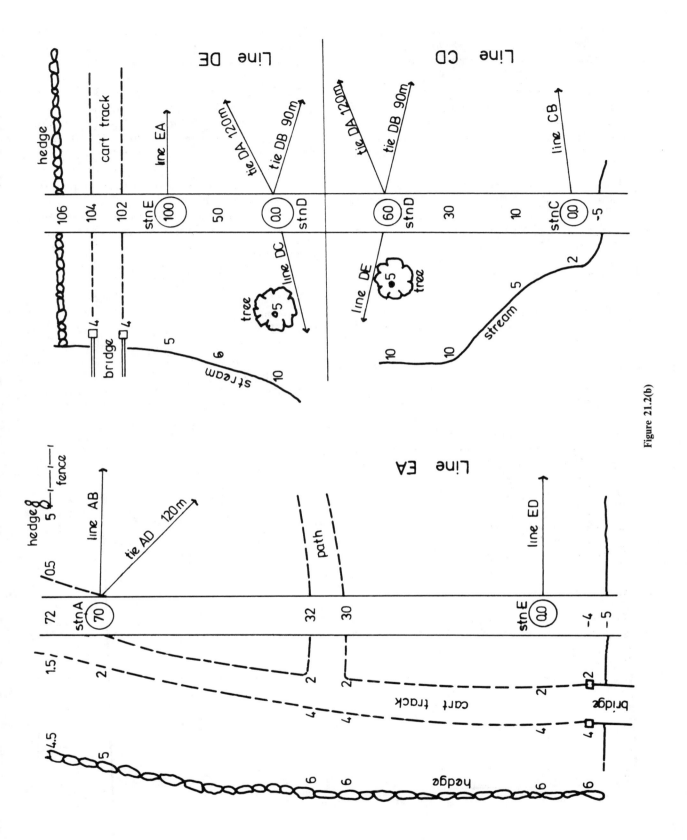

Figure 21.2(b)

(ii) Decide how the survey should be positioned to fit neatly onto the paper provided (size A3).

(iii) Plot the longest line AB marking the stations A and B precisely with a needle.

(iv) Using compasses, describe an arc of 120 m radius from station A on the right-hand side of the line AB.

(v) Using a 90 m radius describe a second arc from station B to cut the previous arc in station D.

(vi) Similarly draw arcs AE and DE to intersect at station E and arcs BC and DC to intersect at station C. This completes the framework.

(vii) Scale off the various chainages along the framework lines and scale the offsets to left or right of the lines as appropriate.

(viii) Add a north point and scale.

Model answer
(i) See Figs 21.3(a) and (b).

Question 3
Figure 21.4 is the plan of Wern Farm drawn to a scale of 1:1250.

(a) Indicate on the plan the positions of the stations which would enable the area to be surveyed by linear measurement. The two stations A and B must be incorporated.

(b) Using a scale rule and the figure as equivalent to the use of a chain on the area, draw up an example of the field book entries for line A – B.

(c) (i) If the ground between A and B was sloping at 8° 30′ from the horizontal, what would be the true length for plotting if the measured length down the slope was 125.200 m?

(ii) How would you correct graphically the position of the features for plotting, from a field book entry with measurements taken down a slope?

(City and Guilds of London Institute Construction Technicians Certificate)

Solution method

(a) The following factors should be borne in mind when locating the stations:
(i) The stations should form well-conditioned triangles.
(ii) They should be close to the boundaries so that the lengths of offsets are short.
(iii) The lines should not run through the buildings.

(b) Draw lightly, offset lines at right angles to the line AB to all points of detail which are to be surveyed; hence scale the chainages and offset lengths and record the measurements.

(c) The true length of the line = slope length × cosine 8° 30′.

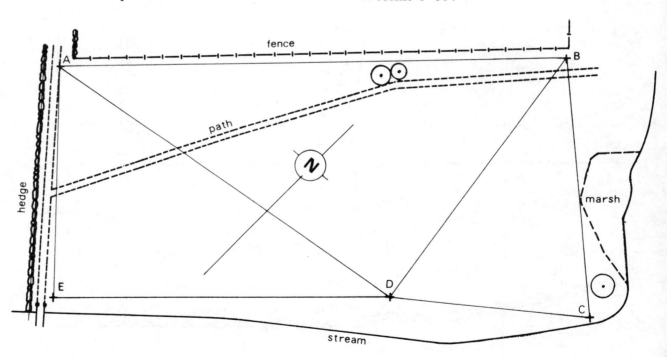

Figure 21.3(b)

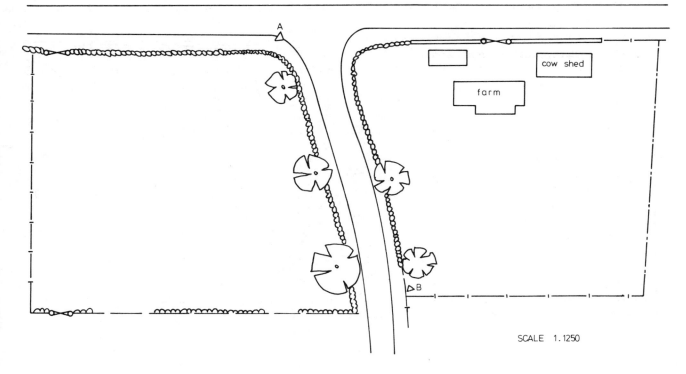

Figure 21.4

Model answer

(a) See Fig. 21.5.

(b) See Fig. 21.6.

(c) (i) Length of line = measured length
$\times$ cosine 8° 30′
= 125.2 × 0.9890
= 123.8 m

(ii) The details along the line AB would be plotted in their correct positions as follows:

1. At a shallow angle to the line AB draw a random AB′ line 125.2 m long. Join B to B′.
2. Along this line AB′ mark the field chainages of the various points of detail, e.g., at points x′ and y′.
3. Draw lines parallel to BB′ through the various chainages points to cut line AB in their correct plan positions at x and y. See Fig. 21.7.

Question 4

Levels taken along the centre line of a proposed roadway at 25 m intervals gave the readings shown in the table below.

The reduced level of formation at A is to be 39.20 m. The road is to have a rising gradient of 1 in 40 between A and B, and 1 in 16 between B and C.

Reduce all the levels, and indicate the depth of cut or fill at each chainage point.

BS	IS	FS	RL	Distance	Remarks
2.30			40.18	0	A
	2.01			25	
	1.61			50	
	1.31			75	
3.95		1.16		100	B
	2.69			125	
3.09		0.33		150	
		0.51		175	C

(IQS Land Surveying)

Solution method

(i) Reduce the levels.

(ii) Given the formation level 'A', calculate the formation levels of the roadway at 25 m intervals from 0 m to 100 m on a rising gradient of 1 in 40, e.g., at 100 m FL'B' = (FL'A' + 100/40) m.

(iii) From the calculated level of B above, calculate the formation levels of the roadway at 25 m intervals from 100 m to 175 m on a rising gradient of 1 in 16, e.g., at 175 FL'C' = (FL'B' + 75/16) m.

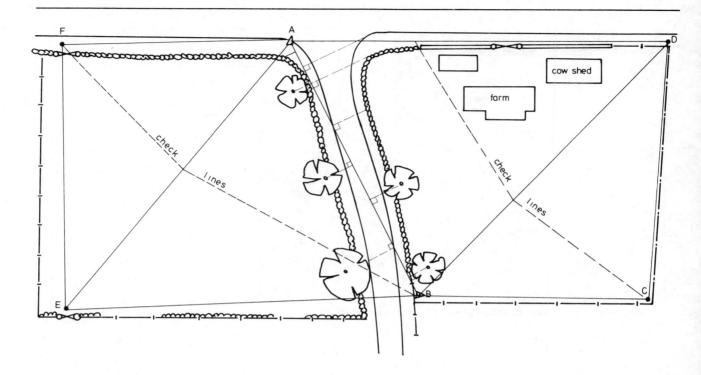

Figure 21.5

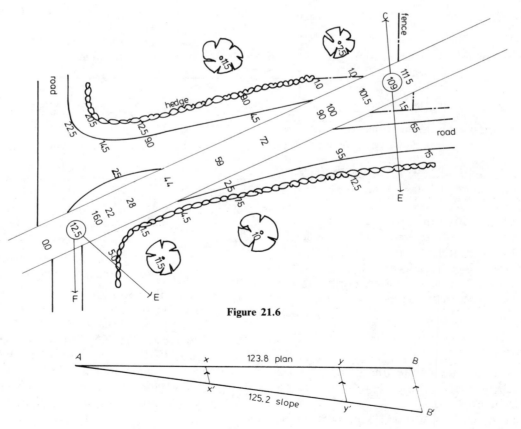

Figure 21.6

Figure 21.7

(iv) Calculate the difference between the formation level and the existing ground level at each chainage point. If the formation level is higher than ground level, there will be filling and, if the formation level is lower than ground level, there will be cutting.

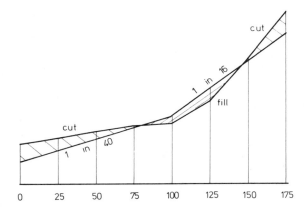

Figure 21.8

Model answer
(i)

BS	IS	FS	Rise	Fall	Reduced level	Distance	Remarks
2.30					40.18	0	A
	2.01		0.29		40.47	25	
	1.61		0.40		40.87	50	
	1.31		0.30		41.17	75	
3.95		1.16	0.15		41.32	100	B
	2.69			1.26	42.58	125	
3.09		0.33	2.36		44.94	150	
		0.51	2.58		47.52	175	C
9.34		2.00	7.34		47.52		
−2.00					−40.18		
7.34					7.34		

(ii) From a point A the roadway is to rise at a gradient of 1 in 40 over the first 100 m. Therefore:

$$\text{Rise over 25 m} = 25/40 \text{ m} = 0.625 \text{ m}$$

Formation levels and depths of cut or fill are as follows:

Chainage (m)	Rise (m)	Formation level (m)	Existing level (m)	Cut (m)	Fill (m)
0 A		39.20	40.18	0.98	—
25	+0.63	39.83	40.47	0.64	—
50	+0.62	40.45	40.87	0.42	—
75	+0.63	41.08	41.17	0.09	—
100 B	+0.62	41.70	41.32	—	0.38

(iii) Similarly, the roadway is to rise at a gradient of 1 in 16 over the last 75 m. Therefore

$$\text{Rise over 25 m} = 25/16 \text{ m} = 1.563 \text{ m}.$$

(iv)

Chainage (m)	Rise (m)	Formation level (m)	Existing level (m)	Cut (m)	Fill (m)
100 B		41.70	41.32	—	0.38
125	+1.56	43.26	42.58	—	0.68
150	+1.57	44.83	44.94	0.11	—
175 C	+1.56	46.39	47.52	1.13	—

Question 5

The following extract is from the level book of the recordings made for a proposed drain run from a building to a sewer connection:

Distance (m)	Reduced level (m)	Remarks
0	109.480	MH cover existing
0	107.950	invert level existing
20	109.710	ground level
40	109.820	do.
60	109.890	do.
80	110.050	do.
100	110.440	do.
120	110.780	do.
140	111.110	do.
160	111.760	do.
180	112.250	do.
200	112.550	do.
		end proposed drain
200	111.450	invert proposed drain.

A 100 mm pipe is specified laid on a 150 mm thick concrete bed.

To ensure economy in excavation work, the drain gradients are to be within the range 1 in 40 maximum to 1 in 80 minimum with a minimum ground cover of 1 m to top of pipe (ignore pipe thickness).

(a) Draw (to scale—1:1000 horizontal; 1:100 vertical) the longitudinal section of the proposed drain which would be suitable to meet the above specification.

(b) By calculation, check the depth of excavation at any gradient change.

(City and Guilds of London Institute Construction Technicians Certificate)

Solution method
(i) Choose a datum for the section and draw a horizontal line to represent it. A suitable datum in this case is 105.0 m.

(ii) Plot the chainages at 1:1000 scale from 0 m to 200 m along this line, and erect a perpendicular at each point.

(iii) Scale the reduced levels above the datum at 1:100 scale and join the points by straight lines to form a section.

(iv) Since the pipe can approach no closer than 1 m to the surface, draw a second line parallel to the surface at a distance of 1 m below it to represent the sub-surface.

(v) Mark the invert level of the pipe at 0 m and 200 m chainages, and by adding 100 m to these invert levels, mark the top of the pipe.

(vi) Scale the gradients from the plan; hence calculate the level of the top of the pipe or any change of gradient.

(vii) Subtract the level of the top of the pipe from the surface level to obtain the depth of excavation.

Model answer

(i)–(v) See Fig. 21.9

(vi)–(vii) From the figure, the most economical gradients for the drain are 1 in 80 from chainage 0 m to chainage 80 m; 1 in 60 from chainage 80 m to chainage 140 m; 1 in 40 from chainage 140 m to chainage 200 m.

Invert level chainage 0 m	= 107.95
Thickness of pipe	= +0.10
∴ Level of top of pipe	= 108.05 m
Gradient from 0 m to 80 m	= 1 in 80 rising
Rise over 80 m	= 1.00 m
∴ Level of top of pipe at 80 m	= 108.05 + 1.00
	= 109.05 m
Surface level at 80 m	= 110.05 m
∴ Cover at 80 m	= 110.05
	− 109.05
	= 1.00
Level of top of pipe at 80 m	= 109.05 m
Gradient from 80 m to 140 m	= 1 in 60
Rise over 60 m	= 1.00 m
∴ Level of top of pipe at 140 m	= 109.05 + 1.00
	= 110.05 m
Surface level at 140 m	= 111.11 m
∴ Cover at 140 m	= 111.11
	−110.05
	= 1.06 m
Check: level of top of pipe at 140 m	= 110.05 m
Gradient from 140 m to 200 m	= 1 in 40
Rise over 60 m	= 1.50 m
∴ Level of top of pipe at 200 m	= 110.05 + 1.50
	= 111.55 m
Invert level at 200 m	= 111.45
Pipe thickness	= +0.10

∴ Level of top of pipe at 200 m   = 111.55 m

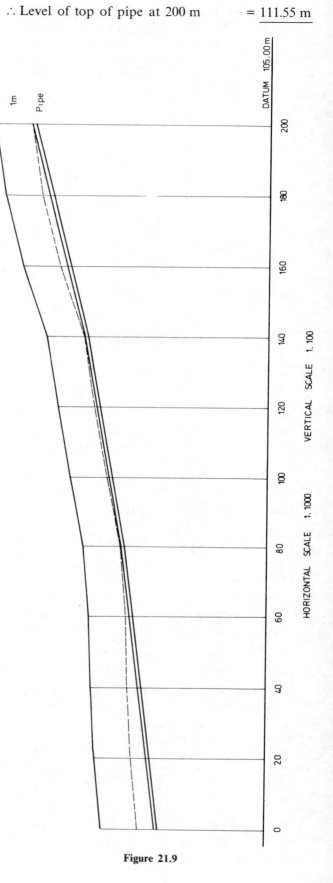

Figure 21.9

Question 6

(a) A tilting level which is to be used on site was checked by the two-peg test method and the following staff readings were obtained:

Staff position	Staff reading	Distance (m)	Instrument position
Peg A	1.428	60	mid-way between A & B
Peg B	1.178		
Peg B	1.354	60	very near to Peg B
Peg A	1.634		

Calculate the amount of instrument error.

(b) A level after checking is found to have an elevated collimation error of 4 minutes and is subsequently used to set out a series of profiles. The horizontal distances to each profile from the instrument position were accurately measured, level book details being as follows:

BS	IS	FS	HOC	RL	Distance (m)	Remarks
2.120				92.260	20	TBM
	1.844				30	Rail 1
	1.192				50	Rail 2
	0.094				60	Rail 3
		2.080			70	Rail 4

Establish the correct reduced level of each profile rail.

(City and Guilds of London Institute Construction Technicians Certificate)

Solution method

(a) (i) Make a sketch from the given information (Fig. 21.10(a))

(ii) Calculate the true difference in level between A and B from the first set of readings. This is the true difference in level, irrespective of whether the instrument is in error.

(iii) Calculate the apparent difference in level between A and B from the second set of readings.

(iv) Calculate the instrumental collimation error over 60 metres. Error = (iii) − (ii) above.

(b) Since the line of collimation of the instrument is elevated by 4 minutes, every staff reading is too great by a varying amount, the amount depending on the distance of the staff from the instrument (Fig. 21.10(b)).

$$\text{Correction} = -(\text{distance} \times \tan 0° \, 04')$$

or more conveniently

$$\text{Correction} = -\,\text{distance} \times \left(\frac{04}{3438}\right) \text{m}$$

$$= -\left(\frac{\text{distance}}{860}\right) \text{m}$$

(i) Correct each staff reading using the formulae above.

(ii) Book the correct readings in a levelling table and reduce the levels.

Model answer

(a) (i) See Fig. 21.10(a)

(ii) True difference in level A to B

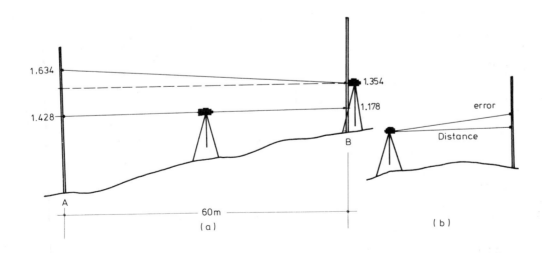

Figure 21.10

$$= 1.428 - 1.178$$
$$= \underline{0.250 \text{ m}} \text{ (B is higher than A)}$$

(iii) Apparent difference in level A to B

$$= 1.634 - 1.354$$
$$= \underline{0.280 \text{ m}}$$

(iv) Since the apparent difference in level is greater than the true difference, the line of collimation must be elevated.

Error = (0.280 − 0.250 m) over 60 m
$$= + 0.030 \text{ m over 60 m}$$
$$= + \underline{1 \text{ mm per 2 m}}$$

(b) (i) Correction to staff readings $= -\left(\dfrac{\text{distance}}{860}\right)$ m

∴ Correction over 20 m = −0.023 m
30 m = −0.035 m
50 m = −0.058 m
60 m = −0.070 m
70 m = −0.081 m

(ii) Correct readings are
BS 2.120 − 0.023 = 2.097 m
IS 1.844 − 0.035 = 1.809 m
IS 1.192 − 0.058 = 1.134 m
IS 0.094 − 0.070 = 0.024 m
FS 2.080 − 0.081 = 1.999 m

Correct reduced levels are as follows:

BS	IS	FS	HOC	RL	Remarks	
2.097			94.357	92.260	TBM	20 m
	1.809			92.548	Rail 1	30 m
	1.134			93.223	Rail 2	50 m
	0.024			94.333	Rail 3	60 m
		1.999		92.358	Rail 4	70 m
2.097	1.999			92.358		
−1.999				−92.260		
0.098				0.098		

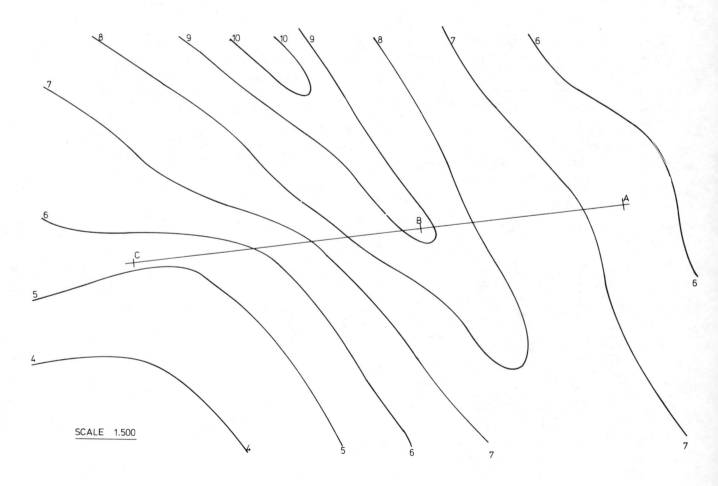

Figure 21.11

Question 7

Draw cross-sections at A, B, and C of an embankment that is to cross the area shown on Fig. 21.11.

The embankment is to be 12 m wide and has a level top with a RL of 15 m. The sides slope at 1 horizontal to 2 vertical.

Choose suitable vertical and horizontal scales for your cross-section.

(RICS Const. Meas. & Val. I)

Solution method

(i) Draw the roadway 12 m wide at a scale of 1:500 (Fig. 21.12).

(ii) Draw embankment contours representing vertical intervals of 1 m. The contours will be parallel to the sides of the roadway at horizontal intervals of 0.5 m, since the sides slope at 1 horizontal to 2 vertical.

(iii) Mark the points where embankment contours and ground contours of similar value coincide.

(iv) Join these points to form the tail of the embankment.

(v) Decide the scales for drawing the cross-sections. It is wise to draw cross-sections on a natural scale whenever possible, and in this case such a scale would be suitable. Alternatively, the vertical scale could be exaggerated and the sections plotted on a horizontal scale of 1:500 and a vertical scale of 1:250.

(vi) Lay a piece of paper across the line of the section at C and mark the positions of the 4, 5, 6, and 7 m contours, and the top and bottom of the embankment (Fig. 21.13).

(vii) On the plotting sheet, draw a horizontal line to represent the datum of zero metres and mark along it the points in (vi) above.

(viii) Draw verticals at these points and scale off the appropriate values of the contours.

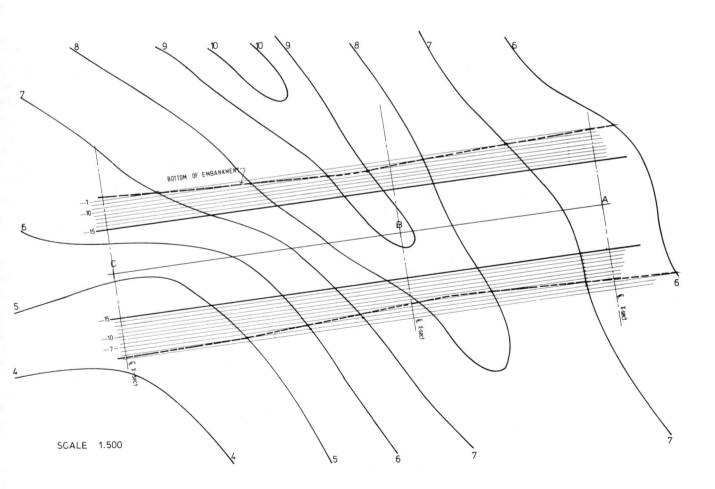

Figure 21.12

(ix) Join these points to form the ground surface.

(x) Scale the formation level of the embankment on the appropriate two vertical lines and join them to the points representing the base of the embankment at ground level.

(xi) Repeat steps (vi) to (x) for cross-sections B and A.

Question 8

The internal angles of a closed theodolite traverse, lettered clockwise ABCDEFGA, were measured as shown in Model answer, col. 2. Side AB has a whole circle bearing of 52° 35′. Correct the measured angles by distributing any error, and list the bearing of each side of the traverse.

Point	A	B	C	D
Angle	73° 21′	122° 48′	57° 06′	238° 18′
Point	E	F	G	
Angle	85° 50′	91° 57′	230° 33′	

(RICS Const. Meas. & Val. II)

Solution method

(i) Draw a diagram using the given information (Fig. 21.14).

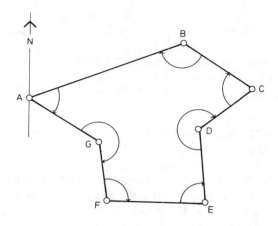

Figure 21.14

(ii) Calculate the sum of the observed values of the angles.

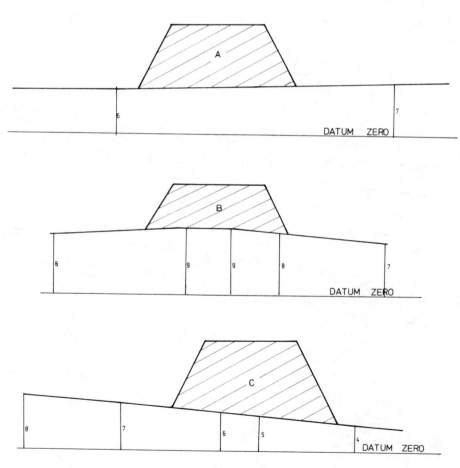

Figure 21.13

(iii) Calculate the sum of the internal angles of a seven-sided polygon [sum = (2n − 4) × 90°].

(iv) Calculate the error of closure. (Error = (iii) − (ii) above hence the angular correction.)

(v) Distribute the error equally among the angles, i.e., divide the error by 7.

(vi) Calculate the adjusted angles [Adj. angle = obs. angle + correction (v)].

(vii) Calculate the bearings of the lines using the adjusted angles (vi).

Model answer

1	2	3	4
	Observed	*Angular*	*Adjusted*
Point	*angle*	*correction*	*angle*
A	73° 21′	+01′	73° 22′
B	122° 48′	+01′	122° 49′
C	57° 06′	+01′	57° 07′
D	238° 18′	+01′	238° 19′
E	85° 50′	+01′	85° 51′
F	91° 57′	+01′	91° 58′
G	230° 33′	+01′	230° 34′
(ii) Σ =	899° 53′	+07′	900° 00′

(iii) Number *n* of internal angles = 7.

$$(2n − 4) × 90° = (10 × 90)°$$
$$= 900° \ 00′ \ 00″$$

(iv) Error of closure = 900° 00′ 00″
　　　　　　　　　　 −899° 53′ 00″
　　　　　　　　　　　 −07′ 00″

∴ correction = +07′ 00″

(v) Angular correction = +(07/7)′
　　　　　　　　　　 = +01′ (Column 3)

(vi) Adjusted angles (column 4).

(vii) Bearing AB　　　= 　52° 35′
　　　+ Â　　　　　　 +73° 22′

∴ Bearing AG　　　= 　125° 57′

Back bearing GA = 　305° 57′
　　+ Ĝ　　　　　　+230° 34′

　　　　　　　　　　536° 31′
　　　　　　　　　　−360°

∴ Bearing GF　　　= +176° 31′

Back bearing FG = 　356° 31′
　　+F̂　　　　　　 +91° 58′

　　　　　　　　　　448° 29′
　　　　　　　　　　−360°

∴ Bearing FE　　　= 　88° 29′

Back bearing EF = 　268° 29′
　　+ Ê　　　　　　 + 85° 51′

∴ Bearing ED　　　= 　354° 20′

Back bearing DE = 　174° 20′
　　+ D̂　　　　　　 +238° 19′

　　　　　　　　　　412° 39′
　　　　　　　　　　−360°

∴ Bearing DC　　　= 　52° 39′

Back bearing CD = 　232° 39′
　　+ Ĉ　　　　　　 + 57° 07′

∴ Bearing CB　　　= 　289° 46′

Back bearing BC = 　109° 46′
　　+ B̂　　　　　　 +122° 49′

∴ Bearing BA　　　= 　232° 35′

Question 9

The following readings on a vertical staff were taken along part of the boundary of a building site (which is to be fenced) by a tacheometer fitted with an anallatic lens and having a multiplying constant of 100.

Inst. Stn	Staff Stn	Stadia hair readings (metres)			Whole circle bearing	Vertical angle
		Top	Mid	Bottom		
A	B	2.100	1.600	1.100	30°	+2°
B	C	2.800	1.600	0.400	60°	+3°
C	D	4.900	3.600	2.300	120°	+5°

Calculate:
(i) the total length of the perimeter ABCDA.
(ii) the reduced level of B, given that the reduced level of the station at A is 76.500 m and the instrument height at A is 1.500 m.

(Scottish Association for National Certificates and Diplomas, Ordinary National Diploma in Building Land Surveying II)

Solution method
(i) Make a sketch from the given information (Fig. 21.15).

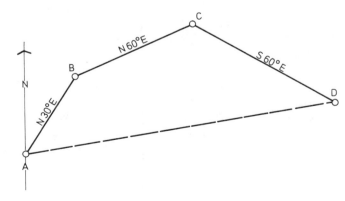

Figure 21.15

(ii) Calculate the plan lengths of lines AB, BC, and CD. Plan length = $mS \cos^2 \theta$.

(iii) From these lengths and the given whole circle bearings, calculate the coordinates of the stations. Assume station A to be the origin.

(iv) From the coordinates of stations A and D, calculate the plan length AD = $\sqrt{(\Delta E^2 + \Delta N^2)}$.

(v) Add the plan lengths AB, BC, CD, and DA to give the required perimeter.

(vi) Calculate the reduced level of B from the calculated plan length AB.
$$RL`B' = RL`A' + i + AB \tan \theta - h$$

Model answer

(ii) Plan length AB = $mS \cos^2 \theta$

$$m = 100$$
$$S = (2.100 - 1.100) \text{ m} = 1.000 \text{ m}$$

Vertical angle $\theta = 2° 00'$

$$\therefore AB = 100 \times 1.000 \times \cos^2 2° 00'$$
$$= 100 \times 1.000 \times 0.9994^2$$
$$= 99.88 \text{ m}$$

Similarly   BC = $100 \times (2.800 - 0.400)$
$$\times \cos^2 3°$$
$$= 100 \times 2.400 \times 0.9986^2$$
$$= 239.33 \text{ m}$$

Similarly   CD = $100 \times (4.900 - 2.300)$
$$\times \cos^2 5°$$
$$= 100 \times 2.600 \times 0.9962^2$$
$$= 258.03 \text{ m}$$

	E	N
(iii) Coords A =	00.00	00.00
D =	480.65	77.16
ΔE =	480.65	ΔN = 77.16

Plan length AD = $\sqrt{(\Delta E^2 + \Delta N^2)}$
$$= \sqrt{(480.65^2 + 77.16^2)}$$
$$= 486.80 \text{ m}$$

(iv) Perimeter ABCDA = (99.88 + 239.33 + 258.03 + 486.80) m
$$= 1084.04 \text{ m}$$

(v) Reduced level A = 76.50 m i = 1.500 m
Plan length AB = 99.88 m h = 1.600 m
AB tan θ   = 99.88 tan 2°
$$= 3.49 \text{ m}$$

(vi) Reduced level 'B' = RL 'A' + i
$$+ AB \tan \theta - h$$
$$= 76.50 + 1.50 + 3.49 - 1.60$$
$$= 79.89 \text{ m}$$

Question 10

What area is bounded by the traverse ABCDEFGA whose metric coordinates are given in the table below:

Point	East coordinate	North coordinate
A	98.3	422.2
B	326.0	940.6
C	678.9	1024.4
D	914.9	1276.0
E	1388.5	812.5
F	860.4	306.7
G	492.3	535.8

(RICS Const. Meas. & Val. II)

Line	Quadrant bearing θ	Plan length L	Plan coords E+	Plan coords W−	$\frac{\Delta E}{\Delta W} = \pm L \sin \theta$ $\frac{\Delta N}{\Delta S} = \pm L \cos \theta$ N+	S−	Total dep.	Total lat.	Station
							00.00	00.00	A
AB	N 30° E	99.88	49.94		86.50		49.94	86.50	B
BC	N 60° E	239.33	207.26		119.67		257.20	206.17	C
CD	S 60° E	258.03	223.45			129.01	480.65	77.16	D
			480.65		206.17 −129.01	129.01	480.65	77.16	
					77.16				

290 Surveying for construction

Solution method

(i) Tabulate the stations and their total coordinates (columns 1–3).
(ii) Multiply the departure (Easting) of every point by the latitude (Northing) of the following point, i.e., $x_1 y_2$, $x_2 y_3$, etc.
(iii) Multiply the departure of every point by the latitude of the preceding point, i.e., $x_2 y_1$, $x_3 y_2$, etc.
(iv) Find the algebraic difference between (ii) and (iii).
(v) Halve this figure to give the area enclosed by the stations of the traverse.

1		2		3	4	5
Stn		Easting m (x)		Northing m (y)	x_1y_2, etc.	x_2y_1, etc.
A	x_1	98.3	y_1	422.2		
B	x_2	326.0	y_2	940.6	92 460.98	137 637.20
C	x_3	678.9	y_3	1024.4	333 954.40	638 573.34
D	x_4	914.9	y_4	1276.0	866 276.40	937 223.56
E	x_5	1388.5	y_5	812.5	743 356.25	1 771 726.00
F	x_6	860.4	y_6	306.7	425 852.95	699 075.00
G	x_7	492.3	y_7	535.8	461 002.32	150 988.41
A	x_1	98.3	y_1	422.2	207 849.06	52 669.14
					3 130 752.36	4 387 892.65
						−3 130 752.36

(iv)   1 257 140.29

(v) ÷ 2   628 570.14 m² = area.

Question 11

Figure 21.16 shows the existing ground levels on a 15 m square grid forming part of a site which is to be excavated to a uniform formation level of 10.00 m above datum.

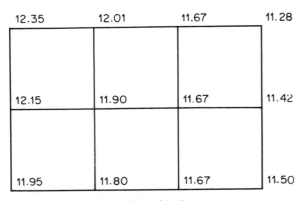

12.35	12.01	11.67	11.28
12.15	11.90	11.67	11.42
11.95	11.80	11.67	11.50

Figure 21.16

Calculate the volume of earth to be excavated assuming vertical sides.

(IQS Land Surveying & Mensuration)

Solution method

(i) Tabulate the stations (column 1) and the depth of excavation required at each (column 2) to achieve formation level.
(ii) Tabulate the number of squares into which each station falls (column 3) and find the sum of this column.
(iii) Multiply the depth of excavation at each station by the number of squares into which it falls (column 4) and find the sum of this column.
(iv) Find the mean depth of excavation of the whole site by dividing the sum of column 4 by the sum of column 3.
(v) Find the volume of excavation by multiplying the total area of the site by the mean depth of excavation.

Model answer

(i)–(iii)

1	2	3	4
Stn	Depth to formation level	No. of squares containing stations	Product (col. 3 × col. 4)
A	2.35	1	2.35
B	2.01	2	4.02
C	1.67	2	3.34
D	1.28	1	1.28
E	2.15	2	4.30
F	1.90	4	7.60
G	1.67	4	6.68
H	1.42	2	2.84
J	1.95	1	1.95
K	1.80	2	3.60
L	1.67	2	3.34
M	1.50	1	1.50
		24	42.80

(iv) Mean depth of excavation = (42.80/24) m
                       = 1.783 m

(v) Total area of site = (30 × 45) m
                           = 1350 m²

Volume of excavation = 1350 × 1.783
                                = 2407.05 m³

Question 12

A vertical parabolic curve is to be designed to connect a downward gradient of 1 in 70 with an upward gradient of 1 in 35. The reduced level of the intersection point is 43.40 m AOD. The length of the vertical curve is to be 140 m. Calculate the reduced levels along the curve at 20 m intervals.

(OND in Building Land Surveying II)

Model answer

(i) See Fig. 21.17.

(ii)
Reduced level I	=	43.40 m

Gradient IT is $+$ 1 in 70

∴ Reduced level T = (43.40 + 70/70) = 44.40 m

Gradient IT_1 is $+$ 1 in 35

∴ Reduced level T_1 = (43.40 + 70/35) = 45.40 m

Gradient IE is $-$1 in 70

∴ Reduced level E = (43.40 − 70/70) = 42.40 m

(iii) Difference in level ET_1

$$= (45.40 - 42.40) \qquad = +3.00 \text{ m}$$

∴ At distance 140 m from T,

$$ax^2 = ET_1 \qquad = +3.00 \text{ m}$$

(iv) ∴ a

$$= \frac{3.00}{140^2}$$

$$= 0.153 \times 10^{-3}$$

(vi) Chainage   Tangent level

= (T + bx) where

$b = -1/70$

T 0	
20	−0.286
40	−0.571
60	−0.857
80	−1.143
100	−1.429
120	−1.714
140(T_1)	−2.000

(vii) Grade correction = ax^2

where $a = 0.153 \times 10^{-3}$

(viii) Curve level = (T + bx + ax^2)

	44.40
+0.061	44.18
+0.245	44.07
+0.551	44.09
+0.980	44.24
+1.531	44.50
+2.204	44.89
+3.000	45.40

Question 13

Two straight lines intersecting at point I are to be connected by a circular curve of 400 m radius. The whole circle bearings of the two lines are 138° 24′ and 210° 00′ respectively, and the chainage of point I is 508.50 m. Derive the data required for the setting out of the curve using a 20 m tape and a theodolite.

(IQS Land Surveying and Mensuration)

Solution method

(i) Make a diagram from the given information (Fig. 21.18).

(ii) Calculate the deviation angle between the straights.

(iii) Calculate the tangent lengths.

(iv) Calculate the chainage of the initial tangent point.

(v) Calculate the curve length.

(vi) Calculate the chainage of the final tangent point.

(vii) Assuming that the curve is to be set out from the initial tangent point using full 20 m chords, calculate the chord deflection angles and tangential angles.

(viii) Calculate the final sub-chord deflection angle and tangential angle.

(ix) Draw up a table of chainages and tangential angles.

Model answer

(i) See Fig. 21.18.

(ii) Deviation angle θ between straights.

θ = 210° 00′ − 138° 24′

θ = 71° 36′

(iii) Tangent length IT = IT_1 = R tan θ/2

= 400 tan 35° 48′ m

= 288.49 m

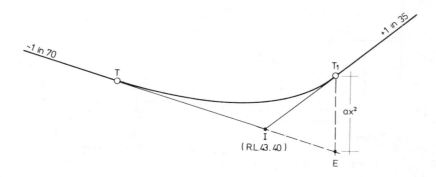

Figure 21.17

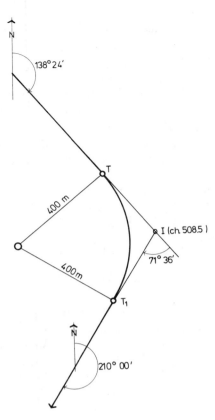

Figure 21.18

(iv) Chainage of initial tangent point T
= chainage of I − tangent length IT
= 508.50 − 288.49
= 220.01 m

(v) Curve length $TT_1 = R \times \theta$ (radians)
= 400 × 1.249 656
= 499.86 m

(vi) Chainage of final tangent point T_1
= chainage of T + curve length
= 220.01 + 499.86
= 719.87 m

(vii) Curve is to be set out using 20 m chords.
Number of full chords = 499.86/20
= 24.993 chords

i.e., 24 full chords = 24 × 20 = 480 m
and 1 final chord = 19.86 m
————
499.86 m

Chord deflection angle $= \left(\dfrac{C}{R} \times 1718.9\right)$ minutes
for full chord

$= \left(\dfrac{20}{400} \times 1718.9\right)$ min.

= 85.945 minutes
= 01° 25′ 56.7″

(viii) Final sub-chord angle $= \dfrac{C}{R} \times 1718.9$ minutes

$= \dfrac{19.86}{400} \times 1718.9$

= 85.343 minutes
= 0° 25′ 20″

(ix)

Chord No.	Chord length (m)	Chord deflection angle	Tangential angle
1	20.0	01° 25′ 57″	01° 25′ 57″
2	20.0	″	02° 51′ 54″
3	″	″	04° 17′ 51″
4	″	″	05° 43′ 47″
5	″	″	07° 09′ 43″
6	″	″	08° 35′ 40″
7	″	″	10° 01′ 37″
8	″	″	11° 27′ 34″
9	″	″	12° 53′ 30″
10	″	″	14° 19′ 27″
11	″	″	15° 45′ 24″
12	″	″	17° 11′ 20″
13	″	″	18° 37′ 17″
14	″	″	20° 03′ 14″
15	″	″	21° 29′ 10″
16	″	″	22° 55′ 07″
17	″	″	24° 21′ 04″
18	″	″	25° 47′ 00″
19	″	″	27° 12′ 57″
20	″	″	28° 38′ 54″
21	″	″	30° 04′ 51″
22	″	″	31° 30′ 47″
23	″	″	32° 56′ 44″
24	″	″	34° 22′ 41″
25	19.86	01° 25′ 20″	35° 48′ 01″

Question 14

Figure 21.19 shows a quadrilateral ABCD and the lengths of those sides that were measured in the

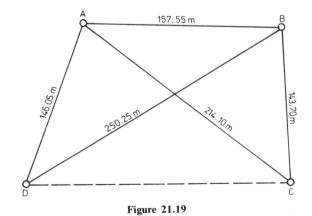

Figure 21.19

Model answers to examination questions 293

course of a chain survey. The side CD was obstructed and could not be measured directly. Calculate the length of CD.

(RICS Const. Meas. & Val. 1975)

Solution method
(i)   In triangle DBA, calculate angle DBA using the Cosine Rule.
(ii) In triangle CBA, calculate angle CBA using the Cosine Rule.
(iii) Calculate angle CBD. CBD = (ii) − (i).
(iv) In triangle DBC, calculate side CD using the Cosine Rule.

Model answer
(i)   In triangle DBA:

$$AB = d = 157.55 \text{ m}$$
$$BD = a = 250.25 \text{ m}$$
$$DA = b = 146.05 \text{ m}$$

Using Cosine Rule:

$$\cos DBA = \frac{a^2 + d^2 - b^2}{2ad}$$

$$= 0.838\,469$$
$$\therefore \text{ angle DBA} = 33° \, 01'$$

(ii) In triangle ABC:

$$BC = a = 143.70 \text{ m}$$
$$AC = b = 214.10 \text{ m}$$
$$BA = c = 157.55 \text{ m}$$

Using Cosine Rule:

$$\cos CBA = \frac{a^2 + c^2 - b^2}{2ac}$$

$$= -0.008\,108$$
$$\therefore \text{ angle CBA} = 90° \, 28'$$

(iii) Angle CBD
$$= 90° \, 28' - 33° \, 01'$$
$$= 57° \, 27'$$

(iv) In triangle DBC

$$DC^2 = DB^2 + BC^2 - 2 \times DB \times BC \cos B$$
$$= 250.25^2 + 143.70^2$$
$$\qquad - (2 \times 250.25 \times 143.70 \times \cos 57° \, 27')$$
$$\therefore DC = 211.14 \text{ m}$$

Question 15

In a survey, the following observations were recorded:

Station	Observing	Bearing	Vertical angle	Distance (m)
A	B	42° 13′	+7° 12′	3654.3
C	B	155° 28′	−4° 42′	2392.5

Note: The distances are the lengths of the lines of sight. Determine:
(i)   the bearing of a line from A to C,
(ii) the horizontal distance AC,
(iii) the height difference between A and C.

(RICS Const. Meas. & Val.)

Solution method
(i)   Draw diagram from given information.
(ii)   Calculate angle ABC from bearings BA and BC.
(iii) Calculate plan lengths BC and BA from given slope lengths and vertical angles.
(iv)   Calculate plan length AC by cosine rule in triangle ABC.
(v)   Calculate angle CAB by sine rule in triangle ABC.
(vi)   Calculate bearing AC from bearing AB and angle CAB.
(vii) Calculate height of C from B, using either plan or slope length BC and vertical angle BC.
(viii) Calculate height of A from B, using either plan or slope length BA and vertical angle BA.
(ix)   Calculate difference in height between C and A.

Model answer

(i)

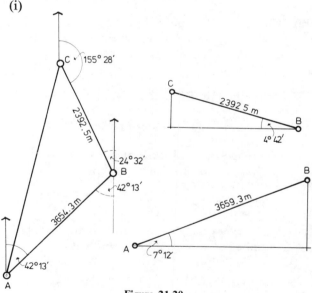

Figure 21.20

(ii) Angle ABC = 180° − (24° 32′ + 42° 13′)
$$= 180° - 66° \, 45'$$
$$= 113° \, 15'$$

(iii) Plan length BC = 2392.5 cos 4° 42′
$$= 2384.45 \text{ m}$$

Plan length BA = 3654.3 cos 7° 12′
$$= 3625.48 \text{ m}$$

(iv)   In $\triangle$ ABC, plan length AC
$$= AB^2 + BC^2 - 2AB \times BC \cos \hat{B}$$
$$= 3625.48^2 + 2384.45^2 - 2 \times 3625.48$$
$$\times 2384.45 \times \cos 113° \; 15'$$
$$= \underline{5065.04 \text{ m}}$$

(v)   In $\triangle$ ABC $\sin \hat{A} = \dfrac{CB \sin \hat{B}}{AC}$
$$= \dfrac{2384.45 \times \sin 113° \; 15'}{5065.04}$$
$$= 0.432\ 535\ 9$$
$$= 25.6286°$$
$$= 25° \; 38'$$

(vi)   Bearing AC = bearing AB − angle CAB
$$= 42° \; 13' - 25° \; 38'$$
$$= \underline{16° \; 35'}$$

(vii)   Elevation C relative to B:
height $= +2392.5 \sin 4° \; 42'$
$$= +196.04 \text{ m}$$

(viii) Elevation A relative to B;
height $= -3654.5 \sin 7° \; 12'$
$$= -458.00 \text{ m}$$

(ix)   Difference in elevation A to C:
height C relative to B = +196.04 m
height A relative to B = −458.00 m
Difference A to C = $\underline{654.04 \text{ m}}$

The following questions are typical of the descriptive questions set in the various construction surveying examinations. The appropriate answers will be found on the pages indicated.

16 Give one example of each of the following types of error and indicate how the surveyor should deal with it:
(i) gross   (ii) cumulative   (iii) random
(iv) instrumental   (v) permissible
(RICS Const. Meas. & Val.)
Answer—pages 25, 26, 76, 77, 166.

17 The permanent adjustments of a theodolite which is to be used for measuring horizontal angles only are based on the relationships of its four axes. Name those axes, show their relationships, and outline the tests to check them.
(IQS Land Surveying and Mensuration)
Answer—pages 152–157.

18 It is required to carry out a chain and level survey of an undeveloped site of about 2 hectares.
(i) List the equipment which would be required to carry out the survey in detail.
(ii) What factors should be considered when selecting the station positions?

(iii) List FIVE causes of error when chain surveying.
(IOB Site Surveying)
Answer—pages: (i) 20–24 (ii) 28–29 (iii) 25–26.

19 Write short illustrated notes on four of the following topics:
(i) Well-conditioned triangles.
(ii) Ordnance datum.
(iii) Tacheometry.
(RICS Const. Meas. & Val.)
Answer—pages: (i) 28 (ii) 53 (iii) 178.

20 Demonstrate with sketches and notes how a chain line may be continued past different types of obstruction.
(RICS Const. Meas. & Val.)
Answer—pages 34–39.

21 (a) Sketch contour patterns to illustrate the following ground characteristics:
(i) a uniform slope
(ii) a concave slope.
(b) While carrying out a contour survey by a direct method the following readings were taken off on a staff:
(i) 4.120 m on a TBM (reduced level 141.260 m).
(ii) 1.200 m on a foresight on Peg A.
The level was then moved to a new position and a backsight reading of 2.640 m taken on the staff held on Peg A. Determine the staff readings required to locate points on the 142 m, 144 m, and 146 m contours.
(c) Describe, with the aid of a sketch, a method for setting out a circular curve by means of a chain and tape. Show how the method is modified when there is an initial sub-chord.
(OND in Building Land Surveying II)
Answer—pages: (a) 96 (b) 97–98 (c) 230–231.

22 An area of approximately 200 m × 200 m of comparatively level ground is to be the site of a large building. Pegs have to be inserted every 10 metres to form a grid over the whole area. To enable contours to be interpolated on the plan, the position of each peg with its ground level must be determined. A temporary bench mark is to be fixed on the site, and this must be related to Ordnance Datum. Describe how to set out the pegs and carry out the levelling, illustrating your answers with sketches.
(RICS Part 1, Elementary Surveying)
Answer—pages 98–100.

23 Write short notes on any THREE of the following, illustrating your answer with sketches where appropriate:
(a) the method for checking and eliminating collimation error in a tilting level;

(b) the effect of unintentional tilt of the staff on fixed-hair tacheometric observations;

(c) a method of obtaining continuous chainage along a line where an obstacle prevents both chaining and ranging;

(d) the causes and effects of local attraction on observation taken during a 'free-needle' compass traverse and a method of correcting these effects.

(OND, Building Land Surveying II)

Answer—pages: (a) 79 (b) 188 (c) 38–39 (d) 121.

24 Describe the more commonly used series of maps produced by the Ordnance Survey and their specific uses.

(RICS Part 1, Elementary Surveying)

Answer—pages 12–14.

25 (a) Describe with the aid of sketches each of the following surveying instruments and how they are used:
(i) cross staff
(ii) prism square
(iii) Abney level

(b) Describe how a reconnaissance survey would be carried out with a view to locating control stations for a chain survey of a small estate.

(OND, Building Land Surveying II)

Answer—pages (a) 35, 36, 24 (b) 28.

26 Produce a labelled sketch of a surveyor's level and annotate with the function of the principal controls.

(RICS Const. Meas. & Val. 1977)

Answer—pages 53, 56.

27 With the aid of sketches, describe how the structural framework and concrete construction for the various depths of a framed building would be set and controlled accurately in level and verticality.

(CGLI Construction Technicians Certificate)

Answer—pages 243–248.

28 Explain with the aid of sketches the following surveying terms:
(i) magnetic north
(ii) quadrant (reduced) bearing
(iii) forward bearing and back bearing
(iv) face left and face right

(IOB Site Surveying)

Answer—pages (i) 110 (ii) 115 (iii) 114 (iv) 149.

29 (a) Explain the term 'local attraction' as used in surveying and describe the procedure to be adopted in the field to eliminate this.

(b) A plan is to be plotted at a scale of 1:500.

Given that the smallest plottable dimension is 0.5 mm on the plan, calculate the accuracy to which the offsets should be measured in the field.

(c) List FOUR points which should be borne in mind when selecting stations in a chain survey, and show with the aid of a sketch how the chainage would be carried over a river that is wider than the length of the chain.

(OND, Building Land Surveying II)

Answer—pages (a) 121–122 (b) 30 (c) 28–29.

30 Describe a method of stadia tacheometry suitable for obtaining sufficient information to draw up a contour plan of a large field.

(OND, Building Land Surveying)

Answer—pages 190–191.

31 A series of levels was run with a dumpy level from an Ordnance Survey bench mark (OBM) to a site to establish the reduced level of the TBM and then from the TBM to another OBM.

(a) Describe how the level will be checked and corrected on site if:
(i) The bubble axis is not at right angles to the vertical axis of the instrument.
(ii) The line of collimation is not parallel to the bubble axis.

(b) Determine the correct staff reading of the following length of sight: 41.025 m. Staff reading 3.842: when the level was in error by a depression of 2 minutes.

(CGLI Construction Technicians Certificate)

Answer—pages: (a) 77–78 (b) 285–286.

32 (a) With the aid of sketches, describe fully:
(i) a method for setting-out a right angle using a chain and tape;
(ii) a hand instrument for setting-out a right angle

(b) List the essential steps to be taken in setting up a subtense bar and making it ready for use.

(c) List the steps to be taken in observing and booking a single staff reading using either a dumpy or a tilting level.

(OND, Building Land Surveying II)

Answer—pages: (a) 34–35 (b) 192–194 (c) 60–61.

33 Describe how to achieve maximum accuracy when using a vernier theodolite to measure horizontal angles. Explain why the techniques are effective i.e., what errors do they eliminate?

(RICS Const. Meas. & Val.)

Answer—pages: 149–150, 157.

22. Computer programs and spreadsheets

Computer programs

Throughout this textbook the use of computer programs has been advocated for the solution of repetitive surveying calculations. This chapter gathers together the programs identified in Chapters 6, 11, 13, 14, and 19.

It must be emphasized at the outset that this textbook does not presume to teach computing; after all a computing textbook does not teach surveying.

The alternative to writing simple programs is to purchase them. Once the reader has experimented with the programs in this chapter and realised the saving in time and effort which accrue, he may well wish to purchase some of the excellent advanced programs currently available. On the other hand, the use of the following programs will certainly add interest to any surveying course and perhaps spur the reader to greater things.

The chapter assumes that the reader has virtually no knowledge of computing. The programs are presented as tools to enable the reader to solve basic surveying problems, simply by typing them into any microcomputer.

Each microcomputer has its own idiosyncracies but there is a large measure of commonality with all of them. The following programs were used on an AMSTRAD PCW 8512 computer and printer. Other microcomputers may require different printer instructions in which case the reader should amend any LPRINT statement to enable the particular printer to be used.

1. Calculation of coordinates using a BASIC program

The coordinate calculations of Chapter 11 may be carried out using BASIC Programs 1 to 5.

Using the $\boxed{P \rightarrow R}$ conversion of the scientific calculator, the length and bearing of a line were input by hand. The calculator then computed the partial coordinates and displayed the answer.

A simple BASIC computer program works on the same principle. The program is typed into the computer *precisely* as shown in following programs which compute the data from Example 1, Chapter 11, in a series of progressive steps.

Example 1 (Chapter 11, page 126) Calculate the coordinates of an open traverse ABCDE given the following data:

line	AB	BC	CD	DE
WCB	30°	110°	225°	295°
length (m)	50.0	70.0	82.0	31.2

Program 1

```
10  INPUT p
20  INPUT b
30  x=p*SIN(b*3.141593/180)
40  y=p*COS(b*3.141593/180)
50  PRINT x
60  PRINT y
70  GOTO 10
```

Explanation

```
10  p is plan length of line
20  b is bearing of line
30  Computer calc.; b is changed to radians
40  Computer calc.; b is changed to radians
50  Answer x is part. dep.
60  Answer y is part. lat.
70  Computer goes back to line 10 to receive
    instructions for the next line of the survey
```

User instructions
1. Type: RUN
2. Computer response: ? type 50.0
3. Computer response: ? type 30
4. Computer prints x, y on screen like this: 25.0
                                              43.0

Repeat for next line

This simple Program 1 may be enhanced, refined and enlarged. Program 2 adds the station designation and prints the results on one line.

Program 2

```
10   INPUT n$
20   INPUT p
30   INPUT b
40   x=p*SIN(b*3.141593/180)
50   y=p*COS(b*3.141593/180)
60   PRINT n$,x,y
70   GOTO 10
```

Explanation

```
10   n$ is the station reference of any station
20   p is plan length of line
30   b is bearing of line
40   Computer calc.
50   Computer calc.
60   Results on one line on screen
70   Return to start
```

User instructions

1. Type: RUN
2. Computer response: ? type B
3. Computer response: ? type 50.0
4. Computer response: ? type 30
5. Computer prints N$, x, y on screen like this:
   B 25.0 43.4
   Repeat for next line

Program 3 shows the input information entered on one line. A prompt has been added to tell the operator what information is required. The results have headings and are printed on a line printer using the command LPRINT (Command varies with type of computer).

Program 3

```
10   PRINT "enter stn.ref.,length,brg.of line"
20   INPUT n$,p,b
30   x=p*SIN(b*3.141593/180)
40   y=p*COS(b*3.141593/180)
50   LPRINT "STN.REF.","PART.DEP.","PART.LAT."
60   LPRINT n$,x,y
70   GOTO 10
```

Explanation

```
10   Prompt
20   Station ref., length, bearing of line
30   Computer calc.
40   Computer calc.
50   Headings printed on line printer
60   Results printed below headings
70   Return to start
```

User instructions

1. Type: RUN
2. Computer response: Enter stn ref., length and bearing of line—type B, 50, 30
3. Computer prints headings and results

STN.REF.	PART.DEP.	PART.LAT.
B	25	43.30127
STN.REF.	PART.DEP.	PART.LAT.
C	65.77848	−23,94143

Table 22.1

In Program 3 the results heading is repeated for each line. Program 4 shows how this is amended to give only one heading.

Program 4

```
10   LPRINT "stn.ref.","part.dep.","part.lat."
20   PRINT "enter stn.ref,length,bearing"
30   INPUT n$,p,b
40   x=p.*SIN(b*3.14159/180)
50   y=p.*cos(b*3.14159/180)
60   LPRINT n$,x,y
70   GOTO 20
```

Explanation

```
10   Headings for results
20   Prompt
30   Stn ref., length, bearing of line
40   Computer calc.
50   Computer calc.
60   Results printed below headings
70   Return to start
```

User instructions

1. Type: RUN
2. Computer response: Prints headings on printer
   Prints prompt on screen—type B, 50, 30
3. Computer ·prints results on printer below headings
   Repeat for next line

STN.REF.	PART.DEP.	PART.LAT.
B	25	43.30127
C	65.77848	−23.94143
D	−57.98257	−57.98294
E	−28.27686	13.18558

Table 22.2

In each of these four programs, only the partial coordinates were calculated. When the total coordinates are to be computed, a deeper knowledge of BASIC programming is required, particularly the technique of looping. A loop is a repetitive part of a program. In any traverse the loop is as follows:

'Start on total coordinates of a point X; calculate the partial coordinates of line XY; add those to the coordinates of X to produce the coordinates of Y'. Repeat for next point.

In Program 5, arrangements are made to enter and list the total coordinates of the starting point on lines 10–40. Lines 80–140 form the loop (k) which is to be repeated from 1 to n times. Lines 50–70 prompt the operator to enter the number of times (n) that the loop is to be repeated.

Program 5

```
10   PRINT "enter stn ref, dep.,lat.,of starting pt.
20   INPUT n$,D,L
30   LPRINT "STN.REF.","DEPARTURE","LATITUDE"
40   LPRINT n$,D,L
50   PRINT "Enter no. of traverse lines"
60   INPUT n
70   FOR k=1 TO n
80   PRINT "Enter stn.ref.,length,brg.of line"
90   INPUT m$,p,b
100  x=p*SIN(b*3.141593/180)
110  y=p*COS(b*3.141593/180)
120  D=(D+x)
130  L=(L+y)
140  LPRINT m$,D,L
150  NEXT k
```

Explanation

10	Prompt
20	Stn ref., dep. lat., of starting point
30	Headings for results
40	Coords of starting point
50	Prompt
60	n is number of lines
70	Loop—repeat calc. of n survey lines
80	Prompt
90	Station ref., length, bearing
100	Computer calc.
110	Computer calc.
120	D is total dep. of stn B (= dep. A + part. dep. line AB)
130	L is total lat. of stn B(= lat. A + part. lat. line AB)
140	Results of stn B on printer
150	Return to line 80

User instruction

1. Type: RUN
2. Computer response: Enter stn ref. etc.:—type B, 0.0, 0.0
3. Computer response:
   Prints STN REF.   DEP.   LAT.
   Prints B,   0,   0
   Prints 'enter no. of lines' on screen – type 4
4. Computer response: Enter stn ref., length and bearing—type B, 50, 30
5. Computer response: Prints total coordinates of stn B on printer below those of A thus:
   B 25.0 43.3.

STN.REF.	DEP.	LAT.
A	0	0
B	24.99998	43.30128
C	90.77851	19.35996
D	32.79594	−38.62298
E	4.519081	−25.4374

Table 22.3

Example 2 (Chapter 11, page 132) Calculate the coordinates of an open traverse ABCDE using a BASIC program given the following data:

Line	AB	BC	CD	DE
WCB	35°	94°	58°	17°
Length (m)	79.0	59.0	52.0	44.0

Solution

STN.REF.	DEPARTURE	LATITUDE
A	0	0
B	45.31255	64.71301
C	104.1688	60.59737
D	148.2673	88.15317
E	161.1317	130.2306

Table 22.4

2. Calculation of bearings

In Chapter 13, pages 164–168, the bearings of the lines of a traverse were calculated using the conventional rules of bearings computation.

The traverse data are as follows:

Traverse ABCDEF — Starting bearing (forward) line AB = 65° 34′ 20″

Measured angles   ABC   110° 05′ 20″
                   BCD   219° 50′ 40″
                   CDE   134° 52′ 50″
                   DEF   250° 58′ 30″

The bearings of the traverse lines are computed in Program 6 (Table 22.5).

Program 6

```
10   REM BEARINGS CALCULATION
20   PRINT "enter ref. of start line"
30   INPUT m$
40   PRINT "enter FORWARD bearing of start line(d,m,s)"
50   INPUT d1,m1,s1
60   LPRINT "LINE","FORWARD WCB."
70   LPRINT m$,d1;m1;s1
80   b=d1+m1/60+s1/3600
90   PRINT "enter no. of lines to be calculated"
100  INPUT n
110  FOR k=I TO n
120  PRINT "enter ref. of next line"
130  INPUT n$
```

Program 6 cont'd

```
140   PRINT "enter measured or adjusted angle(d,m,s)"
150   INPUT d2,m2,s2
160   a=d2+m2/60+s2/3600
170   b=b+a
180   IF b>540 THEN 210
190   IF b>180 THEN 230
200   IF b<180 THEN 250
210   b=b−540
220   GOTO 260
230   b=b−180
240   GOTO 260
250   b=b+180
260   b1=(b−INT(b))*60
270   b2=(b1−INT(b1))*60
280   LPRINT n$,INT(b);INT(b1);INT(b2)
290   NEXT k
```

Explanation

10	Remark
20	Prompt
30	m$ is reference of line
40	Prompt
50	d,m,s—degrees, minutes, seconds
60	Final headings on printer
70	Printout of starting bearing details
80	Starting bearing decimalized
90	Prompt
100	n is no. of lines to be calculated
110	k is the loop
120	Prompt
130	n$ is reference line
140	Prompt
150	Angle in degrees, minutes, seconds
160	Angle decimalized
170	b is bearing of next line
180	If b exceeds 540° go to line 210
190	If b exceeds 180° go to line 230
200	If b is less than 180° go to line 250
210	Following from line 180—subtract 540° from b
220	Instruction to computer—omit lines 230–250
230	Following from line 190—subtract 180° from b
240	Instruction to computer—omit line 250
250	Following from line 200—add 180° to b
260–270	b converted to deg. min. sec.
280	Print next line reference and bearing on printer
290	Repeat calculation from line 110

User instructions

1. Type **RUN**
2. Computer response: Enter starting line reference —type AB
3. Computer response: Enter starting bearing—type 65,34,20
4. Computer response: (a) Printer: LINE WCB
   AB 65 34 20
   (b) Screen: Enter ref. next line—type BC
5. Computer response: Enter measured angle—type 110,05,20
6. Computer response: BC 355 39 40

Solution

LINE	FORWARD WCB.		
AB	65	34	20
BC	355	39	40
CD	35	30	20
DE	350	23	10
EF	61	21	40

Table 22.5

3. Computation of tacheometric data using BASIC computer programs

(a) *Tangential tacheometry*

Table 14.1, partly reproduced below, shows tacheometric data obtained on a survey of three stations B, C, and D from an instrument position A. The relevant first principles calculations have been carried out on pages 178–180.

Station A   Height of instrument   1.400
            Red. lev. of station   72.200

Target station	Staff reading	Intercept S	Vert. angle
B	1.000		+4° 33' 10"
B	3.000	2.000	+5° 05' 50"
C	1.000		+6° 14'
C	1.945	0.945	+6° 48'
D	1.395		−0° 18'
D	3.395	2.000	+0° 30'
Col. 1	2	3	4

The following BASIC Program 7 should be typed precisely as shown into any modern microcomputer and the instructions followed to produce a printout of the results.

Program 7

```
10    PRINT "enter stn.ref. and red.lev. of starting point."
20    INPUT n$,l
30    PRINT "enter inst.height"
40    INPUT IH
50    LPRINT "INST.STN. :-",n$,"RED.LEVEL :-",l
60    LPRINT "_____"
70    LPRINT "STN.REF.","HORZ.DIST.","RED.LEVEL"
80    LPRINT"************************************************************"
90    PRINT "enter no. of target points"
100   INPUT n
110   FOR k=1 TO n
120   PRINT "enter target.stn.ref."
130   INPUT m$
140   PRINT "enter top staff reading, vert.angle as (d,m,s)"
150   INPUT r1,d1,m1,s1
160   a1=(d1+m1/60+s1/3600)*3.141593/180
170   PRINT "enter lower staff reading, vert.angle as (d,m,s)"
180   INPUT r2,d2,m2,s2
190   a2=(d2+m2/60+s2/3600)*3.141593/180
200   i=r1−r2
210   h=i/(TAN(a1)−TAN(a2))
220   v=h*TAN(a2)
230   l1=ih+v−r2+l
240   LPRINT m$,h,l1
250   NEXT k
```

Explanation and notes

10	Prompt
20	'n$' is instrument station name: 'l' is its reduced level
30	Prompt
40	IH is instrument height
50	Print on printer—4 items
60	Underline these items
70	Print on printer—3 items
80	Underline these items with a series of asterisks
90	Prompt
100	'n' is the number of points observed
110	'k' is a loop (a repeat calculation) from 110 to 250
120	Prompt
130	'm$' is any station name
140	Prompt
150	'r1' is top staff reading; d1, m1, s1' is the vertical angle
160	Vertical angle decimalized and changed to radians
170	Prompt
180	'r2' is bottom staff reading: 'd2, m2, s2' is the vertical angle
190	Vertical angle decimalized and changed to radians

200	'i' is intercept = (top staff − bottom staff) reading	
210	'h' is the horizontal distance	
220	'v' is the vertical distance from theodolite centre to top staff reading	
230	'll' is the reduced level of observed point	
240	Print on printer—name of point, horiz. dist. and reduced level	
250	Return to line 110	

User instructions

1. Type: RUN
   Screen response : Enter stn ref. and red. lev. of starting point
2. Type : A, 72.20
   Screen response : Enter inst. height
3. Type : 1.40
   Printer response : INST STN:—A   RED. LEVEL:— 72.20

   STN. REF.   HORZ DIST   RED LEVEL
   ****

   Screen response : Enter no. of target points
4. Type : 3
   Screen response : Enter target stn ref.
5. Type : B
   Screen response : Enter top staff reading, vertical angle
6. Type : 3.00, 5, 5, 50
   Screen response : Enter bottom staff reading, vertical angle
7. Type : 1.00, 4, 33, 10
   Printer response : B   208.98   89.24
   Screen response : Enter target stn ref.
8. Type : C and continue from 6. above.

Solution

INST.STN. :– A		RED.LEVEL :– 72.2
STN.REF.	HORZ.DIST.	RED. LEVEL
*************** ***********		************* ******
B	208.9836	89.24107
C	94.5039	82.91926
D	143.4858	71.4487

Table 22.6

(b) *Stadia tacheometry*

Table 14.4 (page 190) shows tacheometric observations computed by calculator. Program 8 shows the computer calculation.

The relevant field data are as follows:

Instrument station A
Height of instrument 1.390   Reduced level 116.210

Target stn	Horiz. circle	Vert. angle	Stadia Top bottom	'S'	Mid
RO	00° 00'				
			2.040		
B	12° 30'	+5° 40'	1.600	0.440	1.820
			1.670		
C	34° 15'	+2° 26'	1.110	0.560	1.390
			1.380		
D	63° 26'	−8° 10'	1.000	0.380	1.190

Program 8

```
10   LPRINT "STADIA TACHEOMETRY (VERT.STAFF) PROGRAM"
20   LPRINT "                                          "
30   LPRINT "                                          "
40   PRINT "enter stn.ref.and red.lev.of starting point"
50   INPUT n$,1
60   PRINT "enter inst.height"
70   INPUT ih
80   LPRINT "INST.STN. :-",n$,"RED.LEV. :-",l
90   LPRINT "****************************************************************"
100   LPRINT "                                          "
110   LPRINT "                                          "
120   PRINT "enter no. of target points"
130   INPUT n
140   FOR k=1 TO n
150   PRINT "enter stn.ref."
160   INPUT m$
170   PRINT "enter stadia readings-highest, mid,lowest"
180   INPUT r1,r2,r3
190   PRINT "enter vertical angle as(d,m,s or −d,−m,−s)
200   INPUT d,m,s
210   a=(d+m/60+s/3600)*3.141593/180
220   i=r1−r3
230   h=100*i*COS(a)*COS(a)
240   v=h*TAN(a)
250   l1=ih+v−r2+l
260   LPRINT m$,h,l1
270   NEXT k
```

Explanation

10	Title
20, 30	Blank lines on printout
40	Prompt
50	'n$' is instrument station: 1 is its reduced level
60	Prompt
70	'ih' is instrument height
80	Print on printer—4 items

90	Underline these items with a series of asterisks
100, 110	Blank lines on printout
120	Prompt
130	'n' is number of points observed
140	'k' is a loop (a repeat calculation) from 150 to 270
150	Prompt
160	'm$' is any station name
170	Prompt
180	'r1' is highest staff reading, 'r2' is mid reading, 'r3' is lowest reading
190	Prompt
200	'd, m, s' is the vertical angle
210	Vertical angle decimalized and changed to radians
220	'i' is stadia intercept (highest − lowest staff reading)
230	'h' is horizontal distance
240	'v' is vertical distance from theodolite centre to mid staff reading
250	'll' is reduced level of observed point
260	Print on printer—name of point, horiz.dist. and reduced level
270	Repeat calculation from line 150

User instructions

1. Type: RUN
   Screen response   : Enter stn ref. and red. lev. of starting point

2. Type   : A, 116.21
   Screen response   : Enter inst. height

3. Type   : 1.39
   Printer response   : INST STN:— A   RED LEV:— 116.21
                          ************************* ******************
   Screen response   : Enter no. of targets

4. Type   : 3
   Screen response   : Enter stn ref.

5. Type   : B
   Screen response   : Enter stadia readings—highest, mid, lowest

6. Type   : 2.040, 1.820, 1.600
   Screen response   : Enter vertical angle (d, m, s or −d, −m, −s)

7. Type   : 5, 40, 00
   Printer response   : B   43.57   120.10
   Screen response   : Enter stn ref.

8. Type   : C   etc.

Solution

STADIA TACHEOMETRY (VERT.STAFF) PROGRAM

INST.STN. :-A   RED.LEV. :- 116.21

****

B	43.57101	120.1034
C	55.89906	118.5854
D	37,2332	111.0667

Table 22.7

4. Cross-sectional areas and volumes

In Chapter 15, page 212, cross-sectional areas of an embankment were calculated in Example 12 (data reproduced below). Program 9 computes that example. Program 10 extends the example and shows how the volume of material in the embankment is computed.

Example 3 The reduced ground level and formation level of an embankment at 0 m, 30 m, and 60 m chainages are shown below:

Chainage (m)	0	30	60
RL (m)	35.10	36.20	35.80
FL (m)	38.20	38.40	38.60

Given that the formation width of the top of the embankment is 6.00 m, that the transverse ground slope is horizontal, and that the embankment sides slope at 1 unit vertically to 2 units horizontally, calculate the cross-sectional areas at the various chainages.

Program 9

```
10   REM CROSS SECTIONAL AREAS.
20   PRINT "Enter formation width, w"
30   INPUT w
40   PRINT "side slope gradient is 1 in s, enter s"
50   INPUT s
60   LPRINT "Chainage(m)","Area(sq.m)"
70   PRINT "enter no. of cross sections, n"
80   INPUT n
90   FOR k = 1 TO n
100    PRINT "for each section, enter chn.,red.lev.,form,lev
110    INPUT d,r,f
120    c=(r-f)
130    b=ABS(r-f)
140    a=(w+(s*b))*c
150    LPRINT INT(d*100+0.5)/100.INT(a*100+0.5)/100
160    NEXT k
```

Explanation

10	Remark—Title
20–60	Is the part of the program which is constant for each section
20	Prompt
30	w is the formation width

40	Prompt
50	s is side slope gradient
60	Print on printer the headings 'chainage' and 'area'
70–90	Loop details
70	Prompt
80	n is the number of times the loop is repeated
90	k is the loop (lines 100–160) repeated n times
100	Prompt
110	d is the chainage, r is the red.lev., f is the form. lev. of the section
120	c is central cut (positive) or fill (negative)
130	b is the positive value of c called the absolute value
140	a is cross-sectional area
150	Print on printer, below chainage and area, the respective values
160	Repeat lines 110–150 for next cross-section

User instructions

Type: RUN

Screen response	: Enter formation width	Type : 6.0
Screen response	: Enter gradient s	Type : 1
Printer response	: Chainage (m) Area (sq.m)	
Screen response	: Enter no. of cross-sections	Type : 3
Screen response	: Enter chainage	Type : 0
Screen response	: Enter reduced level	Type : 35.10
Screen response	: Enter formation level	Type : 38.20
Printer response	: −37.82 (−ve indicates fill)	
Screen response	: Enter chainage	Type : 10 etc.

Solution

Chainage (m)	Area (sq.m)
0	−37.82
30	−22.88
60	−32.48

Table 22.8

Program 10—Volumes

```
10   REM SIMPSON'S RULE-VOLUME FROM X.SECTS
20   PRINT "enter distance x between sections"
30   INPUT x
40   LET t=0
50   LPRINT "CHN.","AREA","MULT FACT.","AREA*M"
60   PRINT "enter form.width,w"
70   INPUT w
80   PRINT "side slope grade is 1 in s,enter s"
90   INPUT s
100   PRINT "enter no. of x.sects,n"
110   INPUT n
120   FOR k=1 TO n
130   PRINT "for each sect. enter chn.,red.lev.,form.lev., d,r,f"
```

Program 10 cont'd

```
140    INPUT d,r,f
150    PRINT "enter mult.factor for SIMPSONS rule i.e.1,4 or 2"
160    INPUT m
170    c = (r − f)
180    b = ABS(c)
190    a = (w+(s*b))*c
200    y = a*m
210    t = t+y
220    LPRINT d, INT(a*100+0.5)/100,m,INT(y*100+0.5)/100
230    NEXT k
240    v = t*(x/3)
250    LPRINT
260    LPRINT"******************************************"
270    LPRINT "VOLUME=",INT(v*100+0.5)/100,"cub.m."
```

Explanation

All variables in Program 10 have the meanings assigned to them in Program 9. Additionally:

line 30   x is the distance between the cross-sections

line 40   t is the cumulative total of the cross-sectional areas beginning with t = zero

line 160   m is Simpson's rule multiplying factor, namely 1 for first and last, 2 for odd and 4 for even sections

line 220   Prints the areas, a, to two decimal places

line 250   Prints a blank line

line 270   Prints the volume, v, to two decimal places

After the program has been entered, type RUN, The reader should now be capable of following the prompts presented by the program to obtain the volume of the embankment.

Solution

CHN.	AREA	MULT FACT.	AREA*M
0	−37.82	1	−37.82
30	−22.88	4	−91.52
60	−32.48	1	−32.48

****

VOLUME=   −1618.2   cub.m.

Table 22.9

5. Radiation surveys—coordinates

In Chapter 19, page 266, the coordinates of a radiation survey were calculated in Example 2. Program 11 shows the same survey calculated by computer.

The program is written for a theodolite measuring vertical angles (00° in index position of the vertical

circle). If the theodolite measures zenith angles (90°
in index position) substitute for lines 180, 230, and
240, the following statements:

180	PRINT "enter zenith angle (d,m,s)"
230	h = s * sinA
240	v = s * cosA

Program 11

```
10   REM TRI-DIMENSIONAL COORDINATES
20   PRINT "enter inst.stn.ref.& inst.height (n$,i)"
30   INPUT n$,i
40   PRINT "enter east., north, red.lev.,of inst.stn (e,n,r1)"
50   INPUT e,n,r1
60   LPRINT "Station","Easting","Northing","Red.level"
70   LPRINT
80   LPRINT"*************************************************************"
90   LPRINT n$,e,n,r1
100   PRINT "enter no. of target stns."
110   INPUT m
120   FOR k=1 TO m
130   PRINT "enter ref.of target"
140   INPUT t$
150   PRINT "enter wcb. of target(d,m,s)"
160   INPUT d1,m1,s1
170   b=(d1+m1/60+s1/3600)*3.1415927/180
180   PRINT "enter vertical angle (d,m,s) or(−d,−m,−s)"
190   INPUT d2,m2,s2
200   a=(d2+m2/60+s2/3600)*3.1415927/180
210   PRINT "enter slope length"
220   INPUT s
230   h=s*COS(a)
240   v=s*SIN(a)
250   d=h*SIN(b)
260   l=h*COS(b)
270   x=e+d
280   y=n+l
290   PRINT "enter target height"
300   INPUT t
310   r2=r1+i+v−t
320   LPRINT t$,INT(x*100+0.5)/100,INT(y*100+0.5)/100,INT(r2*100+0.5)/100
330   NEXT k
```

Explanation

10	Program title
20	Prompt
30	n$ = station ref., i = inst. height
40	Prompt
50	e, n and r1 are the x, y, z coords of inst. station
60	Print on printer the final headings
70	Underline the headings
80	Leave a blank line on printout
90	Print on printer the ref. and x, y, z coords of station

100	Prompt	
110	m is number of target stations	
120	k is the loop calculation from lines 130 to 320 to be repeated n times	
130	Prompt	
140	t$ is reference of target	
150	Prompt	
160	d1, m1, s1 is bearing to first target	
170	b is the bearing in radians	
180	Prompt	
190	d2, m2, s2 is vertical angle	
200	a is the vertical angle in radians	
210	Prompt	
220	s is the slope length to target	
230	h is the horiz length to target	
240	v is the difference in elevation from theodolite axis to target reflector	
250	d is the partial departure of line	
260	l is the partial latitude of line	
270	x is the easting (x coord) of target	
280	y is the northing (y coord) of target	
290	Prompt	
300	t is the target height	
310	r2 is the reduced level of target	
320	Print on printer stn ref., easting, northing, red. level of target to 2 decimal places	
330	Repeat calculation 130 to 320 for next target station	

User instructions
Type: RUN
Following each prompt, enter the survey data. The computer will print out the x, y, z coordinates of each target station.

Solution

Station	Easting	Northing	Red.level
A	100	100	50
1	78.15	106.61	51
2	84.92	110.89	50.41
3	93.97	116.72	49.92
4	87.62	106.59	49.61

Table 22.10

Program 12 calculates the coordinates of radiated points where the field data have been reduced to bearing, horizontal distance, and vertical distance by an automated EDM. All variables have the meanings assigned in Program 11. The survey data are those of Example 3, page 267. The results are shown in Table 22.11.

Program 12

```
10   REM TRI-DIMENSIONAL COORDINATES (Using automated EDM)
20   PRINT "enter inst.stn.ref.& inst.height (n$,i)"
30   INPUT n$,i
40   PRINT "enter east., north, red.lev.,of inst.stn (e,n,r1)"
50   INPUT e,n,r1
60   LPRINT "Station","Easting","Northing","Red.level"
70   LPRINT
80   LPRINT"*********************************************************"
90   LPRINT n$,e,n,r1
100  PRINT "enter no. of target stns."
110  INPUT m
120  FOR k=1 TO m
130  PRINT "enter ref.of target"
140  INPUT t$
150  PRINT "enter wcb.of target(d,m,s)"
160  INPUT d1,m1,s1
170  b=(d1+m1/60+s1/3600)*3.1415927/180
180  PRINT "enter plan length"
190  INPUT h
200  PRINT "enter difference in height"
210  INPUT v
220  d=h*SIN(b)
230  l=h*COS(b)
240  x=e+d
250  y=n+l
```

```
260    PRINT "enter target height"
270    INPUT t
280    r2=r1+i+v−t
290    LPRINT t$,INT(x*100+0.5)/100,INT(y*100+0.5)/100,INT(r2*100+0.5)/100
300    NEXT k
```

Solution

Station	Easting	Northing	Red.level
***			
A	100	100	50
5	96.69	112.39	49.05
6	101.32	116.76	49.25
7	105.35	118.55	49.9
8	108.4	115.75	50.34

Table 22.11

6. Program 13—levels reduction

It is generally agreed that a computer program of levels reduction is one of the more difficult programs. Consequently, it has been deliberately left to the end of this section.

Table 22.12 shows the levelling Table 6.15 of page 73 calculated by Program 13.

Program 13

```
10    REM LEVELS REDUCTION
20    PRINT "enter bench mark ref.(max.8 letters)"
30    INPUT a$
40    PRINT "enter red.lev.of bench mark"
50    INPUT l
60    LPRINT "STATION","RED.LEV."
61    LPRINT"*********************************************************** "
62    LPRINT
70    LPRINT a$,l
80    PRINT "enter B.S.staff reading"
90    INPUT r1
100   h=l+r1
110   PRINT "enter status of next reading:(i for I.S.):(f for F.S.):(f1 for last F.S.)
120   INPUT b$
130   IF b$="i" THEN 160
140   IF b$="f" THEN 230
150   IF b$="f1"THEN 300
160   PRINT "enter stn. ref. (max.8 letters)"
170   INPUT c$
180   PRINT "enter staff reading"
190   INPUT r2
200   l1=h−r2
210   LPRINT c$,l1
220   GOTO 110
230   PRINT "enter stn.ref.(max.8 letters)"
240   INPUT c$
250   PRINT "enter staff reading"
260   INPUT r2
```

```
270   l=h−r2
280   LPRINT c$,l
290   GOTO 80
300   PRINT "enter stn.ref.(max.8 letters)"
310   INPUT c$
320   PRINT "enter staff reading"
330   INPUT r2
340   l=h−r2
350   LPRINT c$,l
360   END
```

STATION	RED.LEV.
********************	
B.M.	35.27
A	35.4
B	36.24
C	35.76
D	35.74
E	35.61
F	36.33
G	36.33
H	36.54
I	36.69
J	36.54
K	36.41
B.M.	35.28

Table 22.12

Spreadsheets

Introduction

A spreadsheet is an electronic worksheet consisting of a set number of columns and rows. Any surveying calculation which is repetitive and can be tabulated can be neatly and speedily computed on a spreadsheet.

A spreadsheet is a commercial software package. Many spreadsheets are currently available including SuperCalc, VisiCalc and Microsoft's Multiplan, all of which are basically similar in layout and execution.

The following reasonably short explanation and instructions will enable the reader to use the Multiplan version and promote an understanding of the other spreadsheets. It must be emphasized that there are differences, none of them fundamental, and the reader should acquaint himself with the appropriate manual before using any other spreadsheet.

Use of Multiplan

In order to operate the system, the computer is switched on, the spreadsheet disk is loaded into drive A and the user's disk into drive B. Using Multiplan, the user types MP followed by RETURN.

The screen displays the spreadsheet (Fig. 22.1) or rather, a small section of the spreadsheet. There are in fact 63 columns and 255 lines in this Multiplan version.

Commands

The command line appears on the bottom of the screen.

The commands which are of most use to the beginner are:
1. Alpha
   Type 'a'—then type any message in the illuminated cell—press return. The message is printed on the screen in the selected cell.
2. Blank
   Type 'b'—press return. The contents of the illuminated cell are immediately blanked. Groups of cells may be blanked.
3. Copy
   Type 'c'. The contents of the illuminated cell may be copied either down, right, or left through any specified number of rows or columns.
4. Value
   Type 'v'—then type any numeric value or formula in the illuminated cell—press return. The value is printed on the screen in the selected cell.
5. Print
   Type 'p'—the screen contents may then be printed in hard copy on the printer.
6. Transfer
   Type 't'—then type 's' for 'save' the program; type 'l' for 'load' an already saved program; type 'c' for 'clear' any program.

Using the spreadsheet

The simplest way to learn about a spreadsheet is to use it. The principles, once mastered, operate for all

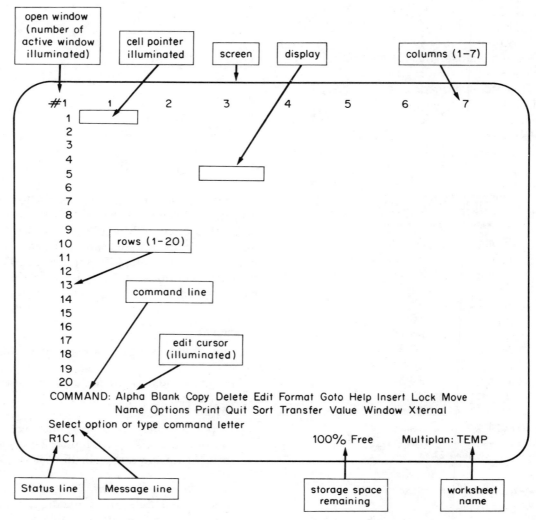

(Copyright © Microsoft Corporation – 'Electronic Worksheet'.)

Figure 22.1

types of surveying programs. Only the formulae are different.

Surveying problems require the input of data, the manipulation of these data via formulae, and the output of the results. The secret of success in the use of spreadsheets for surveying lies in the organization and entering of the formulae.

A simple example will illustrate the fundamentals.

Example 4 Calculate the sum and mean value of three pairs of numbers (x,y) (p,q) (a,b). The sum (s) = (x+y) and the mean value (m) = s/2.
(Data: x = 2, y = 4, p = 10, q = −5, a = 1526.3, b = −3575.5.)
1. Load Multiplan.

2. Select Alpha command.
3. Type the following headings in rows 1–5, cols 1–4.

	Col. 1	Col. 2	Col. 3	Col. 4
R1	Title:	Numbers		
R2	**********************			
R3	First	Second	Sum of	Mean of
R4	Number	Number	Numbers	Numbers
R5	***			
R6				

4. Enter the value of 'x' (i.e. 2) in R6C1 and the value of 'y' (i.e. 4) in R6C2.

5. Cell R6C3 requires the formula s = (x+y).
   (a) Place cursor in the cell R6C3.
   (b) Press command 'value' or press = key.
   (c) Screen response: Enter a formula.
   (d) Place the cursor over x in R6C1.
   (e) Press key + .
   (f) Place the cursor over y in R6C2.
   (g) Press return.
   (h) The formula [RC(−2) + RC(−1)] has been entered, meaning—"the sum equals the cell contents in the same row two columns back added to the cell contents in the same row one cell back".

6. Cell R6C4 requires the formula m = s/2. Place the cursor in the cell and enter the formula [RC(−1)/2] as above.

7. Copy down these formulae for any required number of rows: in this case three.
   (a) Select COPY command.
   (b) Screen response: RIGHT, or LEFT, or DOWN.
   (c) Use the TAB key to move the command cursor over DOWN—type D.
   (d) Screen response: copy DOWN number of cells: ?
   (e) Type 2 RETURN

8. Move the cursor to R6C1—select V then type the data: 2 RETURN.
   Move the cursor to R6C2—type the data: 4 RETURN.

9. The results 6 and 3 will appear automatically in R6C3 and R6C4 respectively.

10. Repeat for data (10, −5) and (1526.3, −3575.5). The spreadsheet, as acted upon by the computer (but not presented on screen to the user) is shown, complete with formulae, in Table 22.13.

1	2	3	4
"Title;"	"NUMBERS"		
"*******"	"*********"	"***************" "*********"	
"First"	"Second"	"Sum of"	"Mean of"
"Number"	"Number"	"numbers"	"numbers"
"*******"	"*********"	"***************" "*********"	
2	4	RC(−2)+RC(−1)	RC(−1)/2
10	−5	RC(−2)+RC(−1)	RC(−1)/2
1526.3	−3575.5	RC(−2)+RC(−1)	RC(−1)/2

Table 22.13

The spreadsheet which appears on screen is shown in Table 22.14.

Title;	NUMBERS		
First Number	Second Number	Sum of numbers	Mean of numbers
2	4	6	3
10	−5	5	2.5
1526.3	−3575.5	−2049.2	−1024.6

Table 22.14

Example 5 (This is Example 1 of Chapter 11, page 126) Calculate the coordinates of an open traverse ABCDE given the following data:

Line	AB	BC	CD	DE
WCB	30°	110°	225°	295°
Length (m)	50.0	70.0	82.0	31.2

Solution
1. Refer to Table 22.15. Select Alpha command and type rows 1 to 6 as they appear in the table.
2. Enter the formulae in the manner of Example 4 (5) in row 8. Row 7 should be left clear of the formulae to enable the total coordinates of origin A to be entered.
3. Enter the survey data in rows 8 to 11 and columns 1 to 3. The results will appear automatically (Table 22.16).

Example 6 The reduced ground levels and specification for a proposed roadway are shown below. Calculate the volume of cut or fill required.

Chainage (m)	20.0	40.0	60.0	80.0
Reduced level (m)	1.90	3.50	6.20	7.30

Roadway specification—gradient 1 in 20 rising
width 5.00 m
side slope gradient 1 in 1

Solution
1. Refer to Table 22.17. Select Alpha command and type rows 1 to 14 as they appear in the table.
2. Enter the formulae in the manner of Example 4 (5) in rows 15 and 16.
3. Enter the survey data in rows 15 to 19. The results will appear automatically (Table 22.18).

Table 22.15

1	2	3	4	5	6	7	8
Spreadsheet	Rect. Co-ords			Amstrad PCW8512		Surv. for Const.	
***	*******						
line	Length	Whole circle	partial departure	partial latitude	total departure	total latitude	station
***	*******	bearing *******	*********	*********	*********	*********	*******
7					0	0	A
8 AB	50	30	RC(-2)*SIN(RC(-1)*3 .14159/180)	RC(-3)*COS(RC(-2)*3 .14159/180)	R(-1)C+RC(-2)	R(-1)C+RC(-2)	B
9 BC	70	110	RC(-2)*SIN(RC(-1)*3 .14159/180)	RC(-3)*COS(RC(-2)*3 .14159/180)	R(-1)C+RC(-2)	R(-1)C+RC(-2)	C
10 CD	82	225	RC(-2)*SIN(RC(-1)*3 .14159/180)	RC(-3)*COS(RC(-2)*3 .14159/180)	R(-1)C+RC(-2)	R(-1)C+RC(-2)	D
11 DE	31.2	295	RC(-2)*SIN(RC(-1)*3 .14159/180)	RC(-3)*COS(RC(-2)*3 .14159/180)	R(-1)C+RC(-2)	R(-1)C+RC(-2)	E

Table 22.16

Spreadsheet	Rect. Co-ords			Amstrad PCW8512		Surv. for Constr.	
************	************			************		************	
line	Length	Whole circle	partial departure	partial latitude	total departure	total latitude	station
************	************	bearing ************	************	************	************	************	*******
**			***	**	0.00	0.00	A
AB	50.00	30.00000	25.00	43.30	25.00	43.30	B
BC	70.00	110.00000	65.78	-23.94	90.78	19.36	C
CD	82.00	225.00000	-57.98	-57.98	32.80	-38.62	D
DE	31.20	295.00000	-28.28	13.19	4.52	-25.44	E

Line No. breakdown:

```
1
2        Cross sectional areas.    This program will calculate areas
3        of trapezoidal shaped sections (5 SECTIONS ONLY.)
4                          *********************
5        ENTER BELOW***                     ANSWER VOL(cub. m)=
6            ***                            ***************
7        1. Formation width,(w)-------          Enter w,------
8        2. Gradient of formation,1 in (n)-----  Enter n,-----
9        3. Side slope gradient 1 in (s)------   Enter s,-----
10       4. Chainage interval (d)-------         Enter d,------
```

Column 9/10 (line 1): R(+5)C(-2)/3*(SUM(R(+10)C(+1):R(+14)C(+1)))

```
11  **** *******  *******  ****************  **************  *******  ***************
```

Sect. No. ***	Chainage	Reduced Level	Formation Level	Central Height	ABS. (c)	Area sq. m	Cut or Fill	Mult fact	Product Col19*Col10
14 **** *******		*******	****************				*******	******	**************
15 1	0	0	0	RC(-1)-RC(-2)	ABS(RC(-1))	(R7C7+R9C7*RC(-1))*RC(-2)	IF(RC(-3)<0,"cut","fill")	1	RC(-3)*RC(-1)
16 2	20	1.9	R(-1)C+(RC(-2)-R(-1)C(-2))/R8C7	RC(-1)-RC(-2)	ABS(RC(-1))	(R7C7+R9C7*RC(-1))*RC(-2)	IF(RC(-3)<0,"cut","fill")	4	RC(-3)*RC(-1)
17 3	40	3.5	R(-1)C+(RC(-2)-R(-1)C(-2))/R8C7	RC(-1)-RC(-2)	ABS(RC(-1))	(R7C7+R9C7*RC(-1))*RC(-2)	IF(RC(-3)<0,"cut","fill")	2	RC(-3)*RC(-1)
18 4	60	6.2	R(-1)C+(RC(-2)-R(-1)C(-2))/R8C7	RC(-1)-RC(-2)	ABS(RC(-1))	(R7C7+R9C7*RC(-1))*RC(-2)	IF(RC(-3)<0,"cut","fill")	4	RC(-3)*RC(-1)
19 5	80	7.3	4	RC(-1)-RC(-2)	ABS(RC(-1))	(R7C7+R9C7*RC(-1))*RC(-2)	IF(RC(-3)<0,"cut","fill")	1	RC(-3)*RC(-1)

Table 22.17

File b;survey5
Cross sectional areas.   This program will calculate areas
of trapezoidal shaped sections (5 SECTIONS ONLY.)
****

ENTER BELOW***   ANSWER Vol(cub. m)=−1153.93
            ***   *******************************
  1. Formation width,(w)-------------------- Enter w,---------------   5.00
  2. Gradient of formation,1 in (n)-------- Enter n,---------------- 20.00
  3. Side slope gradient 1 in (s)---------- Enter s,----------------   1.00
  4. Chainage interval (d)------------------ Enter d,---------------- 20.00
****

Sect. No.	Chainage ***	Reduced Level	Formation Level	Central Height	ABS. (c)	Area sq.m	Cut or Fill	Mult fact	Product Col9*Col10
1	0.00	0.00	0.00	0.00	0.00	0.00	fill	1	0.00
2	20.00	1.90	1.00	−0.90	0.90	−5.31	cut	4	−21.24
3	40.00	3.50	2.00	−1.50	1.50	−9.75	cut	2	−19.50
4	60.00	6.20	3.00	−3.20	3.20	−26.24	cut	4	−104.96
5	80.00	7.30	4.00	−3.30	3.30	−27.39	cut	1	−27.39

Table 22.18

Answers

Chapter 2

1 Descriptive question—page 7.
2 (a) 1:2500. (b) 200 m.
3 1020 m².
4 62 500 kg.
5 87.00 m.
6 12.75 m².
7 Bottom margin 48 mm.
   Side margin 45 mm.
8 9025 m².

Chapter 3

1 To north SK 5366.
   To south SK 5264.
   To west SK 5165.
   To east SK 5365.
2 TQ 3380 SE.
3 See Fig. 3.11.
4 571.5 m², 900 m².
5 1—SK5166 2—SK5266 3—SK5366
   4—SK5165 5—SK5365 6—SK5164
   7—SK5264 8—SK5364.
6 1—SK3657NW 2—SK3657NE 3—SK3757NW 4—SK3757NE
   5—SK3657SW 6—SK3757SW 7—SK3757SE 8—SK3656NW
   9—SK3656NE 10—SK3756NW 11—SK3657NE 12—SK3656SW
   13—SK3656SE 14—SK3756SW 15—SK3756SE.
7 (a) 0.190 ha; 0.469 acres.
   (b) West—road boundary.
      North—road boundary.
      East—boundary at trees.
      South—bank of river.
   (c) Parish boundary—running from north-west
      to south-east—an undefined line across a
      field and through a pond; a line running
      parallel to a hedge and lying 0.91 m from the
      roots of the hedge; a line running along the
      centre road of a roadway; a line running
      along the face of a fence and finishing
      behind the post office.
   (d) 0072.
   (e) 1—Cutting   2—Post Office   3—Embankment
      4—Open brace   5—Boundary merging   6—Bench mark
      7—Signal post   8—Milestone   9—Footpath
      10—Direction of flow of stream.

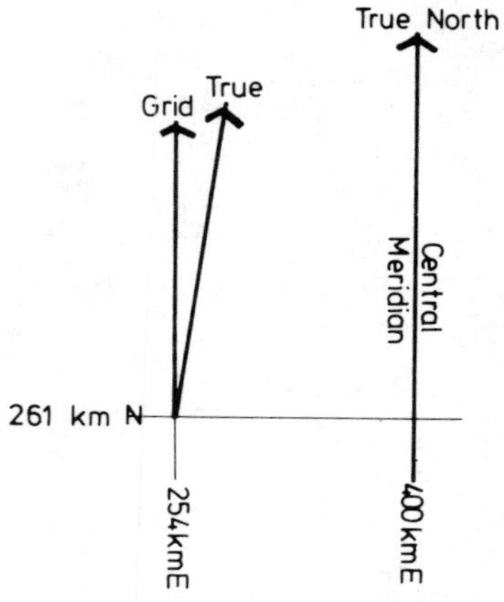

Figure 3.11

Chapter 4

1 (a) Scale of plan—1:2000 offset—max. 10 m.
  (b) Descriptive question—see page 35.
  (c) Descriptive questions—see page 38.
2 Descriptive question—see pages 25–28.
3 (a) See Fig. 4.47.
  (b) See Fig. 4.48
4 Standardization error   = +0.24 m.
  Slope length = 48.46 + 0.24 = 48.70 m.
  Plan length = 48.70 cos 5° = 48.51 m.
5 See Fig. 4.49.
6 Descriptive question:
  (a) See page 29.
  (b) See pages 29–30.
  (c) See pages 25–28.
  (d) See pages 24–25.

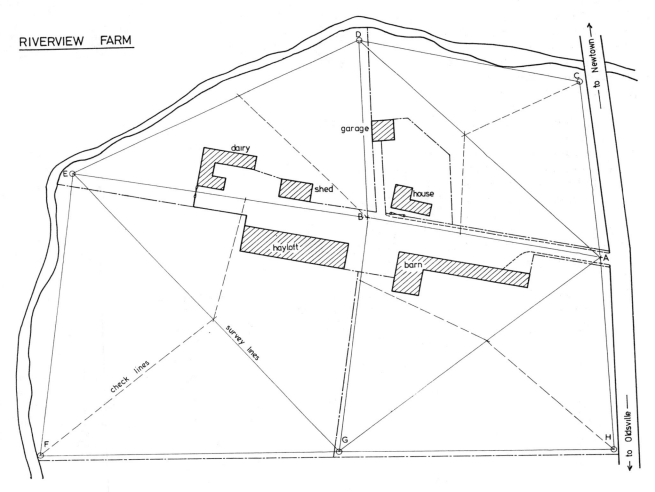

RIVERVIEW FARM

Figure 4.47

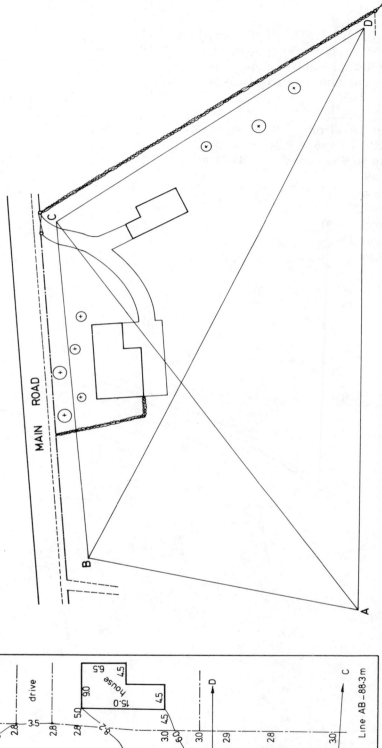

Figure 4.49

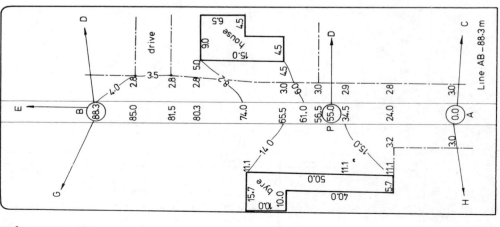

Figure 4.48

Chapter 6

1

BS	IS	FS	HPC	RL	Remarks
1.250			2.685	1.435	BM
	1.285			1.400	Peg A
	1.125			1.560	Peg B
1.555		0.810	3.430	1.875	CP
	−1.400			4.830	Inverted staff
	1.235			2.195	Peg C
1.905		0.665	4.670	2.765	CP
		0.070		4.600	BM
4.710	2.245	1.545		4.600	
−1.545				−1.435	
3.165				3.165	Simple partial check

Sum RLs (except 1st) = 19.225 m

$$2.685 \times 3 = 8.055$$
$$+3.430 \times 3 = 10.290$$
$$+4.670 \times 1 = \underline{4.670}$$
$$23.015$$

Sum IS column 2.245
Sum FS column $\underline{1.545}$
  3.790

$$23.015 - 3.790 = 19.225 \text{ m}$$

2

BS	IS	FS	Rise	Fall	Reduced level	Remarks	Chainage	Formation level	Settlement
3.540					78.675	OBM			
0.410		3.665		0.125	78.550				
0.525		2.245		1.835	76.715				
	2.840			2.315	74.400	Station A	0	74.400	0.000
	2.440		0.400		74.800		30	74.800	0.000
	2.045		0.395		75.195		60	75.200	0.005
2.475		1.655	0.390		75.585		90	75.600	0.015
	2.090		0.385		75.970		120	76.000	0.030
	1.700		0.390		76.360		150	76.400	0.040
	1.315		0.385		76.745		180	76.800	0.055
	0.900		0.415		77.160		210	77.200	0.040
2.465		0.485	0.415		77.575		240	77.600	0.025
	2.055		0.410		77.985		270	78.000	0.015
	1.645		0.410		78.395	Station B	300	78.400	0.005
		2.040		0.395	78.000	TBM			
9.415	10.090	3.995	4.670		78.000				
−10.090		−4.670			−78.675				
−0.675			−0.675		−0.675				

Original level Station A = 74.400. Since the construction gradient of the roadway is 1 in 75 rising from A, the formation levels of the chainage points must increase uniformly by (30/75) = 0.400 m, giving the formation levels in the table above. The maximum settlement occurs at chainage 180 m.

3

BS	IS	FS	HPC	Red. lev.	Remarks
0.872			22.332	21.460	OB mark
0.665		3.980	19.017	18.352	
	2.920			16.097	River level at 'A'
	−1.332			20.349	Soffit of bridge at 'A'
	−1.213			20.329	Soffit of bridge at 'B'
	−1.294			20.311	Soffit of bridge at 'C'
	−1.280			20.297	Soffit of bridge at 'D'
	2.920			16.097	River level at 'D'
4.216		0.597	22.636	18.420	
		1.155		21.481	OB mark
5.753	0.622	5.732		21.481	
−5.732				−21.460	
0.021				0.021	

Sum RLs (except 1st) = 171.733 = [22.332 + (19.017 × 7) + 22.636]
$$- [0.622 + 5.732]$$
$$= 171.733$$

4

BS	IS	FS	Rise	Fall	Reduced levels
3.786					36.642
	1.312		2.474		39.116
	1.960			0.648	38.468
0.872		3.560		1.600	36.868
	3.698			2.826	34.042
	0.670		3.028		37.070
2.238		2.180		1.510	35.560
	1.052		1.186		36.746
2.874		2.806		1.754	34.992
	1.716		1.158		36.150
	0.950		0.766		36.916
		1.412		0.462	36.454
9.770	9.958	9.958	8.612	8.800	36.454
−9.958			−8.800		−36.642
−0.188			= −0.188		= −0.188

BS	IS	FS	Height of collimation	Surface red. level	Grade red. level	Fill	Cut	Remarks
0.824			40.044	39.220				TBM No. 1
	1.628			38.416	38.416	—	—	A
	0.790			39.254	38.816		0.438	20 m from A
	0.383			39.661	39.216		0.445	40 m from A
2.154		1.224	40.974	38.820	39.616	0.796		60 m from A
	2.336			38.638	40.016	1.378		80 m from A
	2.757			38.217	40.416	2.199		100 m from A
2.555		0.461	43.068	40.513				Change point
	2.275			40.793	40.816	0.023		120 m from A
	0.436			42.632	41.216		1.416	140 m from A
	0.227			42.841	41.616		1.225	160 m from A
	0.716			42.352	42.016		0.336	180 m from A
	0.652			42.416	42.416		—	B, 200 m from A
		0.233		42.835				TBM No. 2
5.533	12.200	1.918		42.835				
−1.918				−39.220				
=3.615				= 3.615				

Sum of RLs (except 1st) = 527.388 = $[(40.044 \times 4) + (40.974 \times 3) + (43.068 \times 6)] - [12.200 + 1.918]$
= 527.388

Gradient A to B = (42.416 − 38.416) in 200
= 1 in 50

Rise per 20 m = (1/50) × 20
= 0.400 m

Chapter 7

1 (a) Chainage   0   10   20   30   40   50 (Bridge) 50
   Red. level   90.35   90.96   90.28   90.10   90.21   96.07   91.15
  (b) See Fig. 7.8
  (c) Chainage (m)   0   10   20   30   40   50
     cover (m)   1.45   2.16   1.58   1.50   1.71   2.75
  (d) Clearance 0.42m.
  (e) See Fig. 7.8.

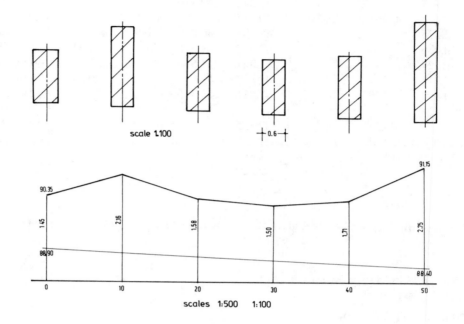

scale 1:100   ⊢ 0.6 ⊣

scales 1:500   1:100

Figure 7.8

2 (a)

Chainage	0 (A)	15	30	45	60	62.5 (B)	75	90	105	120	135	150 (C)
Red. level	57.85	58.09	57.79	58.30	57.85	57.75	58.34	56.89	56.15	56.79	55.53	56.35

  (b) Gradient 1 in 100 falling. Cut at B—0.525 m.
  (c) See Fig. 7.9.
  (d) See Fig. 7.9.

3 (a) (c)

Chn	0	10	20	30	40	50	60	70	80	90
RL	62.912	63.124	63.191	63.727	63.455	63.250	63.831	64.013	64.306	64.973
IL	61.000	61.250	61.500	61.750	62.000	62.250	62.500	62.750	63.000	63.250
Cut	1.912	1.874	1.691	1.977	1.455	1.000	1.331	1.263	1.306	1.723

  (b) (d) See Fig. 7.10.

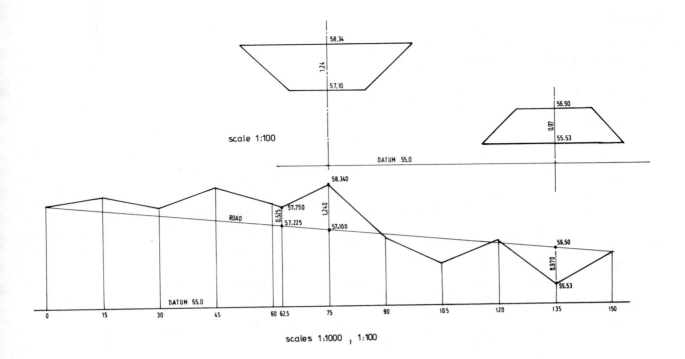

Figure 7.9

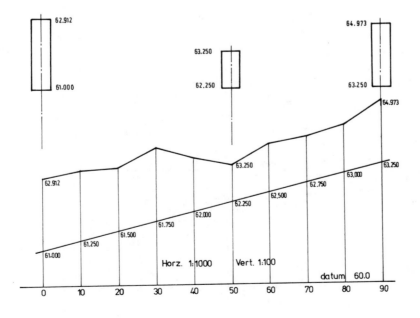

Figure 7.10

Chapter 8

1 Descriptive question—see page 96.

2 (i)

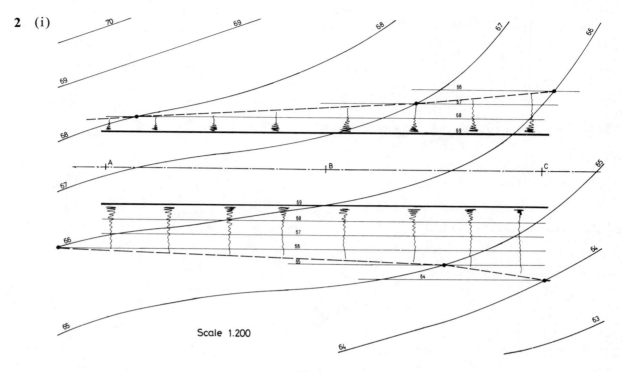

Scale 1.200

Figure 8.25

(ii), (iii)

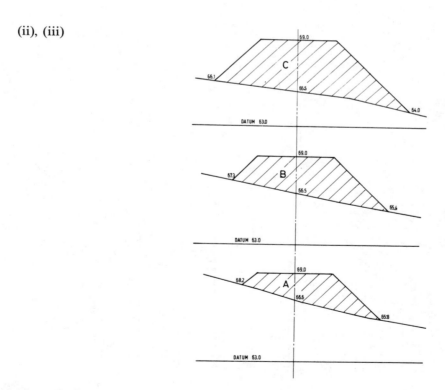

3 0.55 m, 2.550 m.

Figure 8.26

Chapter 9

1 N 48° 50′ W; S 67° 30′ W; N 60° 10′ E;
S 87° 00′ E;   West;
S 12° 10′ E; S 68° 50′ E; S 84° 50′ W;
N 00° 50′ W; S 01° 00′ E

2 329° 50′; 240° 30′; 172° 15; 10° 00′;
90° 00′; 259° 10′; 270° 10′; 115° 30′

3 172° 40′;   00° 30′;   359° 00′;   220° 40′;
S 07° 20′ E; N 00° 30′ E; N 01° 00′ W; S 40° 40′ W

4 181° 00′;   359° 00′;   01° 20′;
S 08° 30′ E; N 10° 30′ W; N 08° 10′ W;
179° 30′;   272° 10′;
S 10° 00′ E; S 82° 40′ W

5 S 68° 20′ W

6 (a) 1960 Declination—12° west
(b) Secular variation—11.25 minutes

7 290°

8 72° 15′ or 252° 15′

9 (a) Descriptive question—see pages 111–113.
(b) Descriptive question—see pages 110–111
and 114.

10 Answer depends on chosen route.

11

26°	88°	233°	315°	8°
8° W	2° E	5° E	5° W	10° W
18°	90°	238°	310°	358°
198°	270°	58°	130°	178°
S18° W	WEST	N 58° E	S 50° E	S 2° E

12

Line	AB	BC	CD	DE
WCB (forward)	106°	117°	51°	329°
WCB (back)	286°	297°	231°	149°

13

Line	AB	BC	CD	DE
WCB (forward)	60°	107°	69°	322.5°

Chapter 10

1 (a) Corrected forward bearings
PQ 22°, QR 94°, RS 166°, ST 235°, TU 261°.
(b) Shortest distance 31 m.

2 See Fig. 10.8.

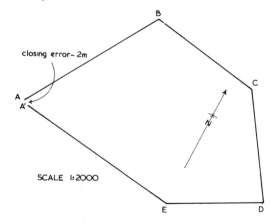

Figure 10.8

3 (a)

Line	Length	WC bearings Forward	Back	Corrections Forward	Back	Corrected WCBs Forward	Back	Declination	True forward WCB
AB	85	68° 00′	247° 00′	−1	0° 00′	67°	247°	−10°	57°
BC	103	146° 30′	326° 30′	0	0° 00′	146.5°	326.5°	−10°	136.5°
CD	58	243° 30′	61° 00′	0	+2° 30′	243.5°	63.5°	−10°	233.5°

(b) See Fig. 10.9.

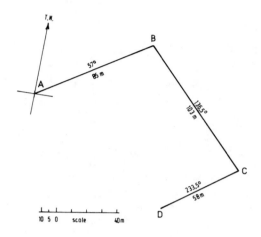

Figure 10.9

4

Line	Bearing forward	Back	Corrections Forward	Back	Corrected bearing Forward	Back
AB	214° 25′	32° 25′	−2°	0	212° 25′	32° 25′
BC	314° 20′	134° 20′	0	0	314° 20′	134° 20′
CD	65° 40′	246° 40′	0	−1°	65° 40′	245° 40′
DE	133° 00′	310° 30′	−1°	+1° 30′	132° 00′	312° 00′
EF	165° 50′	344° 20′	+1° 30′	+3°	167° 20′	347° 20′

Chapter 11

1 (a) Corrected forward bearings are:

AB 110° 30′   BC 17° 00′   CD 86° 15′

DE 39° 30′   EF 45° 00′

(b)

Total departure	Total latitude	Station
0.00	0.00	A
84.30	−31.52	B
108.12	46.42	C
162.51	49.98	D
226.12	127.14	E
321.58	222.60	F

(c) Plotting—Fig. 11.6.

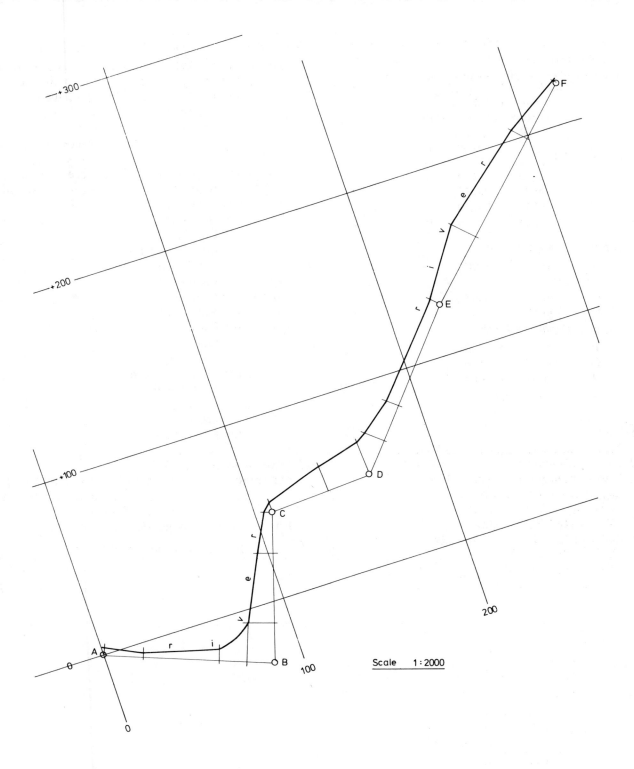

Figure 11.6

Scale   1 : 2000

2 (a) Closing error 1.131 m.

(b) Accuracy 1 in 465.

(c) Corrected final coordinates:

Total departure	Total latitude	Station
0.00	0.00	P
55.58	108.88	Q
129.02	−28.58	R
109.02	−90.28	S
52.71	−110.69	T
0.00	0.00	P

3

Station	Departure	Latitude
P	0.00	0.00
Q	109.67	51.72
R	250.77	−138.48
S	105.51	243.06

4

Station	Departure	Latitude
P	0.00	0.00
Q	246.07	89.96
R	281.07	−61.04
S	39.51	−177.12

Chapter 12

1 Descriptive question—page 140.

2 Descriptive question—pages 152–157.

3 Descriptive question—page 143.

4 Angle ABC   65° 11′ 30″

  BCD   66° 34′ 20″

  CDE 114° 57′ 40″.

5 Descriptive question—page 149.

6

System	Advantages	Disadvantages
Vernier	(a) Cancels effects of eccentricity	(a) Time consuming and laborious
	(b) Gross errors are easily detected	(b) Difficult to read in poor light
		(c) Necessary to read two verniers
Optical scale	(a) Most easily read	(a) Necessary to estimate seconds
	(b) Cheaper instruments	
	(c) Can be illuminated for easy reading	
Optical micrometer	(a) Easily read	None
	(b) Can be illuminated	
	(c) Only one reading necessary	
	(d) Double reading type cancels eccentricity errors.	
	(e) Reads to 1 second	

7 Gross errors—page 157.

Systematic errors—effects of instrumental maladjustments and defects, see pages 152–157.

Random errors—page 153.

8 Angle of depression 00° 49′ 48″.

Chapter 13

1 (a)

	Obs. angle	(b) Corr.	(c) Corrd angle
DAB	89° 16′ 20″	+20″	89° 16′ 40″
ABC	99° 38′ 40″	+20″	99° 39′ 00″
BCD	80° 24′ 20″	+20″	80° 24′ 40″
CDA	90° 39′ 20″	+20″	90° 39′ 40″
	359° 58′ 40″	+1′ 20″	360° 00′ 00″

(d) AD N 45° 36′ 00″ E
   AB S 45° 07′ 20″ E
   BC N 54° 31′ 40″ E
   CD N 45° 03′ 40″ W

2 Closing error $\Delta D = +0.036$ m
   $\Delta L = -0.212$ m
   Distance $= \sqrt{(0.036^2 + 0.212^2)}$
   $= 0.215$ m
   Accuracy $= \underline{1 \text{ in } 2650}$

3

ΣE	ΣW	ΣN	ΣS
136.15	136.00	265.08	265.23
$\Delta D = +0.15$		$\Delta L = -0.15$	

Bowditch's Rule		Line	Traverse Rule	
Departure	Latitude		Departure	Latitude
−0.054	252.464	PQ	—	252.481
−110.794	−113.686	QR	110.821	−113.688
−25.273	−151.477	RS	25.254	−151.467
136.121	12.699	ST	−136.075	12.674

4 (a) Total coordinates

Departure	Latitude	
80.100	138.60	X
00.00	00.00	A
00.00	186.40	B
221.91	260.65	C
392.51	260.65	D
366.77	125.05	E

(b) Length XE = 287.20 m
   Bearing XE = 92° 42′
(c) Angle set out at X = 242° 42′
   Angle set out at E = 261° 57′

5 Total coordinates

Departure	Latitude	Closing error
0.00	0.00	A $= \sqrt{(10.12^2 + 17.25^2)}$
0.00	167.25	B $= \underline{19.99 \text{ m}}$
115.55	364.20	C
230.81	15.24	D Tan bearing closing error
10.12	17.25	A $= \dfrac{10.12}{17.25}$
		$= $ N 30° 23′ 55″ E

Conclusion: line BC is 20 m too long, caused by miscounting the tape lengths.

6 Setting out

Line	Angle	Length
AE	48° 40′ 10″	85.80
AB	287° 15′ 15″	64.20
AC	326° 16′ 40″	89.50

7

Angle	Measured value	Adjustment	Adjusted angle
ABC	283° 31′ 40″	+ 10″	283° 31′ 50″
BCD	329° 06′ 50″	+ 10″	329° 07′ 00″
CDE	90° 47′ 20″	+ 10″	90° 47′ 30″
DEA	299° 43′ 00″	+ 10″	299° 43′ 10″
EAB	256° 50′ 20″	+ 10″	256° 50′ 30″
	1259° 59′ 10″	+ 50″	1260° 00′ 00″

Line	Whole circle bearing	Quadrant bearing
AB	152° 24′ 40″	S 27° 35′ 20″ E
BC	255° 56′ 30″	S 75° 56′ 30″ W
CD	45° 03′ 30″	N 45° 03′ 30″ E
DE	315° 51′ 00″	N 44° 09′ 00″ W
EA	75° 34′ 10″	N 75° 34′ 10″ E

Chapter 14

1 Area of triangle ABC = 2.3003 hectares
Area of triangle ACD = 1.6075 hectares
Total area   = 3.9078 hectares

2 AB = 142.19 m   AC = 41.24 m   BC = 151.57 m

3

	A	B	C	D	E	F
Reduced levels	229.61	231.45	232.98	234.21	233.51	232.12
Chainages	0.00	20.88	39.77	60.65	80.60	100.54
Proposed levels	229.61	230.03	230.41	230.82	231.22	231.62
Cutting	0.00	1.42	2.57	3.39	2.29	0.50

4

Horiz. distance B =   25.99 m
Δ level AB =   +2.75 m
Reduced level B =   101.75 m

Horiz. distance AC =   22.81 m
Δ level AC =   +1.85 m
Reduced level C =   100.82 m

5

AB = 82.6 $\cos^2$ 5° 27′ =   81.85 m
Δ level AB =   7.80 m

BC = 63.8 $\cos^2$ 4° 42′ =   63.37 m
Δ level BC =   −5.91 m

CD = 3.0/(tan 4° 20′ − tan 2° 12′) =   80.30 m
Δ level CD =   3.49 m

Length AD = 81.85 + 63.37 + 80.30 =   225.52 m
Δ level AD = 7.80 − 5.91 + 3.49 =   5.38 m

Gradient AD = 5.38 in 225.52 = 1 in 41.92

6

(i) Horiz. distance AB =   15.14 m
Reduced level B = 102.04 m

(ii) Horiz. distance AB =   15.17 m
Reduced level B = 102.05 m

Reason for minor differences is slight inaccuracies in reading vertical angles, spirit level
settings and staff readings. Also staff may not
be vertical.

Chapter 15

1 80 300 m^2.
2 85 910.4 m^2.
3 319.5805 hectares.
4 (a) 42.5 squares × 25 m^2 = 1062.5 m^2
     (b) Simpson's Rule   = 1062 m^2
     (c) Trapezoidal Rule   = 1047 m^2
     (d) (4.247 revs × 100) = 42.47 × 25 m^2
                                     = 1061.8 m^2
5 (a) 17.5 m^2 (b) 18.68 m^2 (c) 31.68 m^2
6 Area 1 = 30 m^2
     Area 2 = 41.2 m^2 Volume = 2135 m^3
     Area 3 = 18.72 m^2
7 8236 m^3
8 Area of side of trench (trapezoidal rule)

        = 236 m^2

  Volume = 236 × 0.75

        = 177 m^3

9 740.0 m^3.

10 430 m^3 − see Fig. 15.31.

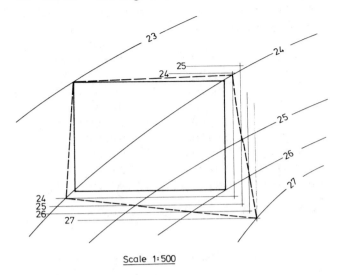

Scale 1:500

Figure 15.31

11 20506 m^3.

Chapter 16

1 (a) Offset at 5 m intervals = 1.65 m, 2.61 m, 2.92 m, 2.61 m, 1.65 m.

(b)

Point	Chainage	Tangential angle (*left*)
T_1	792.20	
1	800.00	00° 33′ 41″
2	820.00	01° 42′ 27″
3	840.00	02° 51′ 13″
4	860.00	03° 59′ 59″
5	880.00	05° 08′ 45″
6 (T_2)	894.92	06° 00′ 03″

2 (a) 08° 30′   (b) 29.73 m   (c) 59.34 m

(d) 167.77 m   227.11 m

(e)

Chord No.	Length	Chainage	Tangential angle
T_1		167.77	
1	12.23	180.00	00° 52′ 33″
2	20.00	200.00	02° 18′ 30″
3	20.00	220.00	03° 44′ 27″
4 (T_2)	7.11	227.11	04° 15′ 00″

3 (a) 329.7 m   (b) 490.0 m   720.2 m
   820.2 m   1223.0 m

4 (a) 162.74.
   (b) 527.62,   830.00

(c) Angles,   1 at 1° 04′ 30″ initial sub-chord
      14 at 1° 44′ 10″
      1 at 0° 52′ 10″ final sub-chord.

5 On AI tangent point is between A and I

   Distance A to TP = 17.29 m

On BI tangent point is outwith BI

   Distance B to TP = −37.71 m

Coords TP1   118.29 mE,   91.15 mN

   TP2   197.94 mE,   109.73 mN

6 (a)   Red level left-hand TP = 64.000 m AOD.
   Red level right-hand TP = 64.500 m AOD.

(b)

Chainage	Curve level	Cut
0	64	0
30	64.42	0.08
60	64.68	0.32
90	64.78	0.32
120	64.72	0.08
150	64.50	0

(c) Highest point—chainage 93.757—Red. lev. 64.781 m AOD.

7 Chainage 1513.3 m.
Red. level 41.34 m AOD.

Chapter 17

1
Bench mark RL	= 67.650 m
Backsight	= 1.200
therefore Height of collimation	= 68.850 m
Invert level A	= 64.350 m
Length of traveller	= 2.000
Sight rail level	= 66.350 m

Staff reading to sight rail A = 68.850 − 66.350
= 2.500 m

Gradient AB = 1 in 100 falling
length   AB = 150 m
therefore fall A to B = 150/100 = 1.500 m

Sight rail level B = 66.350 − 1.500
= 64.850 m

Staff reading to sight rail B = 68.850 − 64.850
= 4.000 m

2 (a)

RL	Remarks
97.240	TBM
97.550	MH5
97.200	MH4
98.100	MH3
98.815	MH2
99.200	MH1

(b)

IL	Remarks
95.600	MH5
96.200	MH4
96.366	MH3
96.550	MH2
96.811	MH1

(c) Sight rails and traveller.

Chapter 18

1

Station	Reduced level (m)
A	156.20
B	173.15
C	192.99
D	167.44
E	177.32

2

	Vert. angle	Length	Δ Height	RL target	
AB	03° 35′ 40″	87.35	+5.48	141.18	B
BC	01° 10′ 20″	153.25	+3.14	144.32	C
CD	05° 35′ 40″	93.57	+9.12	153.44	D

3 BC = 47.36 m.
CD = 30.46 m.

Index